PLACENTA
THE TREE OF LIFE

GENE AND CELL THERAPY SERIES

Series Editors
Anthony Atala & Nancy Templeton

PUBLISHED TITLES

Cellular Therapy for Neurological Injury
Charles S. Jr., Cox

Placenta: The Tree of Life
Ornella Parolini

GENE AND CELL THERAPY

PLACENTA
THE TREE OF LIFE

Edited by
Ornella Parolini

Associate Editor
Antonietta Silini

CRC Press is an imprint of the
Taylor & Francis Group, an **informa** business

CRC Press
Taylor & Francis Group
6000 Broken Sound Parkway NW, Suite 300
Boca Raton, FL 33487-2742

© 2016 by Taylor & Francis Group, LLC
CRC Press is an imprint of Taylor & Francis Group, an Informa business

No claim to original U.S. Government works

Printed on acid-free paper
Version Date: 20160113

International Standard Book Number-13: 978-1-4987-0026-9 (Hardback)

This book contains information obtained from authentic and highly regarded sources. Reasonable efforts have been made to publish reliable data and information, but the author and publisher cannot assume responsibility for the validity of all materials or the consequences of their use. The authors and publishers have attempted to trace the copyright holders of all material reproduced in this publication and apologize to copyright holders if permission to publish in this form has not been obtained. If any copyright material has not been acknowledged please write and let us know so we may rectify in any future reprint.

Except as permitted under U.S. Copyright Law, no part of this book may be reprinted, reproduced, transmitted, or utilized in any form by any electronic, mechanical, or other means, now known or hereafter invented, including photocopying, microfilming, and recording, or in any information storage or retrieval system, without written permission from the publishers.

For permission to photocopy or use material electronically from this work, please access www.copyright.com (http://www.copyright.com/) or contact the Copyright Clearance Center, Inc. (CCC), 222 Rosewood Drive, Danvers, MA 01923, 978-750-8400. CCC is a not-for-profit organization that provides licenses and registration for a variety of users. For organizations that have been granted a photocopy license by the CCC, a separate system of payment has been arranged.

Trademark Notice: Product or corporate names may be trademarks or registered trademarks, and are used only for identification and explanation without intent to infringe.

Visit the Taylor & Francis Web site at
http://www.taylorandfrancis.com

and the CRC Press Web site at
http://www.crcpress.com

Contents

Series Preface vii
Preface ix
Editor xi
Contributors xiii

Chapter 1   Structure and Development of the Human Placenta 1

                    Joanna L. James and Lawrence W. Chamley

Chapter 2   The Role of Mesenchymal Stem Cells in the Functions
                    and Pathologies of the Human Placenta 13

                    Gina Diamanta Kusuma, Padma Murthi, and Bill Kalionis

Chapter 3   The Roles of the Human Placenta in Fetal-Maternal Tolerance 39

                    Jelmer R. Prins and Sicco A. Scherjon

Chapter 4   The Human Placenta in Wound Healing: Historical and
                    Current Approaches 49

                    Carmen García Insausti, José María Moraleda,
                    Gregorio Castellanos, and Francisco José Nicolás

Chapter 5   Cell Populations Isolated from Amnion, Chorion,
                    and Wharton's Jelly of Human Placenta 69

                    Francesco Alviano, Roberta Costa, and Laura Bonsi

Chapter 6   The Immunomodulatory Features of Mesenchymal Stromal
                    Cells Derived from Wharton's Jelly, Amniotic Membrane,
                    and Chorionic Villi: *In Vitro* and *In Vivo* Data 91

                    Marta Magatti, Mohamed H. Abumaree, Antonietta R. Silini,
                    Rita Anzalone, Salvatore Saieva, Eleonora Russo,
                    Maria Elena Trapani, Giampiero La Rocca, and Ornella Parolini

Chapter 7   Use of Placenta-Derived Cells in Neurological Disorders 129

                    Christopher Lawton, Maya Elias, Diego Lozano, Hung Nguyen,
                    Stephanny Reyes, Jaclyn Hoover, and Cesario V. Borlongan

v

Chapter 8 Use of Amnion Epithelial Cells in Metabolic Liver Disorders.............143

Roberto Gramignoli, Fabio Marongiu, and Stephen C. Strom

Chapter 9 The Use of Placenta-Derived Cells in Autoimmune Disorders 161

Antonietta R. Silini, Ornella Parolini, and Mario Delgado

Chapter 10 The Use of Placenta-Derived Cells in Inflammatory
and Fibrotic Disorders... 181

*Euan M. Wallace, Anna Cargnoni, Rebecca Lim, Alex Hodge,
and William Sievert*

Chapter 11 From Bench to Bedside: Strategy, Regulations, and Good
Manufacturing Practice Procedures... 197

Christian Gabriel

Chapter 12 Applications of Placenta-Derived Cells in Veterinary Medicine..... 217

*Barbara Barboni, Valentina Russo, Paolo Berardinelli,
Aurelio Muttini, and Mauro Mattioli*

Index...285

Series Preface

Gene and cell therapies can provide useful therapeutics for numerous diseases and disorders, particularly for those that have no other effective treatments. This book series covers all current topics in gene and cell therapies and their supporting disciplines, from basic research discoveries to clinical applications. Because these fields are continually evolving to produce advanced treatments based on scientific breakthroughs, each book in the series provides a timely in-depth coverage of its focused topic.

The contributors to this second book of the series include a remarkable group of authors who expertly cover topics related to the placenta as a source of cells for therapy. The volume covers important basic areas related to the placenta, such as its development and function, as well as the sources of cells present, like amnion and chorion. Major topics are covered, such the history of the use of placenta for wound healing and the current state of the use of placental cells for neurological disorders, liver disorders, autoimmune diseases, and immunomodulation. Applications of placental derived therapies in veterinary medicine are also covered. This volume, dedicated to the placenta, provides a comprehensive overview of the current state of the field and its future directions.

We would like to thank the volume editor, Dr. Ornella Parolini and the authors, all experts in their fields, for their valuable contributions. We also would like to thank CRC senior acquisitions editor, Dr. C. R. Crumly and the CRC Press editorial and production staff for their efforts and dedication to the Gene and Cell Therapy series.

Anthony Atala, MD
Wake Forest Institute for Regenerative Medicine
and
Nancy Templeton, PhD

Series Editors
Gene and Cell Therapy

Preface

I'm sure that we all can remember one of our first drawings as a child, perhaps that of a tree?

The tree has been one of the most important mythological and religious symbols since ancient times, often regarded as sacred in the ancient world. The image of the tree is also a favorite of many cultures. There are numerous paintings of trees in different religions, such as Judaism, Christianity, Islam, and Hinduism, which indicate prosperity, new beginnings, redemption, and perpetual regeneration.

It is interesting to note that the placenta is tree shaped.

The tree-shaped placenta could be an indication of our origin, of the deep roots which are severed when the umbilical cord is cut, but at which time a new dimension of growth and development begins. As old findings suggest, and seconded by new findings, the placenta's usefulness and importance begin during gestation, when it nourishes the growing fetus and participates in the complex phenomenon of fetomaternal tolerance. This importance continues even after birth when it takes on a whole new dimension and presents itself for applications in regenerative medicine.

The placenta, which is usually thrown away after birth, is likened to a tree; it is constantly giving and receiving. Just like a tree through its branches and trunk, the placenta is a tree of life which collects nourishment from the mother's blood through its branches and passes it to the blood of the fetus. The waste from the fetus is discarded through the umbilical cord to the mother's blood, which in turn absorbs and removes the same. If the placenta is affected, miscarriage, stillbirth, and other life-threatening conditions can occur for both the mother and the fetus. Given its vital role, shockingly little is known about the placenta. The National Institute of Child Health and Human Development describes the placenta as "the least understood human organ and arguably one of the more important, not only for the health of a woman and her fetus during pregnancy, but also for the lifelong health of both."[1]

Therefore, it is with great enthusiasm that I present to you *Placenta: The Tree of Life*, a combined effort of many of the world's renowned scientists, showcasing this intriguing and mysterious organ, especially its potential in novel therapeutic strategies. This book is written by and dedicated to those who are fascinated by the placenta. It will discuss the structure, functions, and pathologies of the human placenta during fetal development, including its functions from the time of its origin in the lining of the uterus. The book also reviews how the placenta sustains life even after birth, including the quest to understand the mechanisms of action underlying the therapeutic benefit of cells isolated from different placental regions, thus truly representing the tree of life. The book encompasses even the most ancient uses of placental tissues, such as the use of fetal membranes as biomaterial in medicine over a century ago, up to the current good manufacturing practices implemented to obtain cells from human placenta envisaged to treat patients with a variety of diseases.

My goal is to take you on an exciting journey which will spark your interest in this organ, which was once discarded after birth. A glance at the Table of Contents, and perhaps reading several chapters, will explain the title of this book,

Placenta: The Tree of Life. The book discusses the marvel of how something that is even today considered biological waste can instead contribute to our nourishment throughout the years. The multiplicity of new scientific societies interested in the therapeutic properties of the placenta clearly underlines the rising interest in the placenta, and I hope that this book will bring the placenta into the center stage it so well deserves.

REFERENCE

1. Guttmacher, A.E., Maddox, Y.T., and Spong, C.Y. (2014). The Human Placenta Project: placental structure, development, and function in real time. *Placenta* 35 (5): 303–4.

Ornella Parolini

Editor

Ornella Parolini obtained her undergraduate degree in biological science from the University of Milan, Milan, Italy, in 1988, and her doctorate degree in cellular and molecular biotechnology in biomedicine from the University of Brescia, Brescia, Italy. Dr. Parolini worked from 1991 to 1994 in Memphis, Tennessee, including St. Jude's Children's Research Hospital, and from 1995 to 2002 in the University of Vienna, Wien, Austria. During these years, she significantly contributed to the field of primary immunodeficiencies.

After a decade of research in the United States and Austria, Dr. Parolini returned to Brescia, Italy, where she established well-funded research programs in regenerative medicine and was appointed director of the E. Menni Research Center (CREM), Fondazione Poliambulanza, Brescia, Italy. Her strong research accomplishments in immunology brought her to study the human placenta. She considered fetal-maternal tolerance as one of the most fascinating aspects of immunology and envisioned that placental tissues could constitute interesting sources of stem cells ideal not only for their stem cell potential but also for their intrinsic and unique immune characteristics owing to the immunologically challenged environment from which they derive. Since 2002, Dr. Parolini has pioneered research on human placenta–derived stem cells, in particular their unique immunomodulatory properties, proposed to be the basis of the tissue repair mechanisms promoting tissue regeneration. CREM is currently recognized internationally for its research and contributions in this field.

Dr. Parolini is the author of more than 100 publications in peer-reviewed scientific journals and several book chapters, and has patents in the placenta stem cell field. She is member of numerous scientific societies. International Placenta Stem Cell Society (IPLASS) was founded by Dr. Parolini and colleagues in 2009, and she became the first elected president of IPLASS in the same year. In 2014, Dr. Parolini was re-elected as the president for a second term, and thus as of October 2015 she is serving her second term as the president of the society.

Antonietta R. Silini obtained her bachelor of science in microbiology from the University of Maryland, College Park, Maryland, and PhD degree in life sciences from the Mario Negri Institute for Pharmacological Research, Milan, Italy, in collaboration with the Open University of London, UK.

Her past research interests include cancer biology, specifically on understanding the crosstalk between tumor and microenvironment in order to identify targets for therapy.

Her current research interests include investigating placental stem cells and their potential therapeutic effects on different diseases, including cancer. Dr. Silini works at the E. Menni Research Center (CREM), Fondazione Poliambulanza, Brescia, Italy, which is internationally recognized for its research and contributions in this field.

Contributors

Mohamed H. Abumaree
College of Science and Health Professions
King Saud Bin Abdulaziz University for
  Health Sciences
King Abdullah International Medical
  Research Center
National Guard Health Affairs
Riyadh, Saudi Arabia

Francesco Alviano
Department of Experimental, Diagnostic
  and Specialty Medicine
Unit of Histology, Embryology and
  Applied Biology
University of Bologna
Bologna, Italy

Rita Anzalone
Dipartimento di Biomedicina
  Sperimentale e Neuroscienze Cliniche
Università degli Studi di Palermo
Palermo, Italy

Barbara Barboni
Basic and Applied Bioscience Unit
Faculty of Veterinary Medicine
University of Teramo
Teramo, Italy

Paolo Berardinelli
Basic and Applied Bioscience Unit
Faculty of Veterinary Medicine
University of Teramo
Teramo, Italy

Laura Bonsi
Department of Experimental, Diagnostic
  and Specialty Medicine
Unit of Histology, Embryology and
  Applied Biology
University of Bologna
Bologna, Italy

Cesario V. Borlongan
Department of Neurosurgery and
  Brain Repair
University of South Florida College
  of Medicine
Tampa, Florida

Anna Cargnoni
Centro di Ricerca "E. Menni"
Fondazione Poliambulanza-Istituto
  Ospedaliero
Brescia, Italy

Gregorio Castellanos
Surgery Service
Virgen de la Arrixaca University
  Clinical Hospital
Murcia, Spain

Lawrence W. Chamley
Department of Obstetrics and
  Gynaecology
School of Medicine
Faculty of Medical and Health
  Sciences
University of Auckland
Auckland, New Zealand

Roberta Costa
Lab of Pathology and Ultrastructural
  Diagnostic
Department of Biomedical and
  Neuromotor Sciences
University of Bologna
Bologna, Italy

Mario Delgado
Cell Biology and Immunology
Instituto de Parasitologia y Biomedicina
  "Lopez-Neyra"
Granada, Spain

Maya Elias
Department of Neurosurgery and
Brain Repair
University of South Florida College
of Medicine
Tampa, Florida

Christian Gabriel
Red Cross Transfusion Service of
Upper Austria
Linz, Austria

and

Ludwig Boltzmann Institute for
Experimental and Clinical
Traumatology
Austrian Cluster for Tissue Regeneration
Vienna, Austria

Roberto Gramignoli
Division of Pathology
Department of Laboratory Medicine
Karolinska Institutet
Stockholm, Sweden

Alex Hodge
Department of Medicine
School of Clinical Sciences
Monash University
Clayton, Victoria, Australia

Jaclyn Hoover
Department of Neurosurgery and
Brain Repair
University of South Florida College
of Medicine
Tampa, Florida

Carmen García Insausti
Cell Therapy Unit
Virgen de la Arrixaca University
Clinical Hospital
Murcia, Spain

Joanna L. James
Department of Obstetrics and
Gynaecology
School of Medicine
Faculty of Medical and Health
Sciences
University of Auckland
Auckland, New Zealand

Bill Kalionis
Department of Perinatal Medicine
Pregnancy Research Centre and
University of Melbourne
Department of Obstetrics and
Gynaecology
Royal Women's Hospital
Melbourne, Victoria, Australia

Gina Diamanta Kusuma
Department of Perinatal Medicine
Pregnancy Research Centre and
University of Melbourne
Department of Obstetrics and
Gynaecology
Royal Women's Hospital
Melbourne, Victoria, Australia

Christopher Lawton
Department of Neurosurgery and
Brain Repair
University of South Florida College
of Medicine
Tampa, Florida

Rebecca Lim
The Ritchie Centre
Department of Obstetrics and
Gynaecology
School of Clinical Sciences
Monash University
Clayton, Victoria, Australia

Contributors

Diego Lozano
Department of Neurosurgery and
Brain Repair
University of South Florida College
of Medicine
Tampa, Florida

Marta Magatti
Centro di Ricerca "E. Menni"
Fondazione Poliambulanza-Istituto
Ospedaliero
Brescia, Italy

Fabio Marongiu
Department of Biomedical Sciences
Experimental Medicine Unit
University of Cagliari
Cagliari, Italy

Mauro Mattioli
Basic and Applied Bioscience Unit
Faculty of Veterinary Medicine
University of Teramo
Teramo, Italy

José María Moraleda
Cell Therapy Unit
Virgen de la Arrixaca University
Clinical Hospital
Murcia, Spain

Padma Murthi
Department of Medicine
School of Clinical Sciences
Monash University
and
Department of Perinatal Medicine
Pregnancy Research Centre and
University of Melbourne
Department of Obstetrics and
Gynaecology
Royal Women's Hospital
Melbourne, Victoria, Australia

Aurelio Muttini
Basic and Applied Bioscience Unit
Faculty of Veterinary Medicine
University of Teramo
Teramo, Italy

Hung Nguyen
Department of Neurosurgery and
Brain Repair
University of South Florida College
of Medicine
Tampa, Florida

Francisco José Nicolás
Molecular Oncology and TGFß
Laboratory
Virgen de la Arrixaca University
Clinical Hospital
Murcia, Spain

Ornella Parolini
Centro di Ricerca "E. Menni"
Fondazione Poliambulanza-Istituto
Ospedaliero
Brescia, Italy

Jelmer R. Prins
Department of Obstetrics
University of Groningen
University Medical Center Groningen
Groningen, The Netherlands

Stephanny Reyes
Department of Neurosurgery and
Brain Repair
University of South Florida College
of Medicine
Tampa, Florida

Giampiero La Rocca
Dipartimento di Biomedicina
  Sperimentale e Neuroscienze
  Cliniche
Università degli Studi di Palermo
and
Dipartimento di Medicina e Terapie
  d'Avanguardia
Strategie Biomolecolari e Neuroscienze
Sezione Cellule Staminali e
  Rimodellamento Tissutale
Istituto Euro-Mediterraneo di Scienza e
  Tecnologia
Palermo, Italy

Eleonora Russo
Dipartimento di Medicina e Terapie
  d'Avanguardia
Strategie Biomolecolari e Neuroscienze
Sezione Cellule Staminali e
  Rimodellamento Tissutale
Istituto Euro-Mediterraneo di Scienza e
  Tecnologia
Palermo, Italy

Valentina Russo
Basic and Applied Bioscience Unit
Faculty of Veterinary Medicine
University of Teramo
Teramo, Italy

Salvatore Saieva
Dipartimento di Medicina e Terapie
  d'Avanguardia
Strategie Biomolecolari e Neuroscienze
Sezione Cellule Staminali e
  Rimodellamento Tissutale
Istituto Euro-Mediterraneo di Scienza e
  Tecnologia
Palermo, Italy

Sicco A. Scherjon
Department of Obstetrics
University of Groningen
University Medical Center Groningen
Groningen, The Netherlands

William Sievert
Department of Medicine
School of Clinical Sciences
Monash University
Clayton, Victoria, Australia

Antonietta R. Silini
Centro di Ricerca "E. Menni"
Fondazione Poliambulanza-Istituto
  Ospedaliero
Brescia, Italy

Stephen C. Strom
Division of Pathology
Department of Laboratory Medicine
Karolinska Institutet
Stockholm, Sweden

Maria Elena Trapani
Dipartimento di Medicina e Terapie
  d'Avanguardia
Strategie Biomolecolari e Neuroscienze
Sezione Cellule Staminali e
  Rimodellamento Tissutale
Istituto Euro-Mediterraneo di Scienza e
  Tecnologia
Palermo, Italy

Euan M. Wallace
The Ritchie Centre
Department of Obstetrics and
  Gynaecology
School of Clinical Sciences
Monash University
Clayton, Victoria, Australia

1 Structure and Development of the Human Placenta

Joanna L. James and Lawrence W. Chamley

CONTENTS

Preface 1
1.1 Overview of the Development of the Human Placenta 2
1.2 Preimplantation Development and Lineage Derivation 2
1.3 Decidua 3
1.4 The Early Postimplantation Placenta and Origins of the Fetal
    Membranes, Connecting Stalk, and Wharton's Jelly 4
1.5 Development of the Villous Placenta 4
1.6 Origins of the Fetal Membranes 6
    1.6.1 Amniotic Membrane 6
    1.6.2 Chorionic Membrane 7
    1.6.3 Membrane Fusion 9
1.7 Umbilical Cord and Wharton's Jelly 9
1.8 Changes in the Anatomy of the Placenta during Gestation 9
1.9 Summary 10
References 11

PREFACE

The correct function of the human placenta is key for adequate nutrient and gas exchange between mother and baby, and thus for the overall success of pregnancy. The ability of the placenta to achieve appropriate transfer at the end of pregnancy depends on adequate placental development in early pregnancy. Despite the importance of early placental development, our understanding of the early gestation human placenta is limited by our inability to access early human implantation sites, and by a lack of suitable animal models with which to study the relatively unique process of human implantation. This chapter describes our current knowledge of how the placental cell lineages first arise from the blastocyst at the time of implantation, how the major structures of the placenta—the villi, the extraplacental membranes, and the umbilical cord—are formed during early pregnancy, and how these components function to ensure successful pregnancy.

1.1 OVERVIEW OF THE DEVELOPMENT OF THE HUMAN PLACENTA

The human placenta is a unique organ that bears limited anatomical resemblance to the placentae of common laboratory animals. Thus, it is often not possible to directly translate the functions of cells from animal placentae to equivalent human placental cells with any great confidence. The human placenta has a villous or branching structure and is hemochorial, meaning that the fetal trophoblast of the placenta is in direct contact with the maternal blood during most of pregnancy. The maternal face of the human placenta is lined by a single multinucleated cell layer called the syncytiotrophoblast; the syncytiotrophoblast is in direct contact with the maternal blood and thus forms the major exchange surface between mother and baby. Underlying the syncytiotrophoblast (on its fetal aspect) is a layer of actively proliferating mononuclear cytotrophoblasts that act as precursors and fuse into the overlying syncytiotrophoblast throughout pregnancy, allowing the continued expansion and regeneration of the mitotically inactive syncytiotrophoblast. In early pregnancy, cytotrophoblasts are also able to breach the syncytiotrophoblast and migrate out of the anchoring villi, where they differentiate to form columns of invasive extravillous trophoblasts that penetrate the maternal tissue of the uterine decidua. As they move away from the villi, the extravillous cytotrophoblasts lose their ability to proliferate and differentiate to an invasive phenotype. When this process first begins shortly after implantation, the extravillous trophoblasts move laterally around the implantation site to create a "cytotrophoblast shell" that encompasses the embryo. From this shell, extravillous trophoblasts continue to invade into the decidual stroma where they surround the maternal spiral arteries. A subset of extravillous trophoblasts, called endovascular trophoblasts, breech these vessels and migrate antidromically along the spiral arteries where they play key roles in replacing the endothelial cells that line the spiral arteries (Boyd and Hamilton 1970; Cartwright et al. 2010). Endovascular trophoblasts also act together with the trophoblasts that surround the spiral arteries to remove or dedifferentiate the arterial smooth muscle layer, rendering these vessels tonically inactive (France et al. 1986). These remodeling processes change the nature of blood flow within the spiral arteries to provide a constant and increased supply of blood to the placental surface as gestation progresses. In a normal pregnancy, this process, referred to as the "physiological changes of pregnancy," continues until midgestation when the vessels are modified as far as a third of the length of their myometrial segments (Brosens et al. 1967). In pregnancies complicated by intrauterine growth restriction, preeclampsia, or both, these physiological changes are impaired (Robertson et al. 1967; Khong et al. 1986).

1.2 PREIMPLANTATION DEVELOPMENT AND LINEAGE DERIVATION

Following conception, the first few symmetric cell divisions of each blastomere of the zygote are identical, and all cells within it are thought to be totipotent. As the embryo begins to compact at the 16–32-cell (morula) stage, the blastomeres begin to show the first signs of polarity and differentiation (Nikas et al. 1996). The first clearly identifiable cell lineage differentiation event in human development occurs around 4 days

Structure and Development of the Human Placenta

postfertilization when the developing embryo progresses to the 64-cell stage and forms a blastocyst with a distinct inner cell mass that will develop into the embryo proper, and an outer single cell layer called the trophectoderm that will go on to form the trophoblast lineages of the placenta. The division of the blastocyst into the inner cell mass and trophectoderm is characterized by the upregulation of Cdx2, Gata2, and human chorionic gonadotropins (hCGs)-α and -β in the trophectoderm, and by a downregulation or loss of expression of Oct4 and Nanog, which in murine blastocysts, are then only observed in the inner cell mass (De Paepe et al. 2014). Trophectoderm lineage specification was thought to be irreversible, as demonstrated by the inability of inner cell mass–derived embryonic stem cell lines to colonize trophectoderm in murine blastocyst chimera experiments (Niwa et al. 2000; Rossant 2001). However, this absolute lineage commitment has been queried in other animal models, such as the cow, where Cdx2 does not completely repress the expression of Oct4, and Cdx2 expression is not an absolute requirement for trophectoderm differentiation (Berg et al. 2011; Goissis and Cibelli 2014). Similarly, it has been shown that trophectoderm from human blastocysts can be reaggregated into blastocyst-like structures that contain both trophectoderm and inner cell mass–like cells that express Nanog, indicating that human trophectoderm differentiation may also be reversible at this early stage (De Paepe et al. 2013).

By the time the blastocyst attaches to the endometrial epithelium around 7 days postfertilization, two morphologically and functionally distinct regions of the trophectoderm can be distinguished: (1) the polar trophectoderm that lies proximal to the inner cell mass and (2) the mural trophectoderm that stretches more thinly around the remainder of the blastocyst cavity. The inner cell mass of the preimplantation blastocyst is a pluripotent mass of cells, but it is likely that crosstalk between the inner cell mass and the underlying polar trophectoderm plays key roles in the further development of embryonic and placental cell lineages from both of these early populations. Therefore, it is important to be aware of key differences that exist between human and murine blastocysts and the modes of implantation in these species. In particular, in human blastocysts it is always the polar trophectoderm that adheres to the endometrial epithelium, whereas in the mouse, the mural trophectoderm at the abembryonic pole is adherent. The implications that this opposite orientation may have for differences in the regulation of lineage development between species remain unclear, but it is likely that inner cell mass–derived signaling molecules have quite different effects on trophectoderm-derived cells between mice and humans.

1.3 DECIDUA

In humans, the uterine endometrium is prepared every month to be receptive to an implanting blastocyst. This monthly process is called decidualization, and it occurs in each menstrual cycle, commencing around the time of ovulation. Decidualization in humans is driven primarily by a rise in progesterone levels from the corpus luteum following ovulation. Decidualization is characterized by the rapid proliferation of the epithelial and stromal cells of the endometrium; differentiation of the glandular epithelium into a highly secretory state; and finally, stromal cell differentiation in which the usually fibroblast-like stromal cells become plump and glycogen rich and take on a characteristic polygonal morphology (Salamonsen et al. 2009).

Decidualization also involves the tightly regulated expression of specific adhesion molecules on endometrial epithelial cells that will facilitate the adhesion and attachment phases of implantation. The concurrent timing of this adhesion molecule expression and stromal cell differentiation ensures that the blastocyst is only able to implant into the decidua for a 2- to 4-day period known as the "implantation window." Although ectopic pregnancies demonstrate that implantation, and even successful pregnancy, is possible in the absence of the decidua (Jackson et al. 1980; Martin et al. 1988), the success of decidualization is widely held to play an important role in implantation and the success of normal pregnancies.

1.4 THE EARLY POSTIMPLANTATION PLACENTA AND ORIGINS OF THE FETAL MEMBRANES, CONNECTING STALK, AND WHARTON'S JELLY

The very early stages of human placental development remain somewhat of an enigma, with as few as 15 normal human implantation sites from the previllous period in existence (Boyd and Hamilton 1970; James et al. 2012). Thus, our understanding of the events from implantation to around the fifth week of gestation (3 weeks postfertilization) are largely based on "snapshots" during this time, making it challenging to tease out a true functional understanding of the processes involved (James et al. 2012).

After adhesion, the blastocyst implants within the endometrium by actively invading into the decidua and remodeling the decidual extracellular matrix to allow the rapid expansion of the trophectoderm-derived components of the placenta (Boyd and Hamilton 1970). By day 9, the embryo has burrowed completely into the decidua, and the endometrial epithelium then heals over the implanted embryo. In contrast to other species, such as the ungulates that develop within the uterine cavity, human placental and embryonic development occurs entirely within the decidua.

1.5 DEVELOPMENT OF THE VILLOUS PLACENTA

As the blastocyst implants, the trophectoderm differentiates to form the first placental cell populations. The polar trophectoderm exhibits the greatest capacity for proliferation and invasion in human blastocysts, and by the eighth day postfertilization two primitive trophoblast populations are evident at the invading edge of the blastocyst: a mononuclear cytotrophoblast-like population and a multinucleated "primitive syncytium" (James et al. 2012). The primitive syncytium is thought to play an important role in expansion of the developing embryo by secreting several serine proteases (urokinase-type and tissue-type plasminogen activators), metalloproteinases (MMPs) (primarily MMP-2 and -9), and collagenases. These enzymes digest the deciduas, creating spaces what are referred to as lacunae. Processes of the primitive syncytium (called trabeculae) advance into the lacunae, thus expanding them and occasionally breaching maternal sinusoids (Hertig et al. 1956; Enders 1989). The primitive cytotrophoblasts rapidly proliferate, resulting in a convoluted layer of cells that begins to migrate into invaginations on the fetal aspect of the trabeculae (Boyd and Hamilton 1970). The formation of this bilayer of trophoblasts (around 12 days postfertilization) marks the beginning of the villous stage of placental development, and these structures are referred to as primary villi (Hertig 1968; Boyd and Hamilton 1970). The spaces between the villi (lacunae) are now referred to as the intervillous spaces.

The inner cell mass also undergoes many key differentiation events during this time. Upon first contact with the endometrial epithelium (7 days postfertilization), a layer of hypoblast (primary endoderm) becomes evident along the surface of the inner cell mass adjacent to the blastocyst cavity. By approximately 8 days postfertilization, a primitive amniotic cavity begins to develop within the inner cell mass, lined on one side by amnioblasts that will go on to form the amniotic membrane, and lined on the other side by the epiblast, a thick layer of columnar cells that together with the proliferating hypoblast form the bilaminar embryonic disk (Figure 1.1).

By 9 days postfertilization, the larger cavity in the embryo, now termed the primary umbilical vesicle or yolk sac, has been surrounded by an inner layer of thin extraembryonic endoderm, which forms the exocoelomic membrane, and a loose outer layer of extraembryonic mesodermal cells that are seen beneath the primitive cytotrophoblast layer by around 10 days postfertilization. The exact origin of this extraembryonic mesoderm in human embryos is unclear. Historically, it was thought to be derived from the trophoblast (Boyd and Hamilton 1970), but more recent literature proposes that it is derived from the hypoblast, the primitive streak, or both (Bianchi et al. 1993; Moore et al. 2013; Carlson 2014). As the placental villi continue to develop, cells from this extraembryonic mesoderm appear to invade into

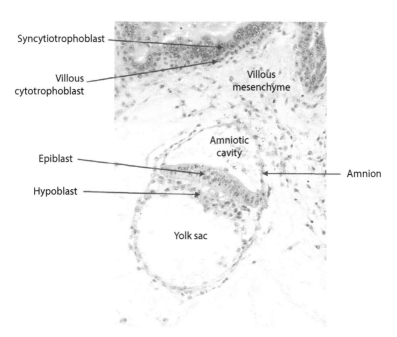

FIGURE 1.1 (See color insert.) Hematoxylin and eosin–stained section through an embryo (Carnegie Collection #7801) presumed to be 13.5 days (Boyd and Hamilton 1970), showing the epiblast and hypoblast (which together comprise the bilaminar embryonic disc) relative to the amnion, the developing placenta (syncytiotrophoblast/cytotrophoblast), and the yolk sac. (Used with the permission of Virtual Human Embryo project [http://virtualhumanembryo.lsuhsc.edu].)

invaginations between the proliferating trophoblast, creating villi with the first semblance of a villous mesenchymal core from 14 days postfertilization (Boyd and Hamilton 1970) (Figure 1.1). The resultant placental villi that are invaded by extraembryonic mesoderm are referred to as secondary villi.

The early villi continue to expand, and at 15 days postfertilization, the first fetal blood vessels are observed as endothelial streaks within the mesenchymal villus core. These vessels are thought to originate from differentiation of the mesenchymal stromal cells within the villus core (Benirschke and Kaufmann 2000; Meraviglia et al. 2012). Indeed, the initial vasculogenesis events in the embryo also arise from the mesodermal lineages, but it is important to note that placental vasculogenesis precedes vasculogenesis in the embryo proper, which is first observed in the extraembryonic yolk sac approximately 3 days later at around 18 days postfertilization (Marcelo et al. 2013). With the formation of vascular structures in the villi, the basic villous structure of the definitive placenta and the establishment of the major cell populations within it are complete by the fourth week of gestation (12–15 days postfertilization). Villi containing a vascularized mesenchymal core are referred to as tertiary villi; these tertiary villi make up the majority of the placenta for the remainder of gestation. At this stage, each villus is surrounded by two prominent layers of trophoblast, the outermost multinucleated syncytiotrophoblast with a more-or-less continuous layer of mononuclear cytotrophoblasts between the syncytiotrophoblast and the mesenchymal core (Figure 1.2). Extravillous trophoblasts have also become evident, protruding in columns from the tips of the newly formed villi and expanding to connect with other extravillous trophoblast outgrowths to form the cytotrophoblast shell that now surrounds the conceptus (Figure 1.2).

1.6 ORIGINS OF THE FETAL MEMBRANES

There are two membranes associated with the human placenta: the amniotic membrane and the chorionic membrane. The amniotic membrane (or amnion) surrounds the embryo/fetus and encloses the fluid-filled amniotic cavity. Lying on the maternal aspect of the amnion is the chorionic membrane (or chorion). Often a third membrane, the decidua, is described. However, the decidua is not a membrane; rather, this tissue is simply the decidualized maternal endometrium that is in close contact with and adherent to the chorion. Indeed, it is because this tissue is shed with the placenta (and during menstruation) that the name decidua (from the Latin *decidere*, meaning "to fall off") is derived.

1.6.1 AMNIOTIC MEMBRANE

The amniotic cavity first appears between the layer of amnioblast (derived from the extraembryonic ectoderm) and the epiblast at around 8 days postfertilization. As the amniotic cavity continues to expand, the amnioblasts form the thin, but tough, amniotic membrane. In the third week postfertilization, folding of the embryonic disc draws this amniotic membrane ventrally over the embryo and fully encloses the embryo within the amniotic sac. Thus, the amniotic membrane forms a bag or balloon that joins to the fetal abdominal skin, via a short transition zone, at

Structure and Development of the Human Placenta

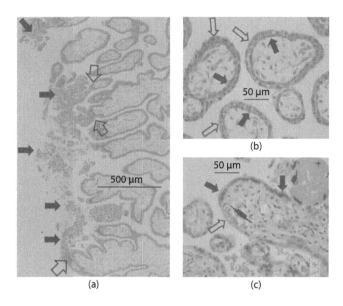

FIGURE 1.2 (See color insert.) (a) Hematoxylin and eosin (H&E)–stained section from a placenta of approximately 5 weeks of gestation showing a region of extravillous trophoblasts from the cytotrophoblast shell (dark arrows). These extravillous trophoblasts emanate from the tips of anchoring villi (hollow arrows). (b) H&E-stained section from a placenta of approximately 6 weeks of gestation. The syncytiotrophoblast (hollow arrows) is relatively homogenous in terms of thickness and distribution of nuclei. On the fetal aspect of the syncytiotrophoblast is a more-or-less continuous layer of villous cytotrophoblasts (solid arrows) beneath which is the sparsely cellular mesenchymal core. (c) H&E-stained section from a placenta at term showing syncytiotrophoblast of varying thickness with clusters of nuclei (syncytial knots, solid arrows) and an adjacent vasculosyncytial membrane lacking nuclei (hollow arrow), allowing a fetal vessel (wide arrow) to come into proximity to the maternal blood in the intervillous space. The mesenchymal core of the villi is more densely cellular, and fetal vessels are more prominent than in early gestation villi (contrast with b).

the umbilicus and covers the surface of the umbilical cord and then runs across and is adherent to the fetal aspect of the placenta (the chorionic plate) before reflecting away from the placenta to encompass the fetus (Figure 1.3). Once fully formed, the amnion is avascular and consists of three layers: (1) the amniotic epithelium (facing the fetus/amniotic cavity), (2) a compact layer, and (3) amniotic mesoderm.

1.6.2 Chorionic Membrane

When the blastocyst implants, the inner cell mass is completely enclosed in a sphere of trophectoderm that will form the placenta, and the development of trabeculae and villi described above is not confined to a single region of the trophectoderm, but rather occurs around the entire sphere. Thus, unlike the disc of placental tissue we are familiar with at term (Figure 1.3), during early gestation the villous placenta forms as a sphere that has villi around its entirety until 6 or 7 weeks of gestation (4–5 weeks postfertilization, Figure 1.4). However from that point, only the villi

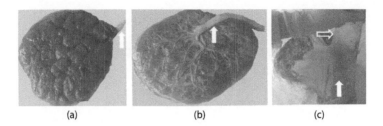

FIGURE 1.3 (See color insert.) A normal term placenta viewed from (a) the maternal aspect, showing the villous surface (the umbilical cord, behind, is indicated by the arrow), and (b) the fetal aspect, with the membranes inverted to expose the pale-colored amnion that covers the entire fetal aspect of the placenta and the umbilical cord (arrow). (c) Although they fuse during gestation, even at term, the amnion (solid arrow) and smooth chorion with adherent decidua (hollow arrow) membranes can be readily separated.

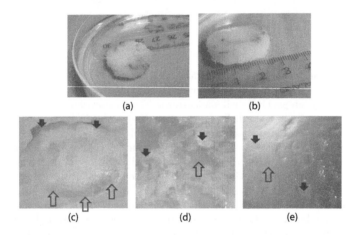

FIGURE 1.4 (See color insert.) The villous placenta initially forms as a sphere encompassing the embryo, but by 6–7 weeks of gestation the villi to the sides and rear of the implantation site begin regressing to form the smooth chorion. (a) A placenta of approximately 6 weeks of gestation, showing villi surrounding the embryo and (b) the same placenta as in (a) showing the surface that is obscured in (a). (c) A placenta of approximately 8 weeks of gestation showing a region where the villi are obviously regressing to form the smooth chorion at the rear and sides of the implantation site (hollow arrows), whereas the disc-like structure that will remain as the definitive placenta (chorion frondosum) can be seen at the top of the image, indicated by the solid arrows. (d and e) Images from a placenta of approximately 7 weeks of gestation showing the regressing villi (solid arrows) and the smooth chorion (hollow arrows).

at the leading edge of the invading embryo continue to grow and form the discoid placenta seen throughout the remainder of gestation (the chorion frondosum). Those villi to the sides and rear of the implantation site rapidly regress, leaving behind a smooth membrane, the chorion laeve or smooth chorion, colloquially referred to as the chorion (Figure 1.4). What causes the regression of the villi of the chorion laeve is unclear, but it has been suggested to be due to exposure to higher levels of oxygen (Jauniaux et al. 2003).

Structure and Development of the Human Placenta

1.6.3 Membrane Fusion

The chorion and the amnion develop separately and are not initially fused, but by the end of the first trimester, expansion of the amniotic cavity has decreased the size of the extraembryonic coelom to the point where the amniotic and chorionic membranes begin to adhere to each other. These membranes remain in this state throughout the remainder of pregnancy, but even at term they can relatively easily be separated (Figure 1.3).

1.7 UMBILICAL CORD AND WHARTON'S JELLY

The umbilical cord connecting the fetus and placenta is formed in the first 8 weeks of gestation. This process begins around day 18 postfertilization, when condensing mesodermal cells begin to form the connecting stalk that joins the embryo to the developing placenta. The umbilical cord begins to develop from this stalk around 22 days postfertilization, when cells from the yolk sac and allantois combine with the connecting stalk and are enveloped by the spreading amnion. Thus, the surface of the umbilical cord comprises amniotic epithelium that is contiguous with the amniotic membrane and fetus. The three vessels (two arteries and one vein) of human umbilical cord are derived from the allantois and yolk sac, although absence of one artery is suggested to be nonpathological and occurs in approximately 1/100 cords (Boyd and Hamilton 1970). Between the vessels is a substance referred to as Wharton's jelly. This jelly consists of a network of myofibroblasts with elastic fibers between which is a jelly-like mixture of mucopolysaccharides (Parry 1970). Contained in Wharton's jelly is a population of mesenchymal stem/stromal cells with the ability to differentiate not only into the three lineages expected of mesenchymal stromal cells but also into myocytes and endothelial, neuronal, and Islet-like cells, among others (reviewed in Moretti et al. 2010).

1.8 CHANGES IN THE ANATOMY OF THE PLACENTA DURING GESTATION

When tertiary villi first form in the newly developing placenta, they are characterized by a continuous covering of syncytiotrophoblast overlying a continuous layer of mononuclear cytotrophoblasts (Figure 1.2). As gestation progresses, changes in the appearance of both of these trophoblast populations occur. First, the number of cytotrophoblasts does not keep pace with the growth of the placenta such that the cytotrophoblast layer appears to diminish, until by term the histological detection of cytotrophoblasts within the placental villi is difficult without the aid of a specific stain (Figure 1.2). Nevertheless, even at term hundreds of millions of cytotrophoblasts are present in the placenta. The discontinuous cytotrophoblast layer allows the placental blood vessels to lie more proximal to the syncytiotrophoblast (Figure 1.2). Changes to the previously uniform dispersion of nuclei within the syncytiotrophoblast and uniform thickness of this cell layer are also observed. With increasing gestational age, distinct clusters of nuclei in the syncytiotrophoblast, called syncytial knots, can be clearly seen (Figure 1.2). Adjacent to these syncytial knots are regions that are devoid of nuclei where the syncytiotrophoblast is thinner. These regions, called vasculosyncytial membranes, are associated with placental blood vessels that may often cause an outward bulging of the villous surface (Figure 1.2). Vasculosyncytial membranes shorten the exchange distance between the maternal and fetal circulations. The syncytiotrophoblast also shows distinct regional expression of many proteins. This patchiness of the syncytiotrophoblast probably reflects either functional differences between regions of this cell or regions of the syncytiotrophoblast of different ages (Abumaree et al. 2006, 2012). The stromal core of late gestation villi is much more densely packed with mesenchymal cells, and fetal vessels are more obvious in the villi than in early gestation.

A key change that occurs across gestation is the level of oxygen to which the villus surface is exposed. Prior to implantation, the trophectoderm of the blastocyst develops in a near anaerobic environment, and during the first weeks postimplantation, the placenta is in a low-oxygen environment (Rodesch et al. 1992). Despite the breach of occasional sinusoids in the decidua by the invading placenta, and extensive remodeling of the spiral arteries by extravillous trophoblasts, maternal blood does not flow to the villous surface until after 10–12 weeks of gestation (Hustin and Schaaps 1987; Foidart et al. 1992; Jauniaux et al. 2001). This lack of flow is because extravillous trophoblasts form plugs in the lumen of spiral arteries that they are remodeling (Boyd and Hamilton 1970). The plugs are loosely cohesive and prevent the flow of maternal red blood cells to the intervillous space such that oxygen levels at the villous surface are physiologically low (in the order of 1–2% compared to approximately 6% in maternal arterial blood) during the first 10 weeks of pregnancy (Rodesch et al. 1992). The trophoblast plugs disperse from approximately 10 weeks of gestation, allowing oxygenated maternal blood to access the intervillous spaces, and the placenta upregulates its antioxidant defenses at this time (Watson et al. 1998). Early exposure of the villous surface to maternal arterial oxygen is suggested to cause miscarriage due to the lack of antioxidant defenses in the early gestation placenta (Jauniaux et al. 2003). Although the trophoblast plugs prevent maternal red cells from accessing the placenta in early pregnancy, they do not completely occlude the spiral arteries since it is clear that extracellular vesicles, including very large multinucleated syncytial nuclear aggregates, derived from the syncytiotrophoblast are deported away from the uterus in the maternal blood from as early as 6 weeks of gestation (Covone et al. 1984).

1.9 SUMMARY

The human placenta is a fascinating organ, the correct function of which is crucial to the success of all pregnancies and to the future health of the offspring. Despite its importance, the human placenta is a poorly studied organ, in part because accessing it during most of gestation is impossible. The placenta is also unique in that once delivered, it is usually discarded (although this is not universally true, e.g., the Maori people of New Zealand retain their placentae for cultural reasons); so the delivered placenta is potentially a rich source of the cells and other products within it. Many placenta-derived cells have unique properties that are described in subsequent chapters.

REFERENCES

Abumaree, M. H., P. R. Stone, and L. W. Chamley. 2006. An in vitro model of human placental trophoblast deportation/shedding. *Mol Hum Reprod* 12 (11): 687–94.

Abumaree, M. H., P. R. Stone, and L. W. Chamley. 2012. Changes in the expression of apoptosis-related proteins in the life cycle of human villous trophoblast. *Reprod Sci* 19 (6): 597–606.

Benirschke, K., and P. Kaufmann. 2000. *Pathology of the Human Placenta*. 4th ed. New York: Springer-Verlag.

Berg, D. K., C. S. Smith, D. J. Pearton, D. N. Wells, R. Broadhurst, M. Donnison, and P. L. Pfeffer. 2011. Trophectoderm lineage determination in cattle. *Dev Cell* 20 (2): 244–55.

Bianchi, D. W., L. E. Wilkins-Haug, A. C. Enders, and E. D. Hay. 1993. Origin of extraembryonic mesoderm in experimental animals: Relevance to chorionic mosaicism in humans. *Am J Med Genet* 46 (5): 542–50.

Boyd, J. D., and W. J. Hamilton. 1970. *The Human Placenta*. Cambridge: Heffer.

Brosens, I., W. B. Robertson, and H. G. Dixon. 1967. The physiological response of the vessels of the placental bed to normal pregnancy. *J Pathol Bacteriol* 93 (2): 569–79.

Carlson, B. M. 2014. *Human Embryology and Developmental Biology*. Elsevier.

Cartwright, J. E., R. Fraser, K. Leslie, A. E. Wallace, and J. L. James. 2010. Remodelling at the maternal-fetal interface: Relevance to human pregnancy disorders. *Reproduction* 140 (6): 803–13.

Covone, A. E., D. Mutton, P. M. Johnson, and M. Adinolfi. 1984. Trophoblast cells in peripheral blood from pregnant women. *Lancet* 2 (8407): 841–3.

De Paepe, C., G. Cauffman, A. Verloes, J. Sterckx, P. Devroey, H. Tournaye, I. Liebaers, and H. Van de Velde. 2013. Human trophectoderm cells are not yet committed. *Hum Reprod* 28 (3): 740–9.

De Paepe, C., M. Krivega, G. Cauffman, M. Geens, and H. Van de Velde. 2014. Totipotency and lineage segregation in the human embryo. *Mol Hum Reprod* 20 (7): 599–618.

Enders, A. C. 1989. Trophoblast differentiation during the transition from trophoblastic plate to lacunar stage of implantation in the rhesus monkey and human. *Am J Anat* 186 (1): 85–98.

Foidart, J. M., J. Hustin, M. Dubois, and J. P. Schaaps. 1992. The human placenta becomes haemochorial at the 13th week of pregnancy. *Int J Dev Biol* 36 (3): 451–3.

France, J. T., W. Jackson, and J. Keelan. 1986. A radioimmunoassay for plasma dehydroepiandrosterone sulphate incorporating placental steroid sulphatase as a hydrolysing reagent. *J Steroid Biochem* 25 (3): 375–8.

Goissis, M. D., and J. B. Cibelli. 2014. Functional characterization of CDX2 during bovine preimplantation development in vitro. *Mol Reprod Dev* 81 (10): 962–70.

Hertig, A. T. 1968. *Human Trophoblast*. Springfield, IL: Charles C Thomas.

Hertig, A. T., J. Rock, and E. C. Adams. 1956. A description of 34 human ova within the first 17 days of development. *Am J Anat* 98 (3): 435–93.

Hustin, J., and J. P. Schaaps. 1987. Echographic and anatomic studies of the maternotrophoblastic border during the first trimester of pregnancy. *Am J Obst Gynecol* 157: 162–8.

Jackson, P., I. W. Barrowclough, J. T. France, and L. I. Phillips. 1980. A successful pregnancy following total hysterectomy. *Br J Obstet Gynaecol* 87 (5): 353–5.

James, J. L., A. M. Carter, and L. W. Chamley. 2012. Human placentation from nidation to 5 weeks of gestation. Part I: What do we know about formative placental development following implantation? *Placenta* 33 (5): 327–34.

Jauniaux, E., J. Hempstock, N. Greenwold, and G. J. Burton. 2003. Trophoblastic oxidative stress in relation to temporal and regional differences in maternal placental blood flow in normal and abnormal early pregnancies. *Am J Pathol* 162 (1): 115–25.

Jauniaux, E., A. L. Watson, and G. Burton. 2001. Evaluation of respiratory gases and acid-base gradients in human fetal fluids and uteroplacental tissue between 7 and 16 weeks' gestation. *Am J Obst Gynecol* 184 (5): 998–1003.

Khong, T. Y., F. de Wolf, W. B. Robertson, and I. Brosens. 1986. Inadequate maternal vascular response to placentation in pregnancies complicated by pre-eclampsia and by small for gestational age infants. *Br J Obst Gynaecol* 93: 1049–59.

Marcelo, K. L., L. C. Goldie, and K. K. Hirschi. 2013. Regulation of endothelial cell differentiation and specification. *Circ Res* 112 (9): 1272–87.

Martin, J. N., Jr., J. K. Sessums, R. W. Martin, J. A. Pryor, and J. C. Morrison. 1988. Abdominal pregnancy: Current concepts of management. *Obstet Gynecol* 71 (4): 549–57.

Meraviglia, V., M. Vecellio, A. Grasselli, M. Baccarin, A. Farsetti, M. C. Capogrossi, G. Pompilio, et al. 2012. Human chorionic villus mesenchymal stromal cells reveal strong endothelial conversion properties. *Differentiation* 83 (5): 260–70.

Moore, K. L., T. V. N. Persaud, and M. G. Torchia. 2013. *The Developing Human: Clinically Oriented Embryology.* 9th ed. Philadelphia, PA: Elsevier.

Moretti, P., T. Hatlapatka, D. Marten, A. Lavrentieva, I. Majore, R. Hass, and C. Kasper. 2010. Mesenchymal stromal cells derived from human umbilical cord tissues: Primitive cells with potential for clinical and tissue engineering applications. *Adv Biochem Eng Biotechnol* 123: 29–54.

Nikas, G., A. Ao, R. M. Winston, and A. H. Handyside. 1996. Compaction and surface polarity in the human embryo in vitro. *Biol Reprod* 55 (1): 32–7.

Niwa, H., J. Miyazaki, and A. G. Smith. 2000. Quantitative expression of Oct-3/4 defines differentiation, dedifferentiation or self-renewal of ES cells. *Nat Genet* 24 (4): 372–6.

Parry, E. W. 1970. Some electron microscope observations on the mesenchymal structures of full-term umbilical cord. *J Anat* 107 (Pt 3): 505–18.

Robertson, W. B., I. Brosens, and H. G. Dixon. 1967. The pathological response of the vessels of the placental bed to hypertensive pregnancy. *J Pathol Bacteriol* 93 (2): 581–92.

Rodesch, F., P. Simon, C. Donner, and E. Jauniaux. 1992. Oxygen measurements in endometrial and trophoblastic tissues during early pregnancy. *Obstetrics and Gynecology* 80 (2): 283–5.

Rossant, J. 2001. Stem cells in the mammalian blastocyst. *Harvey Lect* 97: 17–40.

Salamonsen, L. A., G. Nie, N. J. Hannan, and E. Dimitriadis. 2009. Society for Reproductive Biology Founders' Lecture 2009. Preparing fertile soil: The importance of endometrial receptivity. *Reprod Fertil Dev* 21 (7): 923–34.

Watson, A. L., J. N. Skepper, E. Jauniaux, and G. J. Burton. 1998. Susceptibility of human placental syncytiotrophoblastic mitochondria to oxygen-mediated damage in relation to gestational age. *J Clin Endocrinol Metabol* 83 (5): 1697–705.

2 The Role of Mesenchymal Stem Cells in the Functions and Pathologies of the Human Placenta

Gina Diamanta Kusuma, Padma Murthi, and Bill Kalionis

CONTENTS

Preface 14
Acronyms 14
2.1 Placental Structure and Function 14
2.2 Pathologies of the Placenta 17
2.3 Stem Cells 18
2.4 Stem Cell Populations in the Placenta 19
    2.4.1 Trophoblast Stem Cells 19
    2.4.2 Other Stem Cell Types 20
    2.4.3 Mesenchymal Stem Cells 20
        2.4.3.1 Definition and Background 20
        2.4.3.2 Chorionic Villous MSCs 24
        2.4.3.3 *Decidua Basalis* MSC 25
2.5 Localization of MSC Niche in the Placenta 26
    2.5.1 The Stem Cell Niche 26
    2.5.2 MSC Niche in the Chorionic Placenta 26
    2.5.3 MSC Niche in the *Decidua Basalis* 27
2.6 MSCs and Their Potential Role in Pregnancy Pathologies 28
    2.6.1 Preeclampsia 28
    2.6.2 The Pathogenesis of PE 28
    2.6.3 The Role of CMSCs in PE 29
    2.6.4 The Role of DMSCs in PE 30
2.7 MSCs as a Potential Treatment for PE 31
2.8 Possible Role of MSCs in Other Placental Pathologies 32
References 32

PREFACE

The proper function of the human placenta is essential for all stages of pregnancy and for the successful outcome of a healthy baby. Nurturing the fetus and removing fetal waste products require the placenta to carry out a wide range of functions. The increasing demands on the placenta to support the rapidly growing fetal placenta are met by establishing and maintaining an interface with the mother's uterus and underlying tissues, and by modifying vessels in the maternal circulation to increase blood flow to the placenta. Given the complex morphological and physiological changes that occur in the placenta during its formation, and the varied functions that the placenta performs, it is not surprising that disruption of these processes results in significant placental pathologies. Clinically, the most important of these pathologies is the hypertensive disorder preeclampsia (PE). Despite the placenta's complex and critical roles, there are major gaps in our understanding of the biology and molecular mechanisms involved in normal and pathological placental development. Here, we highlight recent studies providing evidence for stem cells as important new players in the formation of the placenta, the establishment of the fetal-maternal interface, and the pathology of PE.

ACRONYMS

AF	amniotic fluid
BrdU	bromodeoxyuridine
CMSC	chorionic mesenchymal stem cell
DMSC	decidual mesenchymal stem cell
ESC	embryonic stem cell
EVT	extravillous trophoblast
FGR	fetal growth restriction
HLA	human leukocyte antigen
HUVEC	human umbilical vein endothelial cell
ICM	inner cell mass
IL	interleukin
MHC	major histocompatibility complex
MIF	migration inhibitory factor
miR	microRNA
MSC	mesenchymal stem cell
PE	preeclampsia
PDGFR	platelet-derived growth factor receptor
TNF	tumor necrosis factor
VCT	villous cytotrophoblast
VEGF	vascular endothelial growth factor
VSMC	vascular smooth muscle cell

2.1 PLACENTAL STRUCTURE AND FUNCTION

The placenta is a specialized organ essential for the growth of the fetus. As described in Chapter 1, human placental development involves complex and coordinated processes that involve a variety of extraembryonic cell types. In addition, the placenta is not an independent organ since its development and proper function rely on establishing an interface with the mother's uterus and underlying tissues.

The human placenta handles a wide range of functions that change during gestation to cope with the demands of the growing fetus. The main functions of the placenta are to deliver nutrients to the developing fetus, transport gases, and eliminate waste products, in addition to being involved in endocrine secretion, evasion of immune rejection, and detoxification (Moore and Persaud 1998; Gude et al. 2004). Maternal nutrients supplied by the placenta include oxygen, amino acids, water, carbohydrates, vitamins, minerals, and lipids. These nutrients are delivered to the fetus through the placental vasculature and umbilical cord vessels. Various fetal waste products, including carbon dioxide, are removed. The placenta metabolizes many substances that are released into the maternal circulation, fetal circulation, or both. The placenta also protects the fetus by metabolizing particular xenobiotic molecules and by defending the fetus from various infections and maternal diseases. Another important function of the placenta is to provide an immunological barrier to rejection of the fetus. The endocrine function of the placenta is also very important since hormones released by the placenta into both the maternal and fetal circulations influence pregnancy by regulating metabolism, fetal growth, parturition, and many other functions.

The placenta depends on an extensive vascularized capillary network for its role in fetal nutrition, respiration, and hormone synthesis. Vasculogenesis (vessel formation *de novo*) and angiogenesis (formation of new vessels from preexisting vessels) are essential for normal placentation and effective fetal-maternal exchange (Shaman et al. 2013). By the end of gestation, the placenta develops an extensive capillary network with a large surface area to facilitate diffusion exchange between the fetal and maternal circulations.

The interface with maternal tissues involves intimate interactions between migrating or invading extravillous trophoblast cells from the chorion with the maternal stroma and vasculature of the decidua and myometrium. This fetal-maternal interface plays a crucial role in important pathologies of the placenta as described below.

The terminal chorionic villi are the functional units of the mature placenta. Placental villi are enveloped by a continuous layer of multinucleated syncytiotrophoblast, which carries out a variety of essential functions. Beneath the syncytiotrophoblast is a layer of mononuclear villous cytotrophoblast (VCT) cells that differentiate and fuse into the overlying syncytiotrophoblast, allowing the syncytiotrophoblast to replenish and expand throughout placental development (Boyd and Hamilton 1970; Loke and King 1995). Most of the VCTs are committed to differentiate into syncytiotrophoblast (Huppertz et al. 2001). During placental development, the VCTs also differentiate into specialized functional cell types, called extravillous trophoblasts (EVT). There are several distinct EVT subtypes that are classified based on differences in function, and these subtypes include endovascular trophoblast (invade and remodel maternal vessels) and terminally differentiated trophoblast giant cells that invade into the uterine myometrium. EVTs form columns of cells: proliferating cells are present in the proximal regions of the column, while in the distal portions of the columns the EVTs undergo differentiation, detach from the column, and either migrate or invade into the tissues and vasculature of the maternal *decidua basalis* (Lyall 2005), the underlying endometrium, and the inner third of the myometrium (interstitial invasion) (Pijnenborg et al. 1981a) (see Figure 2.1a). Endovascular EVT is a subtype of EVTs that migrate along the vessel walls of the maternal uterine vasculature (endovascular invasion) (Pijnenborg et al. 1981b).

These processes anchor the placenta to the uterine wall, but they also establish the definitive uteroplacental circulation and place fetal trophoblasts in direct contact

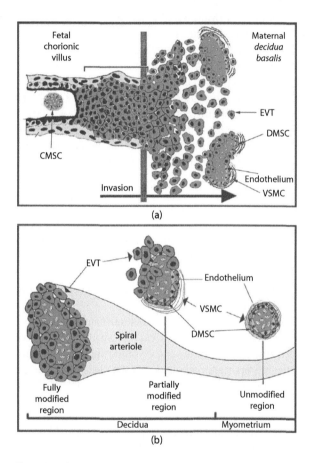

FIGURE 2.1 (**See color insert.**) Uterine spiral arteriole modifications during trophoblast invasion. (a) Diagram of a uterine spiral arteriole during trophoblast invasion. The vascular chorionic mesenchymal stem cell (CMSC) niche in the chorionic villus is shown. Extravillous trophoblasts (EVTs) are shown migrating and invading into the maternal *decidua basalis*. Two partially modified spiral arterioles are shown. The decidual mesenchymal stem cells (DMSCs) in the vascular niche (shown in purple) as well as endothelial cells (green) and vascular smooth muscle cells (VSMCs) are replaced by EVTs. Invading and migrating EVTs are present in the lumen and occupy the vessel wall. (b) Spiral arteriole in various phases of modification or transformation by EVTs. In the unmodified spiral arteriole, DMSCs are present in the vascular niche, and endothelium and VSMCs are present. These cell types are replaced by invading and migrating EVTs in the fully modified arteriole. (Image [a] modified from Lindheimer, M.D. et al., 2009, *Chesley's Hypertensive Disorders in Pregnancy*, 4th ed, Academic Press, Amsterdam. Image [b] modified from Janatpour, M.J. et al., *Dev Genet* 25 (2), 146–57, 2009.)

The Role of Mesenchymal Stem Cells

with maternal blood. Most important is that EVTs remodel the maternal decidual spiral arterioles into large-diameter conduit vessels of low resistance, thereby allowing the placenta to cope with the ever-increasing demands for nutrition and gas exchange from the developing fetus (Brosens et al. 1972; Redman 1991).

2.2 PATHOLOGIES OF THE PLACENTA

Placental development and establishment of the fetal-maternal interface involve multiple fetal and maternal cell types and complex biological processes that are spatially and temporally coordinated. Abnormal placental development and function give rise to many and varied placental pathologies, as described previously (Benirschke et al. 2012). Placental pathologies are placed into broad categories. Defects in placental adherence or penetration (*placenta creta*) are classified according to the depth of invasion of placental EVT cells into the myometrium (Hoozemans et al. 2004). In order from the least to highest degree of EVT invasion into the myometrium are *placenta accreta*, *placenta increta*, and *placenta percreta*. An *abruptio placentae* (also known as placental abruption) occurs when the placental lining separates prematurely from the uterus before delivery (Tikkanen 2011).

The presence of a vascularized allogeneic placenta, the developing fetus, and its membranes in the maternal uterus poses significant challenges for the maternal and fetal immune systems. Pathologies associated with immune response or inflammatory lesions include chorioamnionitis, chronic intervillositis, and chronic deciduitis. The term chorioamnionitis refers to acute inflammation of the placental membranes (amnion and chorion) and is a leading cause of maternal and fetal complications, including preterm birth and neonatal infection (Romero et al. 2002; Ogunyemi et al. 2003). The inflammatory process is generally regarded as a continuum. During the early stage, the neutrophils involved in this inflammatory response are usually maternal (migrating from intervillous sac, decidual vessels, or both) in origin. During later stages, fetal neutrophils (migrating from fetal surface vessels of the chorionic plate or umbilical cord) are involved. Chronic intervillositis is characterized by an intervillous infiltrate of mononuclear cells (monocytes, lymphocytes, and histiocytes) of maternal origin, and the infiltration is frequently associated with either villous or intervillous fibrinoid deposition (Contro et al. 2010). Chronic deciduitis is characterized by abundant lymphocyte infiltration within the decidua (Khong et al. 2000).

Maldevelopment of the placenta incorporates pathologies such as placental villous immaturity, and circumvallate placenta where the fetal membranes grow abnormally on the fetal side and around the edge of the placenta (Benirschke et al. 2012). Obstetric complications associated with obstruction of the cervical os by the placenta or fetal blood vessels are *placenta previa*, where the placenta places either partly or wholly in the lower uterine segment, and *vasa previa*, where the fetal blood vessels are close to, or cross, the lower uterine segment.

Placental pathologies are also associated with neoplastic transformation, the most important of which are trophoblastic neoplasms (Hoffner and Surti 2012).

Fetal and maternal vascular abnormalities include a large group of placental pathologies, such as fetal thrombotic vasculopathy; hypertrophic decidual vasculopathy, where the smooth muscle of the maternal decidual blood vessels is hypertrophic; excess blood vessels in the chorionic villi called chorangiosis; chorangioma; and placental infarctions, which are localized areas of tissue that are dead or dying after deprivation of the blood supply.

The most important clinical pathologies of the placenta are those that result from maternal arterial compromise (i.e., malperfusion) in early pregnancy, which is caused by inadequate EVT invasion and remodeling of the maternal spiral arterioles. Maternal arterial malperfusion is associated with the pregnancy hypertensive disorder of PE, with fetal growth restriction (FGR), and with stillbirth in preterm pregnancies (Khong 2004).

Clinically important placental pathologies affect a significant percentage of human pregnancies, often with serious short- and long-term health consequences to the mother, the baby, or both, and with significant financial and emotional cost to the community. Despite decades of research, our understanding of the molecular mechanisms that underlie the clinically important placental pathologies remains poor.

Tissues and organs, including the placenta, rely on pools of renewable, unspecialized cells called stem cells for their development, growth, and maintenance (Bongso and Richards 2004). Stem cells are required through the life of the organ. These pools of stem cells are a source of differentiating cells necessary for the establishment and growth of fetal tissues and organs. Stem cells are also required throughout adult life to repair and replace damaged cells and to replenish cells that undergo rapid turnover. Most important is that aberrant stem cell function is associated with a variety of human tissue and organ pathologies and with various cancers. As discussed below, increasing evidence shows stem cells of various types have the potential to contribute to placental pathologies.

2.3 STEM CELLS

Stem cells are self-renewing populations of unspecialized cells that undergo symmetric and asymmetric divisions to self-renew or differentiate into various types of mature cells (Marshak et al. 2001; Cai et al. 2004; Zipori 2009). Germ cells, the embryo, the fetus, and adult tissues harbor different stem cells (Mimeault and Batra 2006). Stem cells within tissues are in a quiescent state, but they respond to both intrinsic and extrinsic signals from the microenvironment, and respond by triggering programs for self-renewal or differentiation. This minimal definition of a stem cell cannot always distinguish stem cells from other dividing cells; therefore, criteria that are more stringent are required. Regardless of their origin, stem cells exhibit some "universal" properties. These properties include activated telomerase, high levels of DNA repair enzymes, absence of lineage markers, low uptake of the dye Hoechst 33342, gap junction expression, high aldehyde dehydrogenase activity, and CXCR4 expression (Cai et al. 2004).

Stem cells are classified by their differentiation potential as totipotent, pluripotent, multipotent, bipotent, and unipotent. At the apex of the hierarchy is the zygote: the zygote has the potential to form all fetal and placental cell types (Bongso and Richards 2004). The blastocyst contains a cluster of inner cell mass (ICM) cells that are pluripotent and form cell derivatives of the principal germ layers (ectoderm, mesoderm, and endoderm), but not extraembryonic tissues. Most fetal and adult

The Role of Mesenchymal Stem Cells

organs and tissues harbor multipotent stem cells that differentiate into restricted cell lineages appropriate to their local microenvironment (Alison et al. 2002). At the nadir of the hierarchy are the bipotent and unipotent stem cells, with very restricted differentiation potential (Zipori 2009).

Another classification of stem cells is according to their tissue of origin. Embryonic stem cells (ESCs) derive from the ICM of mammalian blastocysts. Fetal stem cells are derived from various tissues of aborted fetuses. Studies of ESCs and fetal stem cells are hindered by moral and ethical concerns associated with their isolation. Adult stem cells, also called somatic stem cells, are found in most tissues and organs (Zipori 2009).

2.4 STEM CELL POPULATIONS IN THE PLACENTA

Despite the complex structure of the placenta, its myriad of important functions, and the obvious pathologies associated with abnormal development, our understanding of the role of stem cells in normal and abnormal placental development has focused on a single type, the trophoblast stem cell. However, recent studies show the placenta, umbilical cord, fetal membranes, amniotic fluid, and *decidua basalis* contain a variety of stem cell populations with unknown functions: epithelial, endothelial, hematopoietic, and mesenchymal stem cells (MCSs) (Pipino et al. 2013). In this respect, the placenta is like many adult organs and tissues that are known to harbor various populations of stem cells.

2.4.1 TROPHOBLAST STEM CELLS

Trophoblast cells are exclusive to the placenta, and they fall into three differentiated types: VCT, EVT, and the syncytiotrophoblast. Extensive studies highlight the crucial roles that trophoblast cells play in normal and pathological placental development (Benirschke et al. 2012). The current thinking is that VCT is the stem cell population for both syncytiotrophoblast and EVT in early placental development and syncytiotrophoblast in term placentae (Kliman et al. 1986; Loke and King 1995; Morrish et al. 1997; Baczyk et al. 2006). However, another possibility is that there is a distinct population of VCT in the first trimester that gives rise to EVT and syncytiotrophoblast (Simmons and Cross 2005), and a second population of VCT that gives rise to syncytiotrophoblast at term (James et al. 2007).

The events prior to the first trimester of human placental development, during which the identity and role of trophoblast stem cells are established, are not well understood. These cells derive from the trophectoderm, the sphere of cells that envelop the ICM in the preimplantation blastocyst. Although the trophoblast cell lineages derived from the trophectoderm are well understood in the mouse (Simmons and Cross 2005), these lineages are not well understood in humans. Attempts to isolate and characterize the human trophoblast stem cell that drives cytotrophoblast expansion from trophectoderm have not met with success. Trophoblast stem cell–like cells are derived *ex vivo* by using various cell culture methods, but these cells do not accurately model trophoblast stem cell self-renewal and differentiation. A cell population, with properties of a trophoblast stem/progenitor cell, was derived from the chorion of the human placenta. This cell type expressed factors involved in self-renewal or differentiation, and the cells could be differentiated into invasive cytotrophoblasts (i.e., equivalent to EVTs) and into multinucleate syncytiotrophoblast (Genbacev et al. 2011, 2013). Further studies are required to identify trophoblast stem cells at the earliest stages of placental formation since the triggers for important clinical pathologies such as FGR and PE may be activated as early as the implanting blastocyst (Sharkey and Macklon 2013).

2.4.2 OTHER STEM CELL TYPES

The substantial mesenchymal core of the placental villus and the varied functions attributed to mesenchyme-derived cells hinted at stem cell populations other than trophoblast. The mesenchymal core is responsible for paracrine regulation of the overlying trophoblast cell layers, production of the villous extracellular matrix, and regulation of proliferation of endothelial cell precursors during angiogenesis (Guibourdenche et al. 2009). Cells in the mesenchymal core differentiate into support cells such as myofibroblasts, pericytes, and smooth muscle cells that contribute to the maturation of vessels and capillaries in the villi (Huppertz 2008). Researchers seeking alternative stem cells to bone marrow MSCs, which are used for therapeutic purposes, drove the discovery of novel stem cell populations in the mesenchymal core of the chorionic villi. Term placentae are an obvious candidate source of stem cells since the placenta has a large mass and the terminal villi contain a substantial mesenchymal core. Added benefits of human term placentae are their abundance and continuous supply and they are readily obtained by noninvasive means without significant moral or ethical concerns. Unlike most organs, the human placenta is a transient organ that undergoes rapid growth and differentiation throughout its short life span. This attribute is perhaps why the placenta is such an abundant source of stem cells. Stem cells are isolated from not only the chorionic villous stroma of the placenta but also from the fetal membranes (amnion and chorion) and amniotic fluid. The maternal *decidua basalis* component of the fetal-maternal interface and the maternal *decidua parietalis* component of the fetal membranes are sources of stem cells (see Table 2.1). Many studies contributed to the isolation methods and characterization of these stem cell types (Table 2.1). The properties and potential therapeutic applications of stem cells from these tissues are discussed elsewhere in this book. The following discussion focuses in more detail on chorionic and *decidua basalis* MSCs because increasing evidence suggests they play important roles in important placental pathologies.

2.4.3 MESENCHYMAL STEM CELLS

2.4.3.1 De nition and Background

MSCs are multipotent stem cells, originally identified in adult bone marrow, that replicate as undifferentiated cells, but with the potential to differentiate into the mesenchymal lineages if the correct cues are received (Pittenger et al. 1999). Animal models (mouse, zebrafish, and *Drosophila melanogaster*) verified fundamental

TABLE 2.1

Sources of Mesenchymal Stem Cells from the Human Placenta

Mesenchymal Stem Cell Source	Reference	Isolation Protocol	Differentiation Potential	Phenotypic Markers
AF	In 't Anker et al. (2004)	Centrifugation to collect AF cells	Adipogenic, osteogenic	CD90+, CD105+, CD166+, CD49e+, SH3+, SH4+, HLA-ABC+, CD31−, CD34−, CD45−, CD49d−, CD123−, HLA-DR−
	De Coppi et al. (2007)	Immunoselection to isolate c-Kit populations from amniocentesis specimens	Adipogenic, osteogenic, myogenic, endothelial, neural and hepatic	CD29+, CD44+, CD73+, CD90+, CD105+, SSEA-4+, Oct4+, HLA-ABC+, CD45−, CD34−, CD133−, SSEA-3−, Tra-1-81−
	Zhang et al. (2009)	Centrifugation to collect AF cells	Adipogenic, osteogenic, endothelial	CD13+, CD29+, CD44+, CD90+, CD105+, CD34−, CD45−, CD31−, CD14−
Amnion	Portmann-Lanz et al. (2006)	Removal of amnion from the chorion followed by treatment with collagenase	Adipogenic, osteogenic, chondrogenic, myogenic, neurogenic	CD166+, CD105+, CD90+, CD73+, CD49e+, CD44+, CD29+, CD13+, MHC I+, CD14−, CD34−, CD45−, MHC II−
	In 't Anker et al. (2004)	Removal of amnion from the chorion followed by mincing and filtering through a 100-μm nylon filter	Adipogenic, osteogenic	CD90+, CD105+, CD166+, CD49e+, SH3+, SH4+, HLA-ABC+, CD31−, CD34−, CD45−, CD49d−, CD123−, HLA-DR−
	Bacenkova et al. (2011)	Mechanical removal from the amnion followed by treatment with dispase and collagenase and filtering tissue parts through a 40-μm cell strainer	Adipogenic, osteogenic, chondrogenic	CD105+, CD90+, CD44+, CD146+, HLA-ABC+, HLA-DR−, CD34−, CD45−
Chorion	Portmann-Lanz et al. (2006)	Mechanical and enzymatic removal from the surrounding layers followed by treatment with collagenase	Adipogenic, osteogenic, chondrogenic, myogenic, neurogenic	CD166+, CD105+, CD90+, CD73+, CD49e+, CD44+, CD29+, CD13+, MHC I+, CD14−, CD34−, CD45−, MHC II−

(Continued)

TABLE 2.1 *(Continued)*
Sources of Mesenchymal Stem Cells from the Human Placenta

Mesenchymal Stem Cell Source	Reference	Isolation Protocol	Differentiation Potential	Phenotypic Markers
	Soncini et al. (2007)	Mechanical removal from the amnion followed by treatment with dispase, collagenase, and DNase	Adipogenic, osteogenic, chondrogenic	CD13+, CD29+, CD44+, CD54+, CD73+, CD105+, CD166+, CD3−, CD14−, CD34−, CD45−, CD31−
	Bacenkova et al. (2011)	Mechanical removal from the amnion followed by treatment with dispase and filtering tissue parts through a 40-µm cell strainer	Adipogenic, osteogenic, chondrogenic	CD105+, CD90+, CD44+, CD146+, HLA-ABC+, HLA-DR−, CD34−, CD45−
Chorionic villi	Fukuchi et al. (2004)	Mincing of central placental lobules followed by trypsin digestion	Adipogenic, osteogenic	CD44+, CD29+, CD54+, AC133−, CD31−, CD45−
	Igura et al. (2004)	Explant method: cells migrated out from pieces of chorionic villi attached to dishes	Adipogenic, osteogenic, chondrogenic	CD13+, CD44+, CD73+, CD90+, CD105+, HLA-class I+, CD31−, CD34−, CD45−, HLA-DR−
	Poloni et al. (2008)	Chorionic villous sampling from first trimester pregnancies	Adipogenic, osteogenic, chondrogenic, neurogenic	CD90+, CD105+, CD73+, CD44+, CD29+, CD13+, CD45−, CD14−, CD34−, CD117−
Decidua parietalis	In 't Anker et al. (2004)	Decidua was scraped from the chorion followed by mincing and filtering tissue parts through a 100-µm nylon filter	Adipogenic, osteogenic	CD90+, CD105+, CD166+, CD49e+, SH3+, SH4+, HLA-ABC+, CD31−, CD34−, CD45−, CD49d−, CD123−, HLA-DR−
	Macias et al. (2010)	Decidua was separated from the chorion and digested with trypsin	Adipogenic, osteogenic, chondrogenic, myogenic, neurogenic, cardiogenic, pulmonary epithelial	CD105+, CD44+, CD117+, CD29+, CD90+, CD73+, CD13+, CD34−, CD45−, CD133−, BCRP1−

(Continued)

TABLE 2.1 *(Continued)*

Sources of Mesenchymal Stem Cells from the Human Placenta

Mesenchymal Stem Cell Source	Reference	Isolation Protocol	Differentiation Potential	Phenotypic Markers
	Kanematsu et al. (2011)	Decidua was scraped from the chorion followed by treatment with collagenase, dispase, and DNase I	Adipogenic, osteogenic, chondrogenic	CD13+, CD29+, CD44+, CD73+, CD90+, CD166+, HLA-ABC+, CD105+, SSEA-4+, CD14−, CD19−, CD34−, HLA-DR−
Decidua basalis	In 't Anker et al. (2004)	Decidua was dissected from maternal surface of placenta followed by mincing and filtering tissue parts through a 100-μm nylon filter	Adipogenic, osteogenic	CD90+, CD105+, CD166+, CD49e+, SH3+, SH4+, HLA-ABC+, CD31−, CD34−, CD45−, CD49d−, CD123−, HLA-DR−
	Dimitrov et al. (2010)	First trimester decidual tissues were minced carefully followed by treatment with collagenase and filtering through 70-μm nylon filter	Adipogenic, osteogenic, endothelial	HLA-class I+, CD146+, CD29+, CD73+, CD90+, CD45−, CD34−, CD14−, CD19−, CD56−, CD3−
	Huang et al. (2009)	Decidua was dissected and minced followed by treatment with trypsin and collagenase	Adipogenic, osteogenic, chondrogenic	CD29+, CD44+, CD9+, CD105+, CD166+, HLA-ABC+, CD34−, CD40L−, HLA-DR−

AF, amniotic fluid; HLA, human leukocyte antigen; MHC, major histocompatibility complex.

MSC properties, but investigations in humans are limited (Sell 2004; Zipori 2009). Stem cell properties were demonstrated for bone marrow hematopoietic stem cells and MSCs since they repopulate the bone marrow in human transplantation studies (Turksen 2004; Zipori 2009).

The lack of a conclusive definition of an MSC, and the limited ability to show fundamental stem cell properties in human cells, resulted in an abundance of terms in the literature to describe mesenchymal cells with stem cell–like properties from human tissues and organs. These terms include MSC, mesenchymal stromal cell, or multipotent stromal cell. The abbreviation MSC is used to describe this type of human cell. MSCs are isolated from various adult tissues, including cord blood, peripheral blood, adipose tissue, umbilical cord, amniotic fluid, placenta, dental pulp, synovial fluid, skeletal muscle, periosteum, lungs, and cartilage (Erices et al. 2000; Jiang et al. 2002; Shi and Gronthos 2003; Gang et al. 2006; Gucciardo et al. 2009; Macias et al. 2010; Manini et al. 2011; Watt et al. 2013).

2.4.3.2 Chorionic Villous MSCs

The chorionic villi comprise trophoblast cells that envelop a mesenchymal core, which is the source of chorionic mesenchymal stem cells (CMSCs). Fibroblast-like cells with mesenchymal characteristics were isolated and cultured from explanted first trimester placenta tissue (Haigh et al. 1999), but their stem cell properties were not investigated. CMSCs were isolated from chorionic villi during routine chorionic villous sampling (9–12 weeks of gestation), through first trimester explant cultures, but CMSCs are most often isolated from term human placentae (Fukuchi et al. 2004; Igura et al. 2004; Poloni et al. 2008; Castrechini et al. 2010; Abumaree et al. 2013; Liu et al. 2014a). MSCs from the placenta and bone marrow share similarities regarding developmental stage, differentiation potential, proliferation, and expression of cell surface markers (Barlow et al. 2008; Sung et al. 2010).

After delivery of the placenta, it is considered to have two sides: the fetal side of the placenta comprises the umbilical cord insertion region and the chorionic plate, whereas the maternal side of the placenta comprises the bulk of the fetal terminal villi and attached remnants of the basal plate, which includes the maternal *decidua basalis*. Until now, most placental MSC studies focused on fetal CMSCs isolated from either the fetal or maternal side of the term placenta. Isolation methods for CMSCs from term placentae include the explant culture method, enzymatic digestion with placental perfusion, and careful dissection of the terminal villi (Fukuchi et al. 2004; Igura et al. 2004; In 't Anker et al. 2004; Zhang et al. 2006; Castrechini et al. 2010; Nazarov et al. 2012; Abumaree et al. 2013). Different methods to prepare, propagate, and characterize CMSCs make it difficult to compare results between laboratories.

Isolation of CMSCs from the maternal side of the term placenta is problematic. Careful preparation and characterization are necessary to avoid contaminating fetal CMSCs with maternal cells. Avoiding contamination is an important issue since the most common method involves removal of attached *decidua basalis* from the maternal side of the term placenta and then dissection of the underlying terminal villi, followed by mincing and plating onto culture plates. This method of preparing CMSCs is prone to contamination of CMSCs with maternal cells, which probably

The Role of Mesenchymal Stem Cells   25

originate in the maternal *decidua basalis*. Significant problems are associated with the high incidence of maternal cell contamination in human CMSC preparations (Heazlewood et al. 2014). Various groups reported the isolation and characterization of CMSCs, but the fetal origin of the CMSCs was not confirmed (Yen et al. 2005; Brooke et al. 2009; Prather et al. 2009; Portmann-Lanz et al. 2010b; Sung et al. 2010; Tran et al. 2011; Zhu et al. 2014). Without confirmation that CMSC preparations are free of maternal cell contamination, the possibility that maternal cell contaminants could affect the outcome of phenotypic assays, functional assays, or both cannot be excluded. Some studies verified the fetal origin of their isolated CMSCs either by human leukocyte antigen (HLA) typing or fluorescence *in situ* hybridization karyotyping analysis (Fukuchi et al. 2004; Igura et al. 2004; In 't Anker et al. 2004; Castrechini et al. 2012; Nazarov et al. 2012; Abumaree et al. 2013; Liu et al. 2014a).

Parolini et al. (2008, 2010) proposed a set of criteria for defining a placenta-derived MSC. The criteria included are confirmation of the fetal origin for cells isolated from placental chorionic tissue. Criteria for MSCs derived from chorionic villi (i.e., CMSCs) state that the cells should be of fetal origin, with <1% maternal cell contamination. The International Society of Cellular Therapy established minimal criteria to define an MSC. They stipulated that MSCs must adhere to untreated plastic surfaces; express CD105, CD73, and CD90, but lack expression of CD34, CD14, CD19, CD11b, CD79α, or HLA-DR; and differentiate into the mesenchymal lineages (Dominici et al. 2006).

Researchers explored the therapeutic potential of CMSCs in various diseases by using animal models. In a mouse model of lung injury, administration of a mixed population of amniotic epithelial cells, amniotic MSCs, and CMSCs resulted in a reduced lung fibrosis (Cargnoni et al. 2009; Moodley et al. 2013). Intravenous CMSC transplantation in a rat model of ischemic stroke improved functional recovery, whereas CMSC transplantation into diabetic mice resulted in signs of diabetes reversal (Kadam et al. 2010; Kranz et al. 2010). Also, allogeneic injection of CMSCs in a rat model of hind limb ischemia showed improvement in blood flow and capillary density while reducing oxidative stress and endothelial damage (Prather et al. 2009). The results suggested that CMSCs are a reliable supply of therapeutic cells that need no histocompatible tissue matching and are less expensive to prepare and more readily obtained than bone marrow MSCs or adipose MSCs.

2.4.3.3 *Decidua Basalis* MSC

MSCs have been isolated from the *decidua basalis* (i.e., decidual mesenchymal stem cells [DMSCs]) at the fetal-maternal interface of the placenta (also called the basal plate) (In 't Anker et al. 2004; Wulf et al. 2004; Macias et al. 2010; Kanematsu et al. 2011; Kusuma et al. 2015). Several groups reported the isolation of DMSCs from *decidua basalis* region of the placenta by using variations of mincing and enzymatic digestion method (Oliver et al. 1999; In 't Anker et al. 2004; Dimitrov et al. 2010; Indumathi et al. 2013). Early passage DMSCs are heterogeneous, but they become homogeneous and exhibit a characteristic fibroblastic morphology with further passaging in culture (Brooke et al. 2009; Huang et al. 2009). Given the potential heterogeneity of cultures, it is crucial that DMSCs are well-characterized regarding their surface marker expression, parent tissue of origin, and mesenchymal differentiation potential. Lu et al. (2013) showed that DMSCs maintain stable, normal karyotypes; proliferate rapidly in long-term *in vitro* culture; and show immunosuppressive effects. These findings give evidence that the *decidua basalis* is a unique source of MSCs (i.e., DMSCs), with significant potential for therapeutic applications (Lu et al. 2013). Despite many studies of the *in vitro* properties of CMSCs and DMSCs, there is slow progress toward understanding the role of either of these MSC types in normal and pathological placental development.

There are still unanswered questions regarding the origin of DMSCs and their role. The relationship between endometrial MSCs (Gargett et al. 2007) and DMSCs is not defined. In a series of experiments, Oliver et al. (1999) isolated predecidual stromal cells with stem cell–like characteristics from first trimester human *decidua basalis*. They concluded that predecidual stromal cells are involved in defense against infections, and under pathological conditions, they might be responsible for eliminating the conceptus. Dimitrov et al. (2010) reported decidual stromal cells with stem cell properties in the first trimester *decidua basalis* that could differentiate into mesenchymal lineages and that decidualized *in vitro* after stimulation with progesterone and cAMP. The origin of MSCs in human *decidua basalis*, their precise roles in the reproductive process, and their roles in normal placental development need further investigation.

2.5 LOCALIZATION OF MSC NICHE IN THE PLACENTA

2.5.1 THE STEM CELL NICHE

The niche is the specialized local microenvironment where stem cells reside, and its function is to promote stem cell maintenance (Morrison and Spradling 2008). Epigenetic regulation within the stem cell niche in response to environmental cues is also important in determining stem cell function (Pollina and Brunet 2011). The niches that nurture and maintain MSCs in different regions of the placenta, the fetal-maternal interface, and fetal membranes are not well defined. The balance between the quiescent stem state and differentiation state depends on the stem cell niche. Studies of stem cell niches in various tissues show that stem cell behavior is regulated by interactions with supporting cells in their local microenvironment; these interactions are mediated by direct physical contact and secretion of soluble growth factors and cytokines and by extracellular matrix proteins (Fuchs et al. 2004; da Silva Meirelles et al. 2008; Kuhn and Tuan 2009).

2.5.2 MSC NICHE IN THE CHORIONIC PLACENTA

Providing strong evidence for the anatomical location of the stem cell niche *in vivo* is a significant obstacle to understanding placental stem cell biology. The niche gives vital clues as to the potential role of MSCs in placental pathologies. Particular properties are assumed to mark stem cells, such as retaining a bromodeoxyuridine (BrdU) label, a marker for slow-cycling stem cells (Maeshima et al. 2003; Chan and Gargett 2006; Morrison and Spradling 2008; Voog and Jones 2010; Kameyama et al. 2014). This method was applied to cultured chorionic cells and cultured explants of first

The Role of Mesenchymal Stem Cells

trimester chorionic villi (Kennerknecht et al. 1992) to quantify cells that retained label, but the results were not conclusive. Watson et al. (1995) identified BrdU label–retaining cells in explanted 8- to 10-week chorionic villi by *in situ* analysis; cells were detected among the cytotrophoblast, but cells were more abundant in the stroma. However, the inability of the *ex vivo* explant method to maintain placental villous structure and function prevented the confident identification of the MSC niche.

Since *in situ* analysis of label-retaining cells was not a reliable method of identifying stem cells in human placental tissues, an alternate immunohistochemical method using well-characterized MSC surface marker antibodies was used to identify the placental MSC niche. Multiple MSC cell surface markers are necessary since there is no unique marker for MSCs, which complicates the identification of mammalian stem cell niche in tissues. Combinations of MSC cell surface markers FZD-9, STRO-1, 3G5, CD146, CD49a, and α-SMA were used to identify MSCs cultured from bone marrow, periodontal ligament, dental pulp, and human adipose tissue and placenta (Shi and Gronthos 2003; Battula et al. 2008; Lin et al. 2008; Zannettino et al. 2008), and these MSC cell surface markers were later used to identify MSC niches in various human tissues (Shi and Gronthos 2003; Crisan et al. 2008; Lin et al. 2008; Zannettino et al. 2008; Kuhn and Tuan 2009; Castrechini et al. 2010, 2012).

Regarding the chorionic villi of the placenta, immunohistochemical analysis with MSC markers (CD146, VLA-1/CD49a, STRO-1, and 3G5) revealed that CMSCs reside in a vascular niche within the chorionic villi (Castrechini et al. 2010; Liu et al. 2014a). Vascular MSC niches are found in many other organs and tissues (da Silva Meirelles et al. 2008). The vascular niche of CMSCs and their close association with endothelial cells that line the vessel walls suggest CMSCs play a role in angiogenesis and normal vessel function. Human CMSCs enhance angiogenesis by direct differentiation into endothelial cells, through paracrine effects on endothelial cells, or both (Tran et al. 2011). CMSCs (also called human placenta-derived MSCs) express integrins αv, α4, α5, β1, β3, and β5. Differentiation of CMSCs into endothelial-like cells was observed in angiogenesis assays in cell culture as well as *in vivo* and was dependent on α5β1 integrin expression (Lee et al. 2009). Further evidence to support a role for CMSCs in normal vessel function comes from mouse knockouts of the platelet-derived growth factor receptor (PDGFR)-β gene, which encodes a growth factor receptor expressed only in vascular niche, and specifically in MSCs and pericytes. Mice that lack a functional PDGFB or PDGFR-β gene show significant defects in maternal and fetal placental vessel maturation and function (Ohlsson et al. 1999). These animal models hint at possible roles for MSCs in the human placenta and the potential for MSCs to play important roles in placental pathologies that involve the vasculature.

2.5.3 MSC NICHE IN THE *DECIDUA BASALIS*

A vascular DMSC niche was identified by using a combination of MSC cell surface marker antibodies (STRO-1, 3G5, FZD-9, and α-SMA) and immunofluorescence studies (Kusuma et al. 2015). A vascular niche for DMSC in the *decidua basalis* was consistent with another study that identified a vascular niche for *decidua parietalis* MSCs in the choriodecidua of the fetal membranes (Castrechini et al. 2012).

Kusuma et al. (2015) investigated the fate of the DMSC vascular niche in the spiral arterioles of the *decidua basalis* and underlying myometrium. Immunofluorescence analysis with a combination of MSC markers (i.e., STRO-1/FZD-9, α-SMA/FZD-9, and 3G5/FZD-9) showed DMSCs present in the vascular niche around nontransformed spiral arterioles. FZD-9 immunoreactivity was used to detect the DMSC niche in both nontransformed spiral arterioles and spiral arterioles transformed by EVT. FZD-9 immunoreactivity was detected in the vascular niche of nontransformed arterioles. These data provided the first evidence that in fully transformed spiral arterioles the DMSC niche is destroyed or replaced after EVT transformation (Figure 2.1b).

In summary, studies in the chorionic villi of the human placenta and *decidua basalis* show a vascular niche for MSCs, and these results are consistent with vascular niches for MSCs in multiple human organs, including bone marrow, skeletal muscle, pancreas, and adipose tissue (Shi and Gronthos 2003; Crisan et al. 2008; Lin et al. 2008; Zannettino et al. 2008; Kuhn and Tuan 2009; Castrechini et al. 2010, 2012).

2.6 MSCs AND THEIR POTENTIAL ROLE IN PREGNANCY PATHOLOGIES

The role of trophoblast stem cells in normal and pathological placental development is an area of intense research. However, our understanding of the role of other stem cell populations in these processes is rudimentary. Further studies are needed to investigate how MSCs in the niche interact with cell types to which they are closely associated (e.g., endothelial cells and smooth muscle cells) and to determine the effect on MSCs of pathological changes to the niche (Caruso et al. 2012).

In pregnancy disorders such as PE and FGR, defective placentation is thought to play a major role in the pathogenesis. Recent investigations into the functional properties of placenta-derived stem cells give some new insights into our understanding of placenta-related disorders. Most work on the role of MSCs in placental pathologies has focused on their role in PE.

2.6.1 PREECLAMPSIA

PE is a dangerous hypertensive disorder of pregnancy and is a major cause of maternal fetal and neonatal mortality and morbidity (AbouZahr 2003). PE is a multifactorial disorder, where environmental and genetic factors both contribute to the etiology. Although some aspects of the etiology of PE are poorly understood, it is accepted that the disorder is of placental origin and is characterized by impaired placentation, inadequate placental perfusion, and vascular abnormalities.

2.6.2 THE PATHOGENESIS OF PE

Important factors in the molecular pathogenesis of PE include defective placentation, hypoxia, antioxidant depletion, oxidative stress, and angiogenic factor release (Shaman et al. 2013). Maternal endothelial cell dysfunction appears to be central to the pathogenesis of many of the maternal features of PE. Redman (2011) described PE as a two-stage disorder. The feature of the first stage of PE is a defective deep placentation that occurs in the first half of pregnancy and results in a reduced blood supply to the placenta. As described above, a characteristic of normal placentation is the almost complete transformation of the maternal decidual and myometrial segments of spiral arterioles by deeply invading and migrating placental EVTs. In PE, EVT migration and invasion are shallow. The placental bed of PE patients is characterized by fewer fully transformed spiral arterioles (Brosens et al. 2011). The result of an incomplete trophoblast remodeling process is that the spiral arterioles remain closer to the nonpregnant state, where the uterine spiral arterioles are more tortuous, thick walled, and narrow, and they keep their smooth muscle and elastic lamina (Staff et al. 2010).

Reduced vessel dilatation at the end of the spiral arteriole that opens into the intervillous space will most likely result in a higher blood velocity that could damage the villous surface architecture of the chorionic villi and mediate the release of excess numbers of trophoblast microparticles, thereby potentially contributing to the induction of inflammation and systemic maternal endothelial dysfunction (Staff et al. 2010). The maladapted spiral arterioles expose the placenta to hydrostatic and oxidative stresses from utero-perfusion that is both at high pressure and intermittent (Redman 2011). Factors associated with endothelial dysfunction are upregulated in PE; these factors include the soluble receptor for vascular endothelial growth factor (VEGFR-1) also known as sFlt-1 (fms-like tyrosine kinase) and soluble endoglin (Redman and Sargent 2005; Grill et al. 2009; Wang et al. 2009). The activities of these and other factors contribute to the systemic maternal endothelial dysfunction, which manifests in the clinical symptom of PE.

2.6.3 THE ROLE OF CMSCs IN PE

In a cell culture model of PE, oxidative stress in endothelial cells was modeled by treating human umbilical vein endothelial cells (HUVECs) with *tert*-butyl hydroperoxide. Paracrine factors produced by CMSCs protected the endothelium from oxidative stress injury by an antiapoptotic activity mediated by IL6ST/STAT3 and manganese superoxide dismutase activation (Liu et al. 2010). The proximity of CSMCs to endothelial cells in the MSC niche within the chorionic villi (Castrechini et al. 2010) suggests that paracrine effects from CMSCs could protect endothelial cells from oxidative stress in PE. Rolfo et al. (2013) examined growth parameters and cytokine profiles of CMSCs prepared from normal and preeclamptic third trimester chorionic villous tissue. PE-CMSCs showed reduced proliferation and increased cellular senescence compared with CMSCs. Cytokine profiles of conditioned medium from PE-CMSCs and CMSCs showed a variety of proinflammatory cytokines and chemotactic factors were increased in PE-CMSCs. Some important cytokines in the pathogenesis of PE included migration inhibitory factor (MIF), tumor necrosis factor (TNF)-α, VEGF, and key proinflammatory cytokines interleukin (IL)-6 and IL-8. Treatment of chorionic villous explants with conditioned medium from PE-CMSCs reciprocated some of the physiological changes observed in chorionic villi affected by PE, including increased expression of MIF and sFlt-1, reduced production of VEGF, and increased release of free β-hCG. Differential expression of classic MSC surface markers (CD105, CD90, CD73, and CD44) between amnion epithelium MSCs and CMSCs isolated from PE and normal placentae were reported previously (Portmann-Lanz et al. 2010a).

Oxygen levels play a critical role in normal placental development. Pathological pregnancies such as those complicated by PE or FGR are associated with malperfusion of the placenta (Redman 2011). Salomon et al. (2013) determined the effects of oxygen tension on exosome release from first trimester CMSCs. Exosomes are secreted nanovesicles that are mediators of the paracrine effect of MSCs (Collino et al. 2010). Exosomes interact with recipient cells through multiple mechanisms involving surface-expressed ligands and deliver surface receptors, proteins, lipids, mRNA, and distinct patterns of microRNAs (miRs) (Collino et al. 2011). CMSC release of exosomes increased in hypoxic conditions (1 and 3% oxygen) compared with 8% oxygen. Exosomes released from CMSCs under hypoxic conditions increased endothelial cell migration and tube formation. Salomon et al. (2013) proposed that in the early pregnancy placenta under physiological low-oxygen conditions, CMSC-derived exosomes traffic to endothelial cells and promote vasculogenesis and angiogenesis. In the malperfused PE placenta, which is characteristic of both PE and FGR, increased release of CMSC-derived exosomes may be an adaptive response.

2.6.4 THE ROLE OF DMSCs IN PE

The first evidence that PE-DMSCs differ from DMSCs came from Hwang et al. (2010) in an investigation of cytokines levels in normotensive DMSCs and PE-DMSCs. Soluble intracellular adhesion molecule-1 and stromal-derived factor-1 were elevated significantly in normotensive DMSCs compared with PE-DMSCs. These differences were not observed between in amnion-derived MSCs derived from fetal membranes of normotensive and PE pregnancies.

miRs are small noncoding RNA molecules (about 22 nucleotides long) that function by silencing RNA and by posttranscriptional regulation of gene expression. Differences in the miR profiles of PE-DMSCs and DMSCs suggest miRs play important roles in the PE. miR-181a and miR-16 are highly expressed in PE-DMSCs compared with DMSCs. miR-181a is an important regulator of proliferation and immunosuppressive properties of DMSCs (Liu et al. 2012). miR-16 controls the angiogenesis-regulating potential of DMSCs and therefore plays an important role in endothelial function. Microarray analysis confirmed that miR-16 expression was upregulated in PE-DMSCs, suggesting that PE-DMSCs are abnormal and this contributes to the pathogenesis of PE (Wang et al. 2012). In this study, they found overexpression of miR-16 inhibited the proliferation and migration of DMSCs, reduced the ability of HUVECs to form blood vessels, and reduced migration of trophoblast cells. In a subsequent study of miR-494, highly expressed in PE-DMSCs (Zhao et al. 2014), miR-494 was shown to arrest the cell cycle at the G1/S transition in DMSCs. This novel mechanism accounts for the effect of miR-494 on PE-DMSC proliferation and function (Chen et al. 2015).

The data above give evidence that PE-DMSCs are abnormal compared with normotensive DMSCs. A consequence of shallow trophoblast invasion in PE is that

The Role of Mesenchymal Stem Cells

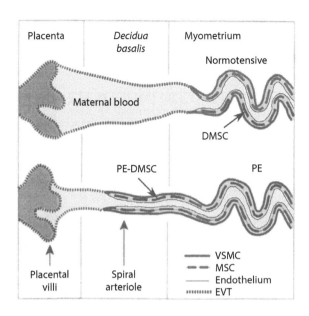

FIGURE 2.2 (See color insert.) Illustration of the proposed increased presence of abnormal decidual mesenchymal stem cells in preeclampsia (PE) as a consequence of shallow spiral arteriole modification. In a normotensive pregnancy, modification or transformation of the spiral arterioles is deep and extends into the inner third of the myometrium. Decidual mesenchymal stem cells (DMSCs) in the vascular niche are normal. Shallow invasion of extravillous trophoblasts (EVTs) in PE results in more unremodeled spiral arteriole vessel walls present in the *decidua basalis* compared with normotensive pregnancy, and these vessels contain abnormal PE-DMSCs. (Image modified from VanWijk, M.J. et al., *Cardiovasc Res* 47 (1), 38–48, 2000.)

more nontransformed spiral arteriole vessel wall is exposed than would be present in a normotensive pregnancy. Kusuma et al. (2015) postulated that the vascular niche contains abnormal PE-DMSCs and that the increased presence of abnormal PE-DMSCs in PE (Figure 2.2) could contribute directly, or indirectly, to the pathogenesis of PE.

2.7 MSCs AS A POTENTIAL TREATMENT FOR PE

Several recent reviews highlighted the potential of MSCs as a therapeutic tool to treat PE (James et al. 2014; Hahn 2015). Liu et al. (2014b) infused intact human DMSCs into T helper 1 (Th1)–induced mice, a model with features of human PE. DMSC infusion resulted in reduced PE-like symptoms through suppression of TNF-α, a regulator of critical factors in the etiology of PE (i.e., sFlt-1 and soluble endoglin). Although these data provide evidence of a reparative effect of human DMSCs, the mouse Th1-induced PE model, and mouse PE models in general, lack critical features of human PE, such as abnormal spiral arteriole remodeling, and they are poor models for human PE. Large animal and primate models of PE are required to verify this intriguing effect of DMSCs. Pregnancy is a transient event in the mother, and the therapeutic use of DMSCs or other MSC types could cause significant problems. Since MSCs have the ability to engraft into tissue, there could be long-term damaging or even pathological effects on the mother, the fetus, or both later in life.

2.8 POSSIBLE ROLE OF MSCs IN OTHER PLACENTAL PATHOLOGIES

Defective deep placentation, which in PE may result in increased numbers of abnormal MSCs around spiral arterioles in the maternal *decidua basalis*, is not unique to PE and is associated with a spectrum of pregnancy complications, including FGR, preterm labor, preterm premature rupture of membranes, late spontaneous abortion, and placental abruption (Brosens et al. 2011). Therefore, abnormal MSCs may play a role in these disorders.

Lei et al. (2015) used a murine model of intrauterine inflammation that induces preterm birth and perinatal brain damage. Administration of human adipose tissue–derived MSCs decreased substantially the rate of preterm birth and perinatal brain damage.

Kim et al. (2015) tested cell growth, cellular senescence, metabolic activity, and the differentiation potential of human MSCs isolated from the umbilical cords of pregnant women with and without gestational diabetes. Umbilical cord MSCs isolated from women with gestational diabetes showed altered metabolic function, premature aging, and poorer cell growth.

Our understanding of recently discovered human stem cell populations and their functions in normal and pathological placental development is at an early stage. Already, exciting new stem cell–based strategies are emerging with the potential to expand our repertoire of treatments for important placental pathologies.

REFERENCES

AbouZahr, C. 2003. Global burden of maternal death and disability. *Br Med Bull* 67: 1–11.
Abumaree, M. H., M. A. Al Jumah, B. Kalionis, et al. 2013. Phenotypic and functional characterization of mesenchymal stem cells from chorionic villi of human term placenta. *Stem Cell Rev* 9 (1): 16–31.
Alison, M. R., R. Poulsom, S. Forbes, and N. A. Wright. 2002. An introduction to stem cells. *J Pathol* 197 (4): 419–23.
Bacenkova, D., J. Rosocha, T. Tothova, L. Rosocha, and M. Sarissky. 2011. Isolation and basic characterization of human term amnion and chorion mesenchymal stromal cells. *Cytotherapy* 13 (9): 1047–56.
Baczyk, D., C. Dunk, B. Huppertz, et al. 2006. Bi-potential behaviour of cytotrophoblasts in first trimester chorionic villi. *Placenta* 27 (4–5): 367–74.
Barlow, S., G. Brooke, K. Chatterjee, et al. 2008. Comparison of human placenta- and bone marrow-derived multipotent mesenchymal stem cells. *Stem Cells Dev* 17 (6): 1095–107.
Battula, V. L., S. Treml, H. Abele, and H. J. Buhring. 2008. Prospective isolation and characterization of mesenchymal stem cells from human placenta using a frizzled-9-specific monoclonal antibody. *Differentiation* 76 (4): 326–36.
Benirschke, K., G. J. Burton, and R. N. Baergen. 2012. *Pathology of the Human Placenta*. Berlin: Springer Verlag.
Bongso, A., and M. Richards. 2004. History and perspective of stem cell research. *Best Pract Res Clin Obstet Gynaecol* 18 (6): 827–42.

The Role of Mesenchymal Stem Cells

Boyd, J. D., and W. J. Hamilton. 1970. *The Human Placenta*. Cambridge, UK: Heffer.

Brooke, G., T. Rossetti, R. Pelekanos, et al. 2009. Manufacturing of human placenta-derived mesenchymal stem cells for clinical trials. *Br J Haematol* 144 (4): 571–9.

Brosens, I., R. Pijnenborg, L. Vercruysse, and R. Romero. 2011. The "great obstetrical syndromes" are associated with disorders of deep placentation. *Am J Obstet Gynecol* 204 (3): 193–201.

Brosens, I. A., W. B. Robertson, and H. G. Dixon. 1972. The role of the spiral arteries in the pathogenesis of preeclampsia. *Obstet Gynecol Annu* 1: 177–91.

Cai, J., M. L. Weiss, and M. S. Rao. 2004. In search of "stemness." *Exp Hematol* 32 (7): 585–98.

Cargnoni, A., L. Gibelli, A. Tosini, et al. 2009. Transplantation of allogeneic and xenogeneic placenta-derived cells reduces bleomycin-induced lung fibrosis. *Cell Transplant* 18 (4): 405–22.

Caruso, M., M. Evangelista, and O. Parolini. 2012. Human term placental cells: Phenotype, properties and new avenues in regenerative medicine. *Int J Mol Cell Med* 1 (2): 64–74.

Castrechini, N. M., P. Murthi, N. M. Gude, et al. 2010. Mesenchymal stem cells in human placental chorionic villi reside in a vascular Niche. *Placenta* 31 (3): 203–12.

Castrechini, N. M., P. Murthi, S. Qin, et al. 2012. Decidua parietalis-derived mesenchymal stromal cells reside in a vascular niche within the choriodecidua. *Reprod Sci* 19 (12): 1302–14.

Chan, R. W., and C. E. Gargett. 2006. Identification of label-retaining cells in mouse endometrium. *Stem Cells* 24 (6): 1529–38.

Chen, S., G. Zhao, H. Miao, et al. 2015. MicroRNA-494 inhibits the growth and angiogenesis-regulating potential of mesenchymal stem cells. *FEBS Lett* 589 (6): 710–17.

Collino, F., S. Bruno, M. C. Deregibus, C. Tetta, and G. Camussi. 2011. MicroRNAs and mesenchymal stem cells. *Vitam Horm* 87: 291–320.

Collino, F., M. C. Deregibus, S. Bruno, et al. 2010. Microvesicles derived from adult human bone marrow and tissue specific mesenchymal stem cells shuttle selected pattern of miRNAs. *PLoS One* 5 (7): e11803.

Contro, E., R. deSouza, and A. Bhide. 2010. Chronic intervillositis of the placenta: A systematic review. *Placenta* 31 (12): 1106–10.

Crisan, M., S. Yap, L. Casteilla, et al. 2008. A perivascular origin for mesenchymal stem cells in multiple human organs. *Cell Stem Cell* 3 (3): 301–13.

da Silva Meirelles, L., A. I. Caplan, and N. B. Nardi. 2008. In search of the in vivo identity of mesenchymal stem cells. *Stem Cells* 26 (9): 2287–99.

De Coppi, P., G. Bartsch, Jr., M. M. Siddiqui, et al. 2007. Isolation of amniotic stem cell lines with potential for therapy. *Nat Biotechnol* 25 (1): 100–6.

Dimitrov, R., D. Kyurkchiev, T. Timeva, et al. 2010. First-trimester human decidua contains a population of mesenchymal stem cells. *Fertil Steril* 93 (1): 210–19.

Dominici, M., K. Le Blanc, I. Mueller, et al. 2006. Minimal criteria for defining multipotent mesenchymal stromal cells. The International Society for Cellular Therapy position statement. *Cytotherapy* 8 (4): 315–17.

Erices, A., P. Conget, and J. J. Minguell. 2000. Mesenchymal progenitor cells in human umbilical cord blood. *Br J Haematol* 109 (1): 235–42.

Fuchs, E., T. Tumbar, and G. Guasch. 2004. Socializing with the neighbors: Stem cells and their niche. *Cell* 116 (6): 769–78.

Fukuchi, Y., H. Nakajima, D. Sugiyama, et al. 2004. Human placenta-derived cells have mesenchymal stem/progenitor cell potential. *Stem Cells* 22 (5): 649–58.

Gang, E. J., J. A. Jeong, S. Han, et al. 2006. In vitro endothelial potential of human UC blood-derived mesenchymal stem cells. *Cytotherapy* 8 (3): 215–27.

Gargett, C. E., R. W. Chan, and K. E. Schwab. 2007. Endometrial stem cells. *Curr Opin Obstet Gynecol* 19 (4): 377–83.

Genbacev, O., M. Donne, M. Kapidzic, et al. 2011. Establishment of human trophoblast progenitor cell lines from the chorion. *Stem Cells* 29 (9): 1427–36.

Genbacev, O., J. D. Lamb, A. Prakobphol, et al. 2013. Human trophoblast progenitors: Where do they reside? *Semin Reprod Med* 31 (1): 56–61.

Grill, S., C. Rusterholz, R. Zanetti-Dallenbach, et al. 2009. Potential markers of preeclampsia—A review. *Reprod Biol Endocrinol* 7: 70.

Gucciardo, L., R. Lories, N. Ochsenbein-Kolble, et al. 2009. Fetal mesenchymal stem cells: Isolation, properties and potential use in perinatology and regenerative medicine. *BJOG* 116 (2): 166–72.

Gude, N. M., C. T. Roberts, B. Kalionis, and R. G. King. 2004. Growth and function of the normal human placenta. *Thromb Res* 114 (5–6): 397–407.

Guibourdenche, J., T. Fournier, A. Malassine, and D. Evain-Brion. 2009. Development and hormonal functions of the human placenta. *Folia Histochem Cytobiol* 47 (5): S35–40.

Hahn, S. 2015. Preeclampsia—Will orphan drug status facilitate innovative biological therapies? *Front Surg* 2: 7.

Haigh, T., C. Chen, C. J. Jones, and J. D. Aplin. 1999. Studies of mesenchymal cells from 1st trimester human placenta: Expression of cytokeratin outside the trophoblast lineage. *Placenta* 20 (8): 615–25.

Heazlewood, C. F., H. Sherrell, J. Ryan, et al. 2014. High incidence of contaminating maternal cell overgrowth in human placental mesenchymal stem/stromal cell cultures: A systematic review. *Stem Cells Transl Med* 3 (11): 1305–11.

Hoffner, L., and U. Surti. 2012. The genetics of gestational trophoblastic disease: A rare complication of pregnancy. *Cancer Genet* 205 (3): 63–77.

Hoozemans, D. A., R. Schats, C. B. Lambalk, R. Homburg, and P. G. Hompes. 2004. Human embryo implantation: Current knowledge and clinical implications in assisted reproductive technology. *Reprod Biomed Online* 9 (6): 692–715.

Huang, Y. C., Z. M. Yang, X. H. Chen, et al. 2009. Isolation of mesenchymal stem cells from human placental decidua basalis and resistance to hypoxia and serum deprivation. *Stem Cell Rev* 5 (3): 247–55.

Huppertz, B. 2008. The anatomy of the normal placenta. *J Clin Pathol* 61 (12): 1296–302.

Huppertz, B., D. S. Tews, and P. Kaufmann. 2001. Apoptosis and syncytial fusion in human placental trophoblast and skeletal muscle. *Int Rev Cytol* 205: 215–53.

Hwang, J. H., M. J. Lee, O. S. Seok, et al. 2010. Cytokine expression in placenta-derived mesenchymal stem cells in patients with pre-eclampsia and normal pregnancies. *Cytokine* 49 (1): 95–101.

Igura, K., X. Zhang, K. Takahashi, et al. 2004. Isolation and characterization of mesenchymal progenitor cells from chorionic villi of human placenta. *Cytotherapy* 6 (6): 543–53.

Indumathi, S., R. Harikrishnan, R. Mishra, et al. 2013. Comparison of feto-maternal organ derived stem cells in facets of immunophenotype, proliferation and differentiation. *Tissue Cell* 45 (6): 434–42.

In 't Anker, P. S., S. A. Scherjon, C. Kleijburg-van der Keur, et al. 2004. Isolation of mesenchymal stem cells of fetal or maternal origin from human placenta. *Stem Cells* 22 (7): 1338–45.

James, J. L., S. Srinivasan, M. Alexander, and L. W. Chamley. 2014. Can we fix it? Evaluating the potential of placental stem cells for the treatment of pregnancy disorders. *Placenta* 35 (2): 77–84.

James, J. L., P. R. Stone, and L. W. Chamley. 2007. The isolation and characterization of a population of extravillous trophoblast progenitors from first trimester human placenta. *Hum Reprod* 22 (8): 2111–19.

Janatpour, M. J., M. F. Utset, J. C. Cross, et al. 1999. A repertoire of differentially expressed transcription factors that offers insight into mechanisms of human cytotrophoblast differentiation. *Dev Genet* 25 (2): 146–57.

The Role of Mesenchymal Stem Cells

Jiang, Y., B. N. Jahagirdar, R. L. Reinhardt, et al. 2002. Pluripotency of mesenchymal stem cells derived from adult marrow. *Nature* 418 (6893): 41–9.

Kadam, S., S. Muthyala, P. Nair, and R. Bhonde. 2010. Human placenta-derived mesenchymal stem cells and islet-like cell clusters generated from these cells as a novel source for stem cell therapy in diabetes. *Rev Diabet Stud* 7 (2): 168–82.

Kameyama, H., S. Kudoh, N. Udaka, et al. 2014. Bromodeoxyuridine (BrdU)-label-retaining cells in mouse terminal bronchioles. *Histol Histopathol* 29 (5): 659–68.

Kanematsu, D., T. Shofuda, A. Yamamoto, et al. 2011. Isolation and cellular properties of mesenchymal cells derived from the decidua of human term placenta. *Differentiation* 82 (2): 77–88.

Kennerknecht, I., S. Baur-Aubele, R. Terinde, and W. Vogel. 1992. Nuclear and chromosomal replication patterns in chorionic villi cells by bromodeoxyuridine labelling and DNA flow cytometry. *Cell Prolif* 25 (4): 321–36.

Khong, T. Y. 2004. Placental vascular development and neonatal outcome. *Semin Neonatol* 9 (4): 255–63.

Khong, T. Y., R. W. Bendon, F. Qureshi, et al. 2000. Chronic deciduitis in the placental basal plate: Definition and interobserver reliability. *Hum Pathol* 31 (3): 292–5.

Kim, J., Y. Piao, Y. K. Pak, et al. 2015. Umbilical cord mesenchymal stromal cells affected by gestational diabetes mellitus display premature aging and mitochondrial dysfunction. *Stem Cells Dev* 24 (5): 575–86.

Kliman, H. J., J. E. Nestler, E. Sermasi, J. M. Sanger, and J. F. Strauss, 3rd. 1986. Purification, characterization, and in vitro differentiation of cytotrophoblasts from human term placentae. *Endocrinology* 118 (4): 1567–82.

Kranz, A., D. C. Wagner, M. Kamprad, et al. 2010. Transplantation of placenta-derived mesenchymal stromal cells upon experimental stroke in rats. *Brain Res* 1315: 128–36.

Kuhn, N. Z., and R. S. Tuan. 2009. Regulation of stemness and stem cell niche of mesenchymal stem cells: Implications in tumorigenesis and metastasis. *J Cell Physiol* 222 (2): 268–77.

Kusuma, G. D., U. Manuelpillai, M. H. Abumaree, et al. 2015. Mesenchymal stem cells reside in a vascular niche in the decidua basalis and are absent in remodelled spiral arterioles. *Placenta* 36 (3): 312–21.

Lee, M. Y., J. P. Huang, Y. Y. Chen, et al. 2009. Angiogenesis in differentiated placental multipotent mesenchymal stromal cells is dependent on integrin alpha5beta1. *PLoS One* 4 (10): e6913.

Lei, J., W. Firdaus, J. M. Rosenzweig, et al. 2015. Murine model: Maternal administration of stem cells for prevention of prematurity. *Am J Obstet Gynecol* 212 (5): 639.e1–10.

Lin, G., M. Garcia, H. Ning, et al. 2008. Defining stem and progenitor cells within adipose tissue. *Stem Cells Dev* 17 (6): 1053–63.

Lindheimer, M. D., J. M. Roberts, and F. G. Cunningham. 2009. *Chesley's Hypertensive Disorders in Pregnancy*. 4th ed. Amsterdam: Academic Press.

Liu, H., P. Murthi, S. Qin, et al. 2014a. A novel combination of homeobox genes is expressed in mesenchymal chorionic stem/stromal cells in first trimester and term pregnancies. *Reprod Sci* 21 (11): 1382–94.

Liu, L., Y. Wang, H. Fan, et al. 2012. MicroRNA-181a regulates local immune balance by inhibiting proliferation and immunosuppressive properties of mesenchymal stem cells. *Stem Cells* 30 (8): 1756–70.

Liu, L., G. Zhao, H. Fan, et al. 2014b. Mesenchymal stem cells ameliorate Th1-induced pre-eclampsia-like symptoms in mice via the suppression of TNF-alpha expression. *PLoS One* 9 (2): e88036.

Liu, S. H., J. P. Huang, R. K. Lee, et al. 2010. Paracrine factors from human placental multipotent mesenchymal stromal cells protect endothelium from oxidative injury via STAT3 and manganese superoxide dismutase activation. *Biol Reprod* 82 (5): 905–13.

Loke, Y.W., and A. King. 1995. *Human Implantation: Cell Biology and Immunology.* Cambridge, UK: Cambridge University Press.

Lu, G., S. Zhu, Y. Ke, X. Jiang, and S. Zhang. 2013. Transplantation-potential-related biological properties of decidua basalis mesenchymal stem cells from maternal human term placenta. *Cell Tissue Res* 352 (2): 301–12.

Lyall, F. 2005. Priming and remodelling of human placental bed spiral arteries during pregnancy—A review. *Placenta* 26 (Suppl A): S31–6.

Macias, M. I., J. Grande, A. Moreno, et al. 2010. Isolation and characterization of true mesenchymal stem cells derived from human term decidua capable of multilineage differentiation into all 3 embryonic layers. *Am J Obstet Gynecol* 203 (5): 495.e9–495.e23.

Maeshima, A., S. Yamashita, and Y. Nojima. 2003. Identification of renal progenitor-like tubular cells that participate in the regeneration processes of the kidney. *J Am Soc Nephrol* 14 (12): 3138–46.

Manini, I., L. Gulino, B. Gava, et al. 2011. Multi-potent progenitors in freshly isolated and cultured human mesenchymal stem cells: A comparison between adipose and dermal tissue. *Cell Tissue Res* 344 (1): 85–95.

Marshak, D. R., R. L. Gardner, and D. Gottlieb. 2001. *Stem Cell Biology, Cold Spring Harbor Monograph.* New York: Cold Spring Harbor Laboratory Press.

Mimeault, M., and S. K. Batra. 2006. Concise review: Recent advances on the significance of stem cells in tissue regeneration and cancer therapies. *Stem Cells* 24 (11): 2319–45.

Moodley, Y., V. Vaghjiani, J. Chan, et al. 2013. Anti-inflammatory effects of adult stem cells in sustained lung injury: A comparative study. *PLoS One* 8 (8): e69299.

Moore, K. L., and T. V. N. Persaud. 1998. *The Developing Human: Clinically Oriented Embryology.* 6th ed. Elsevier Saunders, PA: Philadelphia.

Morrish, D. W., J. Dakour, H. Li, et al. 1997. In vitro cultured human term cytotrophoblast: A model for normal primary epithelial cells demonstrating a spontaneous differentiation programme that requires EGF for extensive development of syncytium. *Placenta* 18 (7): 577–85.

Morrison, S. J., and A. C. Spradling. 2008. Stem cells and niches: Mechanisms that promote stem cell maintenance throughout life. *Cell* 132 (4): 598–611.

Nazarov, I., J. W. Lee, E. Soupene, et al. 2012. Multipotent stromal stem cells from human placenta demonstrate high therapeutic potential. *Stem Cells Transl Med* 1 (5): 359–72.

Ogunyemi, D., M. Murillo, U. Jackson, N. Hunter, and B. Alperson. 2003. The relationship between placental histopathology findings and perinatal outcome in preterm infants. *J Matern Fetal Neonatal Med* 13 (2): 102–9.

Ohlsson, R., P. Falck, M. Hellstrom, et al. 1999. PDGFB regulates the development of the labyrinthine layer of the mouse fetal placenta. *Dev Biol* 212 (1): 124–36.

Oliver, C., M. J. Montes, J. A. Galindo, C. Ruiz, and E. G. Olivares. 1999. Human decidual stromal cells express α-smooth muscle actin and show ultrastructural similarities with myofibroblasts. *Human Reproduction* 14 (6): 1599–605.

Parolini, O., F. Alviano, G. P. Bagnara, et al. 2008. Concise review: Isolation and characterization of cells from human term placenta: Outcome of the first international Workshop on Placenta Derived Stem Cells. *Stem Cells* 26 (2): 300–11.

Parolini, O., F. Alviano, I. Bergwerf, et al. 2010. Toward cell therapy using placenta-derived cells: Disease mechanisms, cell biology, preclinical studies, and regulatory aspects at the round table. *Stem Cells Dev* 19 (2): 143–54.

Pijnenborg, R., J. M. Bland, W. B. Robertson, G. Dixon, and I. Brosens. 1981a. The pattern of interstitial trophoblastic invasion of the myometrium in early human pregnancy. *Placenta* 2 (4): 303–16.

Pijnenborg, R., W. B. Robertson, I. Brosens, and G. Dixon. 1981b. Review article: Trophoblast invasion and the establishment of haemochorial placentation in man and laboratory animals. *Placenta* 2 (1): 71–91.

Pipino, C., P. Shangaris, E. Resca, et al. 2013. Placenta as a reservoir of stem cells: An underutilized resource? *Br Med Bull* 105 (1): 43–68.

Pittenger, M. F., A. M. Mackay, S. C. Beck, et al. 1999. Multilineage potential of adult human mesenchymal stem cells. *Science* 284 (5411): 143–7.

Pollina, E. A., and A. Brunet. 2011. Epigenetic regulation of aging stem cells. *Oncogene* 30 (28): 3105–26.

Poloni, A., V. Rosini, E. Mondini, et al. 2008. Characterization and expansion of mesenchymal progenitor cells from first-trimester chorionic villi of human placenta. *Cytotherapy* 10 (7): 690–7.

Portmann-Lanz, C. B., M. U. Baumann, M. Mueller, et al. 2010. Neurogenic characteristics of placental stem cells in preeclampsia. *Am J Obstet Gynecol* 203 (4): 399.e1–7.

Portmann-Lanz, C. B., A. Schoeberlein, A. Huber, et al. 2006. Placental mesenchymal stem cells as potential autologous graft for pre- and perinatal neuroregeneration. *Am J Obstet Gynecol* 194 (3): 664–73.

Portmann-Lanz, C. B., A. Schoeberlein, R. Portmann, et al. 2010. Turning placenta into brain: Placental mesenchymal stem cells differentiate into neurons and oligodendrocytes. *Am J Obstet Gynecol* 202 (3): 294.e1–294.e11.

Prather, W. R., A. Toren, M. Meiron, et al. 2009. The role of placental-derived adherent stromal cell (PLX-PAD) in the treatment of critical limb ischemia. *Cytotherapy* 11 (4): 427–34.

Redman, C. W. 1991. Current topic: Pre-eclampsia and the placenta. *Placenta* 12 (4): 301–8.

Redman, C. W. 2011. Preeclampsia: A multi-stress disorder. *Rev Med Interne* 32 (Suppl 1): S41–4.

Redman, C. W., and I. L. Sargent. 2005. Latest advances in understanding preeclampsia. *Science* 308 (5728): 1592–4.

Rolfo, A., D. Giuffrida, A. M. Nuzzo, et al. 2013. Pro-inflammatory profile of preeclamptic placental mesenchymal stromal cells: New insights into the etiopathogenesis of preeclampsia. *PLoS One* 8 (3): 19.

Romero, R., J. Espinoza, T. Chaiworapongsa, and K. Kalache. 2002. Infection and prematurity and the role of preventive strategies. *Semin Neonatol* 7 (4): 259–74.

Salomon, C., J. Ryan, L. Sobrevia, et al. 2013. Exosomal signaling during hypoxia mediates microvascular endothelial cell migration and vasculogenesis. *PLoS One* 8 (7): e68451.

Sell, S. 2004. *Stem Cells Handbook*. S. Sell (Ed.). Totowa, NJ: Humana Press.

Shaman, A., B. Premkumar, and A. Agarwal. 2013. Placental vascular morphogenesis and oxidative stress. In A. Agarwal, N. Aziz, and B. Rizk (Eds.), *Studies on Women's Health*. New York, NY: Humana Press, pp. 95–113.

Sharkey, A. M., and N. S. Macklon. 2013. The science of implantation emerges blinking into the light. *Reprod Biomed Online* 27 (5): 453–60.

Shi, S., and S. Gronthos. 2003. Perivascular niche of postnatal mesenchymal stem cells in human bone marrow and dental pulp. *J Bone Miner Res* 18 (4): 696–704.

Simmons, D. G., and J. C. Cross. 2005. Determinants of trophoblast lineage and cell subtype specification in the mouse placenta. *Dev Biol* 284 (1): 12–24.

Soncini, M., E. Vertua, L. Gibelli, et al. 2007. Isolation and characterization of mesenchymal cells from human fetal membranes. *J Tissue Eng Regen Med* 1 (4): 296–305.

Staff, A. C., R. Dechend, and R. Pijnenborg. 2010. Learning from the placenta: Acute atherosis and vascular remodeling in preeclampsia-novel aspects for atherosclerosis and future cardiovascular health. *Hypertension* 56 (6): 1026–34.

Sung, H. J., S. C. Hong, J. H. Yoo, et al. 2010. Stemness evaluation of mesenchymal stem cells from placentas according to developmental stage: Comparison to those from adult bone marrow. *J Korean Med Sci* 25 (10): 1418–26.

Tikkanen, M. 2011. Placental abruption: Epidemiology, risk factors and consequences. *Acta Obstet Gynecol Scand* 90 (2): 140–9.

Tran, T. C., K. Kimura, M. Nagano, et al. 2011. Identification of human placenta-derived mesenchymal stem cells involved in re-endothelialization. *J Cell Physiol* 226 (1): 224–35.

Turksen, K. 2004. *Adult Stem Cells*. Totowa, NJ: Humana Press.

VanWijk, M. J., K. Kublickiene, K. Boer, and E. VanBavel. 2000. Vascular function in pre-eclampsia. *Cardiovasc Res* 47 (1): 38–48.

Voog, J., and D. L. Jones. 2010. Stem cells and the niche: A dynamic duo. *Cell Stem Cell* 6 (2): 103–15.

Wang, A., S. Rana, and S. A. Karumanchi. 2009. Preeclampsia: The role of angiogenic factors in its pathogenesis. *Physiology (Bethesda)* 24 (3): 147–58.

Wang, Y., H. Fan, G. Zhao, et al. 2012. miR-16 inhibits the proliferation and angiogenesis-regulating potential of mesenchymal stem cells in severe pre-eclampsia. *FEBS J* 279 (24): 4510–24.

Watson, A. L., M. E. Palmer, and G. Burton. 1995. Human chorionic gonadotrophin release and tissue viability in placental organ culture. *Hum Reprod* 10 (8): 2159–64.

Watt, S. M., F. Gullo, M. van der Garde, et al. 2013. The angiogenic properties of mesenchymal stem/stromal cells and their therapeutic potential. *Br Med Bull* 108: 25–53.

Wulf, G. G., V. Viereck, B. Hemmerlein, et al. 2004. Mesengenic progenitor cells derived from human placenta. *Tissue Eng* 10 (7–8): 1136–47.

Yen, B. L., H. I. Huang, C. C. Chien, et al. 2005. Isolation of multipotent cells from human term placenta. *Stem Cells* 23 (1): 3–9.

Zannettino, A. C., S. Paton, A. Arthur, et al. 2008. Multipotential human adipose-derived stromal stem cells exhibit a perivascular phenotype in vitro and in vivo. *J Cell Physiol* 214 (2): 413–21.

Zhang, P., J. Baxter, K. Vinod, T. N. Tulenko, and P. J. Di Muzio. 2009. Endothelial differentiation of amniotic fluid-derived stem cells: Synergism of biochemical and shear force stimuli. *Stem Cells Dev* 18 (9): 1299–308.

Zhang, X., A. Mitsuru, K. Igura, et al. 2006. Mesenchymal progenitor cells derived from chorionic villi of human placenta for cartilage tissue engineering. *Biochem Biophys Res Commun* 340 (3): 944–52.

Zhao, G., X. Zhou, S. Chen, et al. 2014. Differential expression of microRNAs in decidua-derived mesenchymal stem cells from patients with pre-eclampsia. *J Biomed Sci* 21: 81.

Zhu, Y., Y. Yang, Y. Zhang, et al. 2014. Placental mesenchymal stem cells of fetal and maternal origins demonstrate different therapeutic potentials. *Stem Cell Res Ther* 5 (2): 48.

Zipori, D. 2009. *Biology of Stem Cells and the Molecular Basis of the Stem State*. New York, NY: Humana Press.

3 The Roles of the Human Placenta in Fetal-Maternal Tolerance

Jelmer R. Prins and Sicco A. Scherjon

CONTENTS

Preface 39
Acronyms 39
3.1 Introduction 40
3.2 The Immune System in General 40
3.3 The Maternal Immune System during Pregnancy 41
3.4 Placenta and Immunity 43
3.5 Fetal Immune Programming 44
3.6 Concluding Remarks 45
References 45

PREFACE

The mechanisms responsible for the maternal immune tolerance of the allogeneic fetus are not fully known. Several mechanisms are thought to play a role in fetal-maternal tolerance; these mechanisms may involve pregnancy-associated cytokines; hormones; environmental factors, such as semen exposure, infection, stress and the placenta. This chapter introduces some general aspects of the immune system and explains some of the immune-regulating mechanisms during pregnancy and thereafter.

ACRONYMS

APC	antigen-presenting cell
DC	dendritic cell
HLA	human leukocyte antigen
IDO	indoleamine 2,3-dioxygenase
MHC	major histocompatibility complex
NK	natural killer
Th	T helper
TRAIL	tumor necrosis factor–related apoptosis-inducing ligand
Treg	regulatory T

3.1 INTRODUCTION

P. B. Medawar was one of the first researchers who tried to explain why the semi-allogeneic fetus is not rejected by the maternal immune system during pregnancy (Billington 2003; Medawar 1953). He explained this unique phenomenon by an anatomical separation between fetus and mother, antigenic immaturity of the fetus, immunological ignorance or inertness of the mother, or a combination of these explanations (Medawar 1953). Because of ongoing research and new knowledge, it is now known that none of the explanations of Medawar are valid. During pregnancy, the maternal immune system balances between rejection and tolerance of the fetus, and an abundance of factors influences this balance (Figure 3.1).

This chapter introduces some general aspects of the immune system and explains some of the immune mechanisms during pregnancy and the role of the placenta in these mechanisms.

3.2 THE IMMUNE SYSTEM IN GENERAL

To be able to fully understand the changes in the maternal immune system during pregnancy, some basic knowledge about the immune system is needed. Therefore, this section describes some general immunology mechanisms.

Traditionally, the immune system is divided into two systems: innate immunity and adaptive immunity (Abbas et al. 2007). The innate immune system provides the first line of defense against pathogens and works mainly through leukocyte recognition of common pathogen structures. All cellular components of the innate immune system are able to recognize microbial structures by specialized pattern recognition receptors (Akira et al. 2006; Akira and Takeda 2004). Adaptive immunity enables adaptation to an infection and increases in magnitude with each successive exposure to a specific microbe (Abbas et al. 2007). The main players of the adaptive immune system are subsets of lymphocytes called T cells and B cells. B cells produce

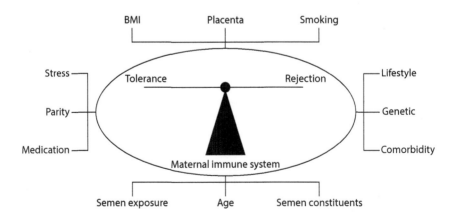

FIGURE 3.1 During pregnancy, the maternal immune system is tolerating an allogeneic fetus. A tremendous amount of factors, endogenous and environmental, skew the maternal immune system toward tolerance or toward rejection of the fetus.

The Roles of the Human Placenta in Fetal-Maternal Tolerance   **41**

antibodies after maturation (LeBien and Tedder 2008) and differentiate into plasma cells or memory B cells. T cells can be classified into helper T cells (CD4+) and cytotoxic T cells (CD8+).

The CD4+ T cell subset consists of different subgroups of cells classified according to their cytokine profile and their function (Zhu and Paul 2008). The main subgroups of CD4+ T cells are T helper (Th)1 cells, Th2 cells, Th17 cells, and regulatory T (Treg) cells. Th1 cells mainly mediate immune responses against intracellular pathogens and are involved in autoimmune responses (Mosmann and Coffman 1989; Mosmann and Sad 1996; Paul and Seder 1994). Th2 cells mainly mediate immune responses against extracellular parasites, such as helminths, and are involved in the pathology of allergic diseases (Mosmann and Coffman 1989; Paul and Seder 1994). Th17 cells mainly mediate immune responses against extracellular bacteria and fungi, and participate in most organ-specific autoimmune diseases (Bettelli et al. 2007; Romagnani et al. 2009). Treg cells are a subset of T cells with immunosuppressive capacities, and are able to regulate immune responses and regulate other immune cell subsets (Sakaguchi 2005; Sakaguchi et al. 1995). Similar to the differentiation of CD4+ cells, CD8+ cells can be divided into subsets based on their function and phenotype: naïve, effector, memory, and regulatory CD8+ cells (Cosmi et al. 2003; Smith and Kumar 2008).

An important process in immunity is antigen processing. Peptides derived from microbes are presented by major histocompatibility complexes (MHCs). MHC class I molecules are expressed on almost all nucleated cells. Normally, MHC class I molecules bind to inhibitory receptors of natural killer (NK) cells, thereby preventing activation of NK cells and thus preventing the killing of "self" cells by NK cells (Bryceson et al. 2011). MHC class II molecules are only expressed on specialized antigen-presenting cells (APCs) and some B cells, and after cytokine stimulation on some endothelial cells and T cells (Abbas et al. 2007). In MHC class I–associated immunity, MHC class I molecules express peptides derived from intracellular antigens. CD8+ cells bind to the MHC class I–expressing cells, become activated, and kill the antigen-expressing cells. MHC class II molecules express peptides mostly obtained by endocytosing extracellular antigens into the cell and then processing them to the MHC class II molecules (Abbas et al. 2007; von Andrian and Mackay 2000). After capturing an antigen, APCs migrate to lymph nodes, where the peptides on MHC II molecules and CD4+ cells bind to the cells. The activated CD4+ cells produce cytokines and induce inflammatory reactions such as activation of macrophages, B cells, or both. During both MHC class I and II responses, memory cells are generated; these cells remain in the lymphoid organs or mucosal tissue to respond to antigen reexposure (von Andrian and Mackay 2000).

3.3 THE MATERNAL IMMUNE SYSTEM DURING PREGNANCY

For years, fetal tolerance was immunologically explained by changes in the Th1/Th2 immune balance. It was thought that during pregnancy there is an immune shift away from Th1 toward Th2 immunity (Raghupathy et al. 1999; Wegmann et al. 1993). Now there is evidence that this shift alone cannot explain maternal tolerance (Svensson et al. 2001; Wilczynski 2005).

During pregnancy, the maternal immune system is continuously challenged by exposure to fetal antigens. The earliest exposure of the maternal immune system to these antigens is via paternal semen; this exposure elicits an inflammatory response in the female reproductive tract (Robertson 2005; Sharkey et al. 2007). The paternal antigens, in combination with seminal immunomodulatory factors such as transforming growth factor and prostaglandin E2, modulate the maternal immune system. Dendritic cells (DCs) become activated by exposure to semen and fetal antigens, and they attract other immune cells toward the uterus (Blois et al. 2007, 2011) (Figure 3.2). DCs play an important role in the maintenance of early pregnancy, especially around implantation, as they have a role in the regulation of angiogenesis (Lee et al. 2011; Pollard 2008). In mouse studies, depletion of uterine DCs resulted in severe impairment of implantation (Plaks et al. 2008). DCs are also involved in the differentiation of a key player in fetal tolerance: Treg cells (Chen et al. 2008). Studies in mouse models have shown that Treg cells are essential for fetal tolerance and that depletion of Treg cells during early pregnancy in mice causes higher rates of fetal resorptions and lower implantation rates (Aluvihare et al. 2004; Guerin et al. 2009;

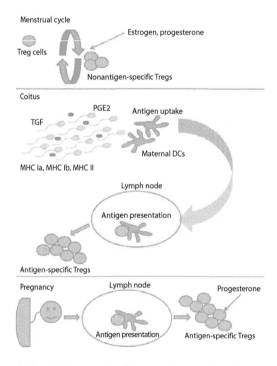

FIGURE 3.2 Regulatory T (Treg) cells are key in fetal-maternal tolerance. In preparation for pregnancy, levels of Treg cells increase in the late follicular phase of the menstrual cycle in a nonantigen-specific manner. Further preparation for pregnancy is achieved at coitus, when seminal immune modulatory factors and fetal antigens result in proliferation of antigen-specific Treg cells. Once pregnancy is established, further expansion of the antigen-specific Treg cells is caused by fetal antigens and immune modulatory cytokines produced by trophoblast cells.

The Roles of the Human Placenta in Fetal-Maternal Tolerance 43

Shima et al. 2010; Zenclussen 2005). Levels of Treg cells are elevated systemically during normal human and mouse pregnancy (Guerin et al. 2009), and levels of Treg cells have already increased by the late follicular phase of the menstrual cycle (Arruvito et al. 2007) (Figure 3.2). Inadequate numbers of Treg cells and functionally limited Treg cells are associated with human reproductive disorders such as infertility, miscarriage, and preeclampsia (Guerin et al. 2009; Sasaki et al. 2004, 2007). It is thought that the balance of Treg and Th17 cells is important for pregnancy outcomes, as higher numbers of Th17 cells are also associated with complicated pregnancy outcomes (Darmochwal-Kolarz et al. 2012; Nakashima et al. 2010; Santner-Nanan et al. 2008; Wang et al. 2010).

The memorizing function of the adaptive immune system is also seen in pregnancy. It is, for example, known that the duration of the sexual relationship (without barrier methods) is inversely related to immune-associated complications of pregnancy such as preeclampsia and fetal growth restriction (Kho et al. 2009; Robillard et al. 1994). More specifically, in a mouse study, fetal-specific CD4+ memory Treg cells have been found to be important for pregnancy success (Rowe et al. 2012). It was shown that memory Treg cells were generated during pregnancy, and during a subsequent pregnancy the memory Treg cell population rapidly re-expanded (Rowe et al. 2012). The importance of these memory Treg cells was further shown by the finding that fetal resorption rates were significantly lower in second pregnancies compared to first pregnancies after partial depletion of maternal Treg cells (Rowe et al. 2012).

It is estimated that in early pregnancy, around 70% of the leukocytes in the decidua are NK cells (Moffett-King 2002; Lash and Bulmer 2011). During pregnancy a specific NK cell subset develops within the decidua, CD56+ decidual (uterine) NK cells. These cells mainly produce cytokines and have no cytotoxic functions (Akira et al. 2006; Hanna et al. 2006; Moffett-King 2002), but they do have an important role in angiogenesis, spiral artery remodeling, and trophoblast differentiation and proliferation (Hanna et al. 2006; Moffett-King 2002). Other cells that are involved in remodeling of spiral arteries and placental development are macrophages and monocytes (Harris 2010). Macrophages in the decidua during healthy pregnancy appear to have a remodeling or regulatory role and are mainly (although not fully) of the M2 subtype (Nagamatsu and Schust 2010). There is evidence that during pregnancy, the balance of macrophage subsets is important for pregnancy success, as a study by Schonkeren et al. (2011) showed an altered balance of macrophage subsets in complicated pregnancies.

3.4 PLACENTA AND IMMUNITY

The decidua is populated with different subsets of leukocytes during pregnancy, and the levels of these subsets vary during pregnancy (Gomez-Lopez et al. 2010; Hunt 1994). In early pregnancy, NK cells and macrophages are the dominant immune cells, making up around 70 and 25% of the decidual leukocytes, respectively. Then, levels of NK cells gradually diminish, and at the end of pregnancy the decidua is mainly populated by granulocytes, T cells, mast cells, and macrophages (Gomez-Lopez et al. 2010).

The placenta itself has several specific immune-protective characteristics. Probably the most described protective mechanism is the absence of classical MHC class I expression on trophoblasts. This absence results in partial prevention of immune attack by maternal T cells (Moffett-King 2002; Munoz-Suano et al. 2011). Normally, the absence of MHC class I molecules would cause NK cells to kill these cells; however, trophoblast cells express human leukocyte antigen (HLA)-C and the nonclassical MHC molecules HLA-E and HLA-G (King et al. 2000a, 2000b; Rajagopalan and Long 1999), which suppress NK cell–mediated killing (Moffett-King 2002). Neutralization of possibly cytotoxic NK and T cells with direct specificity for HLA-C and HLA-E, or indirectly (e.g., via the minor histocompatibility antigens) (Dierselhuis et al. 2014; Perchellet et al. 2013), seems to be important for the acceptance of the semiallogeneic fetus. Incompatibility of maternal killer immunoglobulin-like receptor KIR2 genotype and fetal HLA-C allotype is associated with increased risks of pregnancy complications, such as preeclampsia, miscarriage, and fetal growth restriction (Hiby et al. 2004, 2010, 2014).

Further protection from maternal immune cells in a nonspecific manner is arranged by Fas ligands that are present on trophoblast tissue, indoleamine 2,3-dioxygenase (IDO), complement inhibitor proteins, and tumor necrosis factor–related apoptosis-inducing ligand (TRAIL) (Hunt et al. 1997; Makrigiannakis et al. 2001; Munn et al. 1998; Xu et al. 2000). Activated maternal T cells, which express Fas, undergo apoptosis when binding to the Fas ligands on trophoblasts. The production of IDO by trophoblast cells reduces the availability of tryptophan to T cells, thereby inhibiting the proliferation of T cells (Mellor et al. 2001). TRAIL is expressed on trophoblasts and is known to induce apoptosis when binding to its receptor (Phillips et al. 1999, 2001). The presence and functionality—locally—in the placenta of fetus-induced Treg cells causing a fetoallogeneic, paternal antigen–derived, specific immunotolerance was demonstrated both in murine models and in human (CD4 and CD8 Treg cells) pregnancy (Aluvihare et al. 2004; Tilburgs et al. 2006, 2008, 2010).

3.5 FETAL IMMUNE PROGRAMMING

Cytokines, chemokines, and hormones can cross the fetal-maternal barrier (placenta); therefore, alterations in maternal immune development could likely also lead to fetal immune alterations. Moreover, exogenous factors (e.g., microbial exposure, smoking, medication) could very well cause direct changes in fetal immunity and cause indirect changes via epigenetic processes such as methylation. Although more and more is known about the possible effects of prenatal environmental and maternal factors on the fetal immune system and the possible long-term outcomes, this is still a relatively new research area and much more research is needed (Hsu and Nanan 2014). Several observational studies have found altered fetal immune parameters in immune-associated complications of pregnancies, such as increases in NK cells and activated T cells, indicating an activated inflammatory state of the fetal immune system (Bujold et al. 2003; Darmochwal-Kolarz et al. 2001; Hinz et al. 2010).

Besides effects on the fetal immune system, dysregulation of the maternal immune system could also affect the neurodevelopment of the fetus. The effects of immune imbalances have been reviewed extensively (Filiano et al. 2014). It is known,

The Roles of the Human Placenta in Fetal-Maternal Tolerance   **45**

for example, that cytokines affect neuronal function and behavior (reviewed in Filiano et al. 2014), whereas imbalance of immune molecules and cytokines can produce lifelong changes in neuronal function and behavior (Filiano et al. 2014). As the blood–brain barrier during pregnancy is not fully developed, the fetal brain is even more vulnerable to immune alterations and its effects on neuronal development. Therefore, maternal imbalances between cytokines (as seen in immune-associated complications of pregnancy, or caused by environmental factors such as smoking, stress, and nutrition) could lead to abnormal brain development and increased risk of neurodevelopmental disorders (e.g., schizophrenia and autism) (Filiano et al. 2014; Marques et al. 2015).

3.6 CONCLUDING REMARKS

As described in this chapter, during pregnancy adequate maternal immune regulation is essential for pregnancy outcome since the maternal immune system has to tolerate an allogeneic fetus. The complex mechanisms by which the fetus avoids rejection are incompletely understood. Disturbances in maternal immune tolerance are implicated in several reproductive pathologies; moreover, disturbances in the prenatal immune environment affect growth and health of offspring in later life. To fully understand these relationships, more research is needed.

REFERENCES

Abbas, A. K., A. H. Lichtman, and S. Pillai. 2007. *Cellular and Molecular Immunology.* Elsevier Saunders, PA: Philadelphia.
Akira, S., and K. Takeda. 2004. Toll-like receptor signalling. *Nat Rev Immunol* 4: 499–511.
Akira, S., S. Uematsu, and O. Takeuchi. 2006. Pathogen recognition and innate immunity. *Cell* 124: 783–801.
Aluvihare, V. R., M. Kallikourdis, and A. G. Betz. 2004. Regulatory T cells mediate maternal tolerance to the fetus. *Nat Immunol* 5: 266–71.
Arruvito, L., M. Sanz, A. H. Banham, and L. Fainboim. 2007. Expansion of CD4+CD25+and FOXP3+ regulatory T cells during the follicular phase of the menstrual cycle: Implications for human reproduction. *J Immunol* 178: 2572–8.
Bettelli, E., T. Korn, and V. K. Kuchroo. 2007. Th17: The third member of the effector T cell trilogy. *Curr Opin Immunol* 19: 652–57.
Billington, W. D. 2003. The immunological problem of pregnancy: 50 years with the hope of progress. A tribute to Peter Medawar. *J Reprod Immunol* 60: 1–11.
Blois, S. M., U. Kammerer, C. Alba Soto, et al. 2007. Dendritic cells: Key to fetal tolerance? *Biol Reprod* 77 (4): 590–98.
Blois, S. M., B. F. Klapp, and G. Barrientos. 2011. Decidualization and angiogenesis in early pregnancy: Unravelling the functions of DC and NK cells. *J Reprod Immunol* 88: 86–92.
Bryceson, Y. T., S. C. Chiang, S. Darmanin, et al. 2011. Molecular mechanisms of natural killer cell activation. *J Innate Immun* 3 (3): 216–26.
Bujold, E., T. Chaiworapongsa, R. Romero, et al. 2003. Neonates born to pre-eclamptic mothers have a higher percentage of natural killer cells (CD3-/CD56+16+) in umbilical cord blood than those without pre-eclampsia. *J Matern Fetal Neonatal Med* 14: 305–12.
Chen, W., X. Liang, A. J. Peterson, D. H. Munn, and B. R. Blazar. 2008. The indoleamine 2,3-dioxygenase pathway is essential for human plasmacytoid dendritic cell-induced adaptive T regulatory cell generation. *J Immunol* 181: 5396–404.

Cosmi, L., F. Liotta, E. Lazzeri, et al. 2003. Human CD8+CD25+ thymocytes share phenotypic and functional features with CD4+CD25+ regulatory thymocytes. *Blood* 102: 4107–14.

Darmochwal-Kolarz, D., M. Kludka-Sternik, J. Tabarkiewicz, et al. 2012. The predominance of Th17 lymphocytes and decreased number and function of treg cells in preeclampsia. *J Reprod Immunol* 93: 75–81.

Darmochwal-Kolarz, D., B. Leszczynska-Gorzelak, J. Rolinski, and J. Oleszczuk. 2001. Preeclampsia affects the immunophenotype of neonates. *Immunol Lett* 77: 67–71.

Dierselhuis, M. P., E. Jankowska-Gan, E. Blokland, et al. 2014. HY immune tolerance is common in women without male offspring. *PLoS One* 9: e91274.

Filiano, A. J., S. P. Gadani, and J. Kipnis. 2014. Interactions of innate and adaptive immunity in brain development and function. *Brain Res* 1617: 18–27.

Gomez-Lopez, N., L. J. Guilbert, and D. M. Olson. 2010. Invasion of the leukocytes into the fetal-maternal interface during pregnancy. *J Leukoc Biol* 88: 625–33.

Guerin, L. R., J. R. Prins, and S. A. Robertson. 2009. Regulatory T-cells and immune tolerance in pregnancy: A new target for infertility treatment? *Hum Reprod Update* 15: 517–35.

Hanna, J., D. Goldman-Wohl, Y. Hamani, et al. 2006. Decidual NK cells regulate key developmental processes at the human fetal-maternal interface. *Nat Med* 12: 1065–74.

Harris, L. K. 2010. Review: Trophoblast-vascular cell interactions in early pregnancy: How to remodel a vessel. *Placenta* 31 (Suppl): S93–8.

Hiby, S. E., R. Apps, O. Chazara, et al. 2014. Maternal KIR in combination with paternal HLA-C2 regulate human birth weight. *J Immunol* 192: 5069–73.

Hiby, S. E., R. Apps, A. M. Sharkey, et al. 2010. Maternal activating KIRs protect against human reproductive failure mediated by fetal HLA-C2. *J Clin Invest* 120: 4102–10.

Hiby, S. E., J. J. Walker, K. M. O'Shaughnessy, et al. 2004. Combinations of maternal KIR and fetal HLA-C genes influence the risk of preeclampsia and reproductive success. *J Exp Med* 200: 957–65.

Hinz, D., J. C. Simon, C. Maier-Simon, et al. 2010. Reduced maternal regulatory T cell numbers and increased T helper type 2 cytokine production are associated with elevated levels of immunoglobulin E in cord blood. *Clin Exp Allergy* 40: 419–26.

Hsu, P., and R. Nanan. 2014. Foetal immune programming: Hormones, cytokines, microbes and regulatory T cells. *J Reprod Immunol* 104–105: 2–7.

Hunt, J. S. 1994. Immunologically relevant cells in the uterus. *Biol Reprod* 50: 461–6.

Hunt, J. S., D. Vassmer, T. A. Ferguson, and L. Miller. 1997. Fas ligand is positioned in mouse uterus and placenta to prevent trafficking of activated leukocytes between the mother and the conceptus. *J Immunol* 158: 4122–28.

Kho, E. M., L. M. McCowan, R. A. North, et al. 2009. Duration of sexual relationship and its effect on preeclampsia and small for gestational age perinatal outcome. *J Reprod Immunol* 82: 66–73.

King, A., D. S. Allan, M. Bowen, et al. 2000a. HLA-E is expressed on trophoblast and interacts with CD94/NKG2 receptors on decidual NK cells. *Eur J Immunol* 30: 1623–31.

King, A., S. E. Hiby, L. Gardner, et al. 2000b. Recognition of trophoblast HLA class I molecules by decidual NK cell receptors—A review. *Placenta* 21 (Suppl A): S81–5.

Lash, G. E., and J. N. Bulmer. 2011. Do uterine natural killer (uNK) cells contribute to female reproductive disorders? *J Reprod Immunol* 88: 156–64.

LeBien, T. W., and T. F. Tedder. 2008. B lymphocytes: How they develop and function. *Blood* 112: 1570–80.

Lee, J. Y., M. Lee, and S. K. Lee. 2011. Role of endometrial immune cells in implantation. *Clin Exp Reprod Med* 38: 119–25.

Makrigiannakis, A., E. Zoumakis, S. Kalantaridou, et al. 2001. Corticotropin-releasing hormone promotes blastocyst implantation and early maternal tolerance. *Nat Immunol* 2: 1018–24.

The Roles of the Human Placenta in Fetal-Maternal Tolerance   **47**

Marques, A. H., A. L. Bjorke-Monsen, A. Teixeira, and M. N. Silverman. 2015. Maternal stress, nutrition and physical activity: Impact on immune function, CNS development and psychopathology. *Brain Res* 1617: 28–46.

Medawar, P. B. 1953. Some immunological and endocrinological problems raised by the evolution of viviparity in vertebrates. *Symp Soc Exp Biol* 7: 320–38.

Mellor, A. L., J. Sivakumar, P. Chandler, et al. 2001. Prevention of T cell-driven complement activation and inflammation by tryptophan catabolism during pregnancy. *Nat Immunol* 2: 64–8.

Moffett-King, A. 2002. Natural killer cells and pregnancy. *Nat Rev Immunol* 2: 656–63.

Mosmann, T. R., and R. L. Coffman. 1989. TH1 and TH2 cells: Different patterns of lymphokine secretion lead to different functional properties. *Annu Rev Immunol* 7: 145–73.

Mosmann, T. R., and S. Sad. 1996. The expanding universe of T-cell subsets: Th1, Th2 and more. *Immunol Today* 17: 138–46.

Munn, D. H., M. Zhou, J. T. Attwood, et al. 1998. Prevention of allogeneic fetal rejection by tryptophan catabolism. *Science* 281: 1191–3.

Munoz-Suano, A., A. B. Hamilton, and A. G. Betz. 2011. Gimme shelter: The immune system during pregnancy. *Immunol Rev* 241: 20–38.

Nagamatsu, T., and D. J. Schust. 2010. The immunomodulatory roles of macrophages at the maternal-fetal interface. *Reprod Sci* 17: 209–18.

Nakashima, A., M. Ito, S. Yoneda, et al. 2010. Circulating and decidual Th17 cell levels in healthy pregnancy. *Am J Reprod Immunol* 63: 104–9.

Paul, W. E., and R. A. Seder. 1994. Lymphocyte responses and cytokines. *Cell* 76: 241–51.

Perchellet, A. L., S. Jasti, and M. G. Petroff. 2013. Maternal CD4(+) and CD8(+) T cell tolerance towards a fetal minor histocompatibility antigen in T cell receptor transgenic mice. *Biol Reprod* 89: 102.

Phillips, T. A., J. Ni, and J. S. Hunt. 2001. Death-inducing tumour necrosis factor (TNF) superfamily ligands and receptors are transcribed in human placentae, cytotrophoblasts, placental macrophages and placental cell lines. *Placenta* 22: 663–72.

Phillips, T. A., J. Ni, G. Pan, et al. 1999. TRAIL (Apo-2L) and TRAIL receptors in human placentas: Implications for immune privilege. *J Immunol* 162: 6053–9.

Plaks, V., T. Birnberg, T. Berkutzki, et al. 2008. Uterine DCs are crucial for decidua formation during embryo implantation in mice. *J Clin Invest* 118: 3954–65.

Pollard, J. W. 2008. Uterine DCs are essential for pregnancy. *J Clin Invest* 118: 3832–5.

Raghupathy, R., M. Makhseed, F. Azizieh, et al. 1999. Maternal Th1- and Th2-type reactivity to placental antigens in normal human pregnancy and unexplained recurrent spontaneous abortions. *Cell Immunol* 196: 122–30.

Rajagopalan, S., and E. O. Long. 1999. A human histocompatibility leukocyte antigen (HLA)-G-specific receptor expressed on all natural killer cells. *J Exp Med* 189: 1093–100.

Robertson, S. A. 2005. Seminal plasma and male factor signalling in the female reproductive tract. *Cell Tissue Res* V322: 43–52.

Robillard, P. Y., T. C. Hulsey, J. Perianin, et al. 1994. Association of pregnancy-induced hypertension with duration of sexual cohabitation before conception. *Lancet* 344: 973–5.

Romagnani, S., E. Maggi, F. Liotta, L. Cosmi, and F. Annunziato. 2009. Properties and origin of human Th17 cells. *Mol Immunol* 47 (1): 3–7.

Rowe, J. H., J. M. Ertelt, L. Xin, and S. S. Way. 2012. Pregnancy imprints regulatory memory that sustains anergy to fetal antigen. *Nature* 490: 102–6.

Sakaguchi, S. 2005. Naturally arising Foxp3-expressing CD25+CD4+ regulatory T cells in immunological tolerance to self and non-self. *Nat Immunol* 6: 345–52.

Sakaguchi, S., N. Sakaguchi, M. Asano, M. Itoh, and M. Toda. 1995. Immunologic self-tolerance maintained by activated T cells expressing IL-2 receptor alpha-chains (CD25). Breakdown of a single mechanism of self-tolerance causes various autoimmune diseases. *J Immunol* 155: 1151–64.

Santner-Nanan, B., M. J. Peek, R. Khanam, et al. 2008. Increased Th17 cells and decreased CD4+Foxp3+ T regulatory cells in third trimester pregnancies with pre-eclampsia. *38th Annual Australasian Society of Immunology*, October 2015, Canberra, Australia.

Sasaki, Y., D. Darmochwal-Kolarz, D. Suzuki, et al. 2007. Proportion of peripheral blood and decidual CD4(+) CD25(bright) regulatory T cells in pre-eclampsia. *Clin Exp Immunol* 149: 139–45.

Sasaki, Y., M. Sakai, S. Miyazaki, et al. 2004. Decidual and peripheral blood CD4+CD25+ regulatory T cells in early pregnancy subjects and spontaneous abortion cases. *Mol Hum Reprod* 10: 347–53.

Schonkeren, D., M. L. van der Hoorn, P. Khedoe, et al. 2011. Differential distribution and phenotype of decidual macrophages in preeclamptic versus control pregnancies. *Am J Pathol* 178: 709–17.

Sharkey, D. J., A. M. Macpherson, K. P. Tremellen, and S. A. Robertson. 2007. Seminal plasma differentially regulates inflammatory cytokine gene expression in human cervical and vaginal epithelial cells. *Mol Hum Reprod* 13: 491–501.

Shima, T., Y. Sasaki, M. Itoh, et al. 2010. Regulatory T cells are necessary for implantation and maintenance of early pregnancy but not late pregnancy in allogeneic mice. *J Reprod Immunol* 85: 121–9.

Smith, T. R., and V. Kumar. 2008. Revival of CD8+ treg-mediated suppression. *Trends Immunol* 29: 337–42.

Svensson, L., M. Arvola, M. A. Sallstrom, R. Holmdahl, and R. Mattsson. 2001. The Th2 cytokines IL-4 and IL-10 are not crucial for the completion of allogeneic pregnancy in mice. *J Reprod Immunol* 51: 3–7.

Tilburgs, T., D. L. Roelen, B. J. van der Mast, et al. 2006. Distribution of CD4(+)CD25(bright) and CD8(+)CD28(-) T-cells in decidua and maternal blood during human pregnancy. *Placenta* 27 (Suppl A): S47–53.

Tilburgs, T., D. L. Roelen, B. J. van der Mast, et al. 2008. Evidence for a selective migration of fetus-specific CD4+CD25bright regulatory T cells from the peripheral blood to the decidua in human pregnancy. *J Immunol* 180: 5737–45.

Tilburgs, T., S. A. Scherjon, and F. H. Claas. 2010. Major histocompatibility complex (MHC)-mediated immune regulation of decidual leukocytes at the fetal-maternal interface. *J Reprod Immunol* 85: 58–62.

von Andrian, U. H., and C. R. Mackay. 2000. T-cell function and migration. Two sides of the same coin. *N Engl J Med* 343: 1020–34.

Wang, W. J., C. F. Hao, L. Yi, et al. 2010. Increased prevalence of T helper 17 (Th17) cells in peripheral blood and decidua in unexplained recurrent spontaneous abortion patients. *J Reprod Immunol* 84: 164–70.

Wegmann, T. G., H. Lin, L. Guilbert, and T. R. Mosmann. 1993. Bidirectional cytokine interactions in the maternal-fetal relationship: Is successful pregnancy a TH2 phenomenon? *Immunol Today* 14: 353–6.

Wilczynski, J. R. 2005. Th1/Th2 cytokines balance—Yin and yang of reproductive immunology. *Eur J Obstet Gynecol Reprod Biol* 122: 136–43.

Xu, C., D. Mao, V. M. Holers, et al. 2000. A critical role for murine complement regulator crry in fetomaternal tolerance. *Science* 287: 498–501.

Zenclussen, A. C. 2005. CD4(+)CD25+ T regulatory cells in murine pregnancy. *J Reprod Immunol* 65: 101–10.

Zhu, J., and W. E. Paul. 2008. CD4 T cells: Fates, functions, and faults. *Blood* 112: 1557–69.

4 The Human Placenta in Wound Healing
Historical and Current Approaches

Carmen García Insausti, José María Moraleda,
Gregorio Castellanos, and Francisco José Nicolás

CONTENTS

Preface 50
4.1   Introduction 50
4.2   Acute Wound Healing 51
    4.2.1   Hemostasis 51
    4.2.2   Inflammation 51
    4.2.3   Proliferative Phase 52
        4.2.3.1   Fibroblast Migration 52
        4.2.3.2   Collagen Synthesis 52
        4.2.3.3   Angiogenesis 52
        4.2.3.4   Granulation Tissue Formation 53
        4.2.3.5   Epithelialization 53
    4.2.4   Remodeling Phase 53
4.3   Chronic Wounds 54
4.4   Amniotic Membrane and Wound Healing 56
4.5   The Use of AM in Chronic Wound Healing 57
    4.5.1   Boiled and Dehydrated AM 57
    4.5.2   Fresh AM 58
    4.5.3   Irradiated AM 59
    4.5.4   Cryopreserved AM 59
4.6   Mechanisms Involved in AM-Induced Skin Re-Epithelialization 59
4.7   TGF-β, Chronic Wound Healing, and AM 60
4.8   Concluding Remarks 62
References 63

PREFACE

In this chapter, we review current knowledge of the physiological process of wound healing, a highly coordinated process that is disrupted in chronic wounds. In most chronic wounds, the healing process is stuck in the inflammatory or proliferative phase. This is particularly evident in large, massive wounds with considerable tissue loss. Such wounds become senescent in the process of inflammation or proliferation, losing the ability to epithelialize. Generally, chronic wounds do not respond to current treatments; therefore, they need special interventions. A plethora of wound dressings and devices have become available during the past two decades. However, their performance is not optimal because chronic wounds persist as a serious unresolved medical concern. Since 2007, we have been using amniotic membrane (AM) to treat wounds with considerable tissue loss with good results. AM is a tissue of particular interest as a biological dressing, and it has well-documented re-epithelialization effects that are in part related to its capacity to synthesize and release biologically active factors. The mechanisms involved in AM-induced skin re-epithelialization are largely unknown. Our studies have demonstrated that AM is able to induce epithelialization in large wounds that were unable to epithelialize. AM induces several signaling pathways that are involved in cell migration, proliferation, or both. In addition, AM is able to selectively antagonize the antiproliferative effect of transforming growth factor beta (TGF-β) by modifying the genetic program that TGF-β induces on keratinocytes. The combined effects of AM on keratinocytes, promoting cell proliferation and migration and antagonizing the effect of TGF-β, is the perfect combination that allows chronic wounds to move out of their nonhealing state and progress into epithelialization.

4.1 INTRODUCTION

Skin is a complex organ with different functions; it defends the body against external chemicals and mechanical or physical insults, and it creates a safety barrier against infection. It also participates in water exchange, electrolyte balance, and body temperature regulation (Enoch and Price 2004). When skin structure breaks down due to any cause, these functions are no longer adequately performed, and it is therefore vital to restore its integrity as soon as possible. A wound is defined as a breakage in the epithelial integrity of the skin. However, the disruption could be deeper, extending to the dermis, subcutaneous fat, fascia, muscle, or even the bone.

Wound healing is the biological process responsible for restoring tissue that has suffered a lesion, thereby compromising its morphological integrity. Normal wound healing represents a dynamic and well-ordered biological process involving soluble mediators, blood cells, the extracellular matrix, and parenchymal cells (Martin 1997; Singer and Clark 1999). This process leads to a completely healed wound, usually seen after simple injury. Such a wound is defined as one that has returned to its normal anatomical structure, function, and appearance within a reasonable time, and one that has attained complete skin closure without drainage or dressing requirements. In contrast to these simple wounds, some wounds fail to heal in a timely and orderly manner, resulting in chronic, nonhealing wounds. Chronic wounds can have several causes, including venous, arterial, and neuropathic involvement; pressure;

The Human Placenta in Wound Healing 51

vasculitis; and burns (Enoch and Price 2004). Extended traumatic deep wounds usually also become chronic (Insausti et al. 2010a). It has been reported that important cellular, molecular, and biochemical differences exist between the acute and the chronic wound environment (Enoch and Price 2004).

4.2 ACUTE WOUND HEALING

Healing in acute wounds goes through four phases that overlap in time: hemostasia, inflammation, proliferation, and tissue remodeling (Enoch and Price 2004; Werner and Grose 2003). The regulation of these events is multifactorial.

4.2.1 HEMOSTASIS

Hemostasis is the first stage of wound healing. After tissue injury, the lacerated vessels immediately constrict and thromboplastic tissue products, predominantly from the subendothelium, are exposed. Platelets aggregate and form the initial hemostatic plug. The activation of coagulation factors leads to the formation of thrombin; thrombin converts fibrinogen to fibrin that is subsequently polymerized into a stable clot. As a thrombus is formed, hemostasis in the wound is achieved (Lorenz and Longaker 2003). The clot, consisting of fibrin, fibronectin, vitronectin, von Willebrand factor, and thrombospondin, provides the provisional matrix for cellular migration (Clark et al. 1982; Grinnell et al. 1981). The aggregated platelets degranulate, and the alpha-granules release platelet-derived growth factor (PDGF), insulin-like growth factor 1 (IGF-1), epidermal growth factor (EGF), and TGF-β, all of which are 1) potent chemoattractants for inflammatory cells, 2) activation factors for local fibroblasts and endothelial cells, and 3) vasoconstrictors (Lorenz and Longaker 2003). After the transient vasoconstriction induced by these platelet factors, local small vessels dilate, secondary to the effects of vasoactive amines such as serotonin from the platelet-dense granules, and the coagulation and complement factors. These factors increase blood vessel permeability and exudation of fluid into the extravascular space and cause tissue edema, which is more prominent during the inflammatory phase.

4.2.2 INFLAMMATION

Inflammation is the next phase of healing. It begins with complement activation and the initiation of the classical molecular cascade that leads to infiltration of the wound with granulocytes or polymorphonuclear leukocytes (PMNLs). The complement components also stimulate the release of histamine and leukotrienes from mast cells. Within a short time, the PMNLs begin to adhere to the endothelial cells in the adjacent blood vessels through margination and start to move through the vessel wall by diapedesis (Enoch and Price 2004). The local endothelial cells then break cell-to-cell contact, thereby enhancing the margination of inflammatory cells into the wound site (Roberts and Tabares 1995). An efflux of white blood cells, first neutrophils and later monocytes, and also plasma proteins enter the wound site. The early neutrophil infiltrate scavenges cellular debris, foreign bodies, and bacteria. Activated complement fragments aid in bacterial killing through opsonization. Within 2–3 days, the inflammatory cell population begins to shift to one of monocyte predominance. Circulating monocytes are attracted to the wound site (Clark 1996; Lorenz and Longaker 2003; Wahl and Wahl 1992) by a variety of chemoattractants, such as the complement, clotting components, immunoglobulin G fragments, collagen and elastin breakdown products, and cytokines. These elicited monocytes differentiate into macrophages, and in conjunction with resident macrophages, orchestrate the repair process. Macrophages, besides phagocytosing tissue and bacterial debris, secrete multiple growth factors that activate and attract local endothelial cells, fibroblasts, and keratinocytes to begin their respective repair functions. Macrophages also release proteolytic enzymes such as collagenase that can lead to tissue debridement (Enoch and Price 2004). Macrophages appear to have a pivotal role in the transition between inflammation and repair (Riches 1996; Singer and Clark 1999).

4.2.3 PROLIFERATIVE PHASE

The proliferative phase starts at about day 3 and lasts for 2 weeks after wounding. It is characterized by the replacement of the provisional fibrin–fibronectin matrix with newly formed granulation tissue. The whole phase can be understood through five concomitant events.

4.2.3.1 Fibroblast Migration

The fibroblast migration phase begins with the deposition of the fibrin–fibrinogen matrix and the activation and turnover of local fibroblasts (fibroplasia). These cells replace the clot and produce an extracellular matrix (ECM) as the filling material. The initial fibrin–fibrinogen matrix is populated with platelets and macrophages. These macrophages and the local ECM release growth factors that initiate fibroblast activation. Fibroblasts migrate into the wound by using the newly deposited fibrin–fibronectin matrix as a scaffold. Local fibroblasts become activated and increase protein synthesis in preparation for cell division. As fibroblasts proliferate, they become the prominent cell type by 3–5 days in clean, noninfected wounds (Lorenz and Longaker 2003). After cell division and proliferation, fibroblasts begin synthesis and secretion of ECM products. The initial wound matrix is provisional and is composed of fibrin and a glycosaminoglycan (GAG), hyaluronic acid (Insausti et al. 2010a; Lorenz and Longaker 2003). This acid provides a matrix that enhances cell migration.

4.2.3.2 Collagen Synthesis

As fibroblasts enter and populate the wound, they use hyaluronidase to digest the provisional hyaluronic acid-rich matrix, and larger, sulfated GAGs are subsequently deposited. Concomitantly, collagens are deposited by fibroblasts onto the fibronectin and GAG scaffold in a disorganized array. Collagen types I and III are the major fibrillar collagens making up the ECM and are the major structural proteins both in unwounded and wounded skin. They provide strength and integrity for all tissues in the body and therefore play a particularly vital role in wound repair. Most collagen types are synthesized by fibroblasts; however, it is now known that some types are synthesized by epidermal cells (Lorenz and Longaker 2003; Marinkovich et al. 1993).

4.2.3.3 Angiogenesis

The process of forming new blood vessels occurs concurrently during all stages of the healing process. During the hemostatic phase, TGF-β and PDGF secreted

The Human Placenta in Wound Healing

by the platelets attract macrophages and granulocytes and promote angiogenesis. The macrophages, in particular, play a key role in angiogenesis by releasing several other angiogenic substances such as tumor necrosis factor α (TNF-α) and fibroblastic growth factor b (FGF-b). Angiogenic capillary sprouts invade the fibrin–fibronectin-rich wound clot and organize into a microvascular network throughout the granulation tissue (Enoch and Price 2004; Tonnesen et al. 2000). As collagen accumulates in the granulation tissue to produce scar tissue, the density of blood vessels diminishes. Disturbance of this dynamic process may influence the development of chronic wounds (Enoch and Price 2004; Lorenz and Longaker 2003).

4.2.3.4 Granulation Tissue Formation

Granulation tissue is made up mainly of proliferating fibroblasts, capillaries, and tissue macrophages in a matrix of collagen; GAGs, including hyaluronic acid; and the glycoproteins fibronectin and tenascin (Clark 1989; Kurkinen et al. 1980). Its formation is evident as early as 48 h after wounding and by 96 h fibroblasts become the predominant cell type in this tissue (Enoch and Price 2004).

4.2.3.5 Epithelialization

Epithelialization is the resurfacing of a wound with new epithelium. It consists of both the migration and proliferation of keratinocytes at the periphery of the wound (Santoro and Gaudino 2005). Within hours of the injury, morphological changes in keratinocytes at the wound margin are evident. In skin wounds, the epidermis thickens and marginal basal cells enlarge and migrate over the wound healing defect (Lorenz and Longaker 2003; Santoro and Gaudino 2005). Once these epithelial cells begin migrating, they do not divide until epidermal continuity is restored. New epithelial cells for wound closure are provided by fixed basal cells in a zone near the edge of the wound (Lorenz and Longaker 2003; Woodley 1996). Their daughter cells flatten and migrate over the wound matrix as a sheet (epiboly). Cell adhesion glycoproteins, such as tenascin and fibronectin, provide the "railroad tracks" to facilitate epithelial cell migration over the wound matrix. After the reestablishment of the epithelial layer, keratinocytes and fibroblasts secrete laminin and type IV collagen to form the basement membrane (Lorenz and Longaker 2003; Marinkovich et al. 1993). The keratinocytes then become columnar and divide as the layering of the epidermis is established, thus re-forming a barrier to further contamination and moisture loss. Further epithelial cell growth and differentiation reestablishes the stratified epithelium. The rate of epithelial coverage is increased if the wound does not require debridement, if the basal lamina is intact, and if the wound is kept moist. A dry eschar (scab) slows the rate of epithelialization. Several growth factors, such as EGF, FGF-b, and keratinocyte growth factor (KGF), modulate epithelialization.

4.2.4 REMODELING PHASE

Matrix synthesis and the remodeling phase are initiated concurrently with the development of granulation tissue and continue over prolonged periods. As the matrix matures, fibronectin and hyaluronic acid are broken down and collagen bundles increase in diameter at the same rate as increasing wound tensile strength

(Clark et al. 1995; Enoch and Price 2004; Welch et al. 1990). Wound tensile strength increases rapidly from 1 to 8 weeks postwounding. However, these collagen fibers never regain more than 80% of unwounded skin strength (Enoch and Price 2004). The regulation of the remodeling is poorly understood, but simplistically it can be conceptualized as the balance between synthesis, deposition, and degradation. Lysyl oxidase is the major intermolecular collagen crosslinking enzyme. Collagen crosslinking improves wound tensile strength. The matrix metalloproteinases (MMPs) collagenases, gelatinases, and stromelysins degrade ECM components. The balance of collagen deposition and degradation is in part determined by the regulation of MMP activity. As remodeling of the wound continues, MMP activity decreases and the activity of tissue inhibitors of metalloproteinases increases. TGF-β plays an important role in mediating this interaction, underlining the ability of TGF-β to promote matrix accumulation. Early collagen deposition is highly disorganized, but its subsequent organization is primarily achieved by wound contraction. Wound remodeling occurs when the underlying contractile connective tissue shrinks to bring the wound margins closer together. Contraction occurs through the interactions between fibroblasts and the surrounding ECM. These interactions may be influenced by many extracellular factors, including TGF-β, PDGF, and FGF (Grinnell 1994). With continued remodeling the outgrowth of capillaries is halted, blood flow to the area is reduced, and metabolic activity in the area declines. An acellular, avascular scar is the final result of an acute wound healing process.

4.3 CHRONIC WOUNDS

In severe pathological conditions, wound healing development can be altered, delaying the whole process (Baskovich et al. 2008; Bello and Phillips 2000). For example, in big wounds with a considerable loss of tissue affecting the skin, subcutaneous tissue, fascia, or even muscle, the first phase of filling can lengthen for too long, thereby generating a chronic wound that interferes with the development of normal wound healing (Baskovich et al. 2008; Bello and Phillips 2000). Chronic wounds have been defined as those that have failed to proceed through an orderly and timely process to produce anatomical and functional integrity, or that proceed through the repair process without establishing a sustained anatomical and functional result (Singer and Clark 1999). The normal process of healing could be disrupted at one or more of the points in the phases of hemostasis, inflammation, proliferation, and remodeling (Enoch and Price 2004; Lazarus et al. 1994), due to different factors that could result in impaired vascularization affecting oxygenation, deficient levels of cytokines, fibrotic and desiccated tissues, or growth factors promoting cell senescence and necrosis, among others.

In most chronic wounds, the healing process is stuck in the inflammatory phase or proliferative phase. An extreme or long-lasting inflammatory process can be negative for wound healing because of the production of free radicals in the environment, with their negative consequences upon microbes and healthy cells. This response is evident in chronic wounds and causes a persistent destruction of tissue (Martin and Leibovich 2005). Macrophages, which arrive at a late stage, are capable of cleaning the wound of cell debris and ECM residues such as fibrin and perishing neutrophils. These macrophages produce numerous cytokines, angiogenic factors, and growth factors that have an important role in fibrotic proliferation and angiogenesis. Mastocytes are other important cells coming from the bloodstream, although they are recruited later than neutrophils and macrophages. Mastocytes have a role in the post-inflammatory phase and later mechanism of tissue repair. All these cells secrete a cocktail of growth factors and cytokines with a tissue repair function that directs the behavior of cells responsible for wound closure. In addition, mastocytes produce an inflammation amplification signal that introduces more neutrophils, macrophages, and mastocytes into the wound. Only when inflammation has toned down and the repair is complete will inflammatory cells be dispersed from the wound (Enoch and Price 2004). Wound fluid derived from chronic venous leg ulcers is rich in proinflammatory cytokines such as TNF-α, interleukin (IL)-1β, and TGF-β1 (Harris et al. 1995).

Massive wounds with important tissue loss become senescent in the process of inflammation or proliferation, losing the ability to epithelialize (Baskovich et al. 2008; Bello and Phillips 2000). This failure in the re-epithelialization process requires intervention to provide the epithelial layer for the final sealing of the skin. The wound environment must be optimized prior to treatment. Surgical debridement must be aggressive and down to the level of viable, bleeding tissue. Sharp surgical debridement of all necrotic, nonviable tissues and wound debris from the wound bed is the most effective and rapid way to create a clean, viable environment. Furthermore, removal of bacteria and other unwanted wound contents might expedite wound closure, increase the availability of viable cells, and produce an optimal response to cytokines (Mulder and Vande Berg 2002).

In large-surface and deep wounds, with a substantial loss of soft tissue, in which primary wound closure is neither possible nor practicable, the most important issue is to dress the wound with appropriate materials to keep it free from infection, to reduce or eliminate pain and all potential factors inhibiting natural healing, and to replace or substitute the missing tissue as much as possible. In these cases, autologous skin grafting in the form of split- or full-thickness skin is a standard therapeutic criterion. In some patients, however, alternative therapies are needed because of the existence of comorbidity (Bello and Phillips 2000). Most of these therapies try to stimulate the formation of granulation tissue. Granulation tissue is characterized by a dense population of blood vessels, macrophages, and fibroblasts embedded within a loose provisional matrix of fibronectin, hyaluronic acid, and collagen (Lorenz and Longaker 2003).

Among the measures that could increase granulation tissue is the application of vacuum-assisted closure (VAC), which in the experience of some investigators and our own, stimulates granulation tissue (Bovill et al. 2008; Insausti et al. 2010a; Morykwas et al. 1997). The presence of granulation tissue has been used as a clinical indicator that the wound is ready for skin graft treatment (Lorenz and Longaker 2003) or for the use of other measures that could improve epithelialization (Insausti et al. 2010a).

A plethora of wound dressings and devices have become available over the past two decades. They include synthetic dressings of several types and allo-skin or auto-skin substitutes, although their cost remains too high for routine clinical practice (Greaves et al. 2013; Lorenz and Longaker 2003). New technologies involving growth

56 Placenta: The Tree of Life

factors and bioengineered tissues are relatively recent and have produced relatively good results; nevertheless, like some other dressings they are quite expensive.

4.4 AMNIOTIC MEMBRANE AND WOUND HEALING

Amniotic membrane (AM) is the innermost layer of the placenta. AM has a fetal origin and can be separated from the placenta by blunt dissection. AM is a tissue of particular interest as a biological dressing due to its special structure, biological properties, and immunological characteristics. It has low immunogenicity and well-documented re-epithelialization effects, together with anti-inflammatory, antifibrotic, antimicrobial, and nontumorigenic properties. These pleiotropic functions are related in part to its capacity to synthesize and release biologically active substances, including cytokines and signaling molecules such as TNF-α, TGF-α, TGF-β, basic EGF, EGF, KGF, hepatic growth factor, IL-4, IL-6, IL-8, natural inhibitors of metalloproteases, β-defensins, and prostaglandins, among others (Parolini et al. 2008; Parolini and Soncini 2006; Yang et al. 2006). In addition to these functions, AM may function as a substrate where cells can proliferate and differentiate. Alternatively, AM is a biomaterial that can be easily obtained, processed, and transported (Miki and Strom 2006).

AM use offers considerable advantages compared to skin transplantation. Its application does not produce rejection because it has low immunogenicity and does not induce uncontrolled proliferation (Insausti et al. 2010b). All these effects are related to its capacity for the production and release of biologically active substances (see above).

AM has been used in clinical medicine for more than 100 years. Davis (1910) reported a comprehensive review of 550 cases of skin transplantation to various types of burns and wounds, using natural AM obtained from labor and delivery at the Johns Hopkins University. Sabella (1913) and Stern (1913) separately reported on the use of preserved AM in skin grafting for burns and ulcers. Since then, there have been several reports of the uses of AM in the treatment of wounds of different etiologies and other applications: first in the reconstructive surgery of different tissues and organs, including the mouth, tongue, nasal mucosa, larynx, eardrum, vestibule, bladder, urethra, vagina, and tendons (Brandt et al. 2000; Fishman et al. 1987; Ganatra 2003; Georgy and Aziz 1996; Morton and Dewhurst 1986; Tolhurst and van der Helm 1991; Zohar et al. 1987); second, as a peritoneum substitute in reconstruction procedures of pelvic exenteration surgery; third, in adherence prevention in the abdomen and pelvic surgery; and finally as a covering of onfaloceles and the like (Davis 1910; Ganatra 2003; Sabella 1913; Stern 1913; Trelford and Trelford-Sauder 1979).

In ophthalmology, the use of AM was reported for the first time in 1940 by de Rötth, who used fresh fetal membranes, namely, amnion and chorion, in the ocular surface as a biological dressing in the management of conjunctival alterations (de Rötth 1940). Later, Sorsby and Symons (1946) used preserved AM as a temporary coating in the treatment of acute caustic ocular lesions. Although the results were favorable, for reasons probably related to difficulties in the preparation, preservation, and handling of the AM, its use was abandoned for almost four decades. In 1995, with the reconstitution assays of rabbit corneas with limbic disorder by using human

The Human Placenta in Wound Healing 57

preserved AM by Kim and Tseng (1995), there was a renewed widespread interest in the use of AM in ophthalmology, and many papers were published related to the efficacy of the AM in various ocular surface conditions and in diseases such as epidermolysis bullosa (Gomes et al. 2005; Trelford and Trelford-Sauder 1979; Mermet et al. 2007). Currently, AM is a resource widely used in ophthalmology (Baradaran-Rafii et al. 2007; Dua et al. 2004; Gomes et al. 2005), and to a lesser degree, in the treatment of wounds, burns lesions, chronic ulcers of the legs (Colocho et al. 1974; Hasegawa et al. 2007; Mermet et al. 2007; Singh et al. 2004), and in other surgical and nonsurgical procedures (Brandt et al. 2000; Díaz-Prado et al. 2010; Fishman et al. 1987; Georgy and Aziz 1996; Insausti et al. 2010a; Morton and Dewhurst 1986; Redondo et al. 2011; Sangwan et al. 2006; Tolhurst and van der Helm 1991; Yeager et al. 1985; Zohar et al. 1987).

4.5 THE USE OF AM IN CHRONIC WOUND HEALING

The source of AM for wound healing is donated placenta. AM has been used for wound healing either as intact AM without epithelium removal or as denuded AM without the epithelium (Akle et al. 1985; Wilshaw et al. 2006): In some cases, AM was used fresh and in others cases AM was preserved. Currently, it is known that the use of fresh AM is not practical for clinical use (Zelen et al. 2015). Methods to remove the epithelium or preserve AM are very diverse and exceed the scope of this chapter. However, we describe some papers where the researchers have used different forms of AM as a biological dressing to help the wound healing process of chronic ulcers.

4.5.1 BOILED AND DEHYDRATED AM

One of the earlier reports about the treatment of chronic leg ulcers in the elderly using AM is that of Troensegaard-Hansen (1950). The treatment of seven patients with chronic leg ulcers with AM boiled in saline allowed them to heal in 5–10 weeks, with ulcers remaining healed after the application.

A dehydrated form of human amnion/chorion membrane (dHACM) allograft has been used in several reports for the improvement and cure of chronic wounds such as diabetic neuropathic foot ulcers, venous or arterial insufficiency ulcers, and pressure ulcers. Diabetic lower extremity wounds, for example, remain refractory to standard treatment and care (Zelen et al. 2015). Venous leg ulcers are also slow to heal, with less than a third of them healed by 24 weeks of treatment (Margolis et al. 2004) using standard care.

dHACM has been used in several cases with success. A preliminary case study involved research on three patients that were diagnosed with insulin-dependent or noninsulin-dependent diabetes and were receiving a conservative treatment for a foot ulcer. Advanced therapy with dHACM was used because the wounds did not decrease to less than 50% after 4 weeks of care. In the three cases, the use of dHACM was able to completely heal the wounds (Shah 2014). After these cases, dHACM has been used for the treatment of different types of diabetic ulcers, with very satisfactory results (Zelen 2013). There has also been a prospective randomized comparative study of the application of dHACM allograft in the management of these ulcers

(Zelen et al. 2014b). In these cases, wound recidivism is often a problem, so a study was conducted on patients that had received dHACM and healed in the previous trials. Eighteen patients were followed up for about a year, and all except one of the ulcers were healed after a year, suggesting the durability of the treatment (Zelen et al. 2014a). For these kinds of ulcers, the researchers compared the weekly versus biweekly application of AM in a trial with 40 patients. They concluded that the differences in healing rates for the weekly versus biweekly applications of dHACM showed a clear clinical improvement from weekly application because it induced more than 40% faster healing of wounds (Zelen et al. 2014a). The faster healing with weekly application also has an economic advantage (Zelen et al. 2015).

This kind of dressing has also been used in different types of ulcer cases where no diabetes was involved. For example, another case series described four patients referred for plastic surgery that were treated with dHACM instead. Healing was observed in patients with a variety of wound types, all having in common chronic wounds that were refractory to treatment. Application of this therapy was successful, eliminating the need for surgical intervention (Sheikh et al. 2014). In another retrospective case series of five patients, Forbes and Fetterolf (2012) referred to the use of AM for wounds that fail to heal at least over a month: recalcitrant wounds (Forbes and Fetterolf 2012). The ulcers treated with dHACM showed improved healing accompanied by a change in healing trajectories compared with those documented before application (Forbes and Fetterolf 2012).

Venous leg ulcers represent the largest category of ambulatory wounds in developed countries and are responsible for causing pain and disability in thousands of patients (Zelen et al. 2015). A multicenter randomized controlled clinical trial with 84 patients evaluated the use of dHACM and multilayered compression therapy versus a multilayered compression technique alone in the treatment of venous leg ulcers (Serena et al. 2014). This study showed that venous ulcers treated with dHACM allograft were significantly better at 4 weeks than wounds treated with multilayered compression alone. Also, it has been suggested that dHACM injection may be a viable treatment for a variety of tendon and ligament injuries (Zelen et al. 2015). For example, there has been a clinical trial evaluating the injection of dHACM for the treatment of plantar fasciitis (Zelen et al. 2013). Patients receiving dHACM experienced significantly reduced pain over the study period, whereas this was not observed in the controls.

4.5.2 Fresh AM

Faulk et al. (1980) used fresh amnion applied to chronic leg ulcers and compared it to regular dressings before the application of autologous skin transplantation. They reported that a superior, clean, red, delicate layer of granulation tissue had formed throughout the wound bed after 5 days of treatment with AM. Ward and Bennett (1984) used fresh amnion that was applied for 5 days to venous ulceration before the application of an autograft. Twenty-eight patients were properly followed up with 50% recurrence at 1 year. In a different study, Ward et al. (1989) tested the use of fresh, frozen, tissue culture–maintained, or lyophilized amnion on chronic leg ulcers in 27 patients. They found no statistically significant differences between the groups, only judging that lyophilized AM was the easiest to use and store.

The Human Placenta in Wound Healing

4.5.3 IRRADIATED AM

Singh et al. (2004) used gamma-irradiated AM on 50 patients with ulcers of different etiologies such as leprotic, diabetic, traumatic, gravitational, and superficial burns. They used between two and six applications of AM, and they reported the pain relief and healing of ulcers that were refractory to healing (Singh et al. 2004). Also, Gajiwala and Lobo Gajiwala (2003) treated six cases of superficial pressure sore and eight cases of nonhealing ulcers. In all cases, the treatment was carried out using irradiated human amnion. The amnion provided good biological covering in all the patients. It was easy to handle and apply, and it provided pain relief. Healing duration varied, depending on the extent and depth of the wound, and on the amount of exudate. The superficial bedsores healed with a single application of amnion. Reduced exudate, healthy granulation tissue, and enhanced epithelialization were observed after the application of amnion (Gajiwala and Lobo Gajiwala 2003).

4.5.4 CRYOPRESERVED AM

Keeping AM under −80°C in the presence of cryoprotectant substances preserves all its functions so it may be used as a biological dressing (Kruse et al. 2000; Yang et al. 2006). AM cryopreserved in glycerol has been used for the treatment of venous leg ulcers. In a prospective pilot study, Mermet et al. (2007) used glycerol-cryopreserved AM in 15 patients with chronic venous leg ulcers. In all cases, the researchers registered a significant increase in granulation tissue along with a significant decrease in fibrinous slough. A significant clinical response occurred in 80% of the patients, including complete healing in three of them during the 3-month follow-up period. All patients experienced a significant reduction of ulcer-related pain rapidly after AM transplantation (Mermet et al. 2007).

We reported the use of cryopreserved AM as a wound dressing in two patients with large and deep traumatic wounds. Negative pressure wound therapy was applied until a complete restoration of the soft tissue was observed. Until now, no sign of epithelialization has been observed in either patient. Rather than a skin transplant reconstruction, the application of AM was the decided upon approach. During AM treatment, epithelialization started very soon in both patients and continued until the total epithelialization of wound (Insausti et al. 2010a). Neither of the two cases presented adverse events. AM-induced newly formed skin showed evidence of fibrotic scarring tissue with large and abundant blood vessels in both the deep and superficial dermis.

4.6 MECHANISMS INVOLVED IN AM-INDUCED SKIN RE-EPITHELIALIZATION

The mechanisms involved in AM-induced skin re-epithelialization are largely unknown. It has been shown that AM might exert its wound healing effect by accelerating keratinocyte migration from the wound edge and inducing its differentiation, thereby contributing to the generation of intact epithelium (Lee and Tseng 1997). Niknejad et al. (2008) considered that the stimulatory effect on epithelialization from the wound bed, the wound edge, or both is mediated by growth factors and progenitor cells released by AM. In addition, it has been reported that the maintenance of the integrity of the basement membrane and stromal matrix improves the healing potency of AM and is crucial in promoting rapid re-epithelialization (Kubo et al. 2001).

Insausti et al. (2010a) had previously worked on HaCaT cells, a spontaneously immortalized human keratinocyte cell line, as a model to understand the molecular consequences of AM application on human wounds (Boukamp et al. 1988). This research showed that HaCaT cells exhibited different molecular responses upon stimulation with AM that were attributed to the effects of soluble AM-released factors on HaCaT cells (Insausti et al. 2010a). The application of AM to keratinocytes induced the activation of the phosphorylation of extracellular-signal-regulated kinase (ERK)1/2, c-Jun N-terminal kinase (JNK)1/2, and p38 (Insausti et al. 2010a). The addition of AM-conditioned medium induced similar responses, suggesting a trans effect of AM on the elicitation of these events. In addition, the researchers reported that HaCaT cells stimulated with AM showed an increased expression of *c-JUN*. Members of the AP1 family had been involved in keratinocyte migration and the wound healing process (Angel et al. 2001; Gangnuss et al. 2004; Li et al. 2003; Yates and Rayner 2002). AM induced the phosphorylation of JNK1 and two kinases in HaCaT cells (Insausti et al. 2010a); JNK1 is a positive regulator of c-JUN, contributing to its phosphorylation and stabilization (Ronai 2004; Sabapathy et al. 2004). Finally, the expression of c-Jun in the wounds treated with AM was very strong, and particularly evident at the basal epithelium near the leading edge and at the dermal leading edge or keratinocyte tongue, indicating that c-Jun expression might be an important event for epithelialization occurring at the AM-stimulated wound borders (Insausti et al. 2010a).

4.7   TGF-β, CHRONIC WOUND HEALING, AND AM

Wound fluid derived from chronic venous leg ulcers is rich in proinflammatory cytokines, such as TNF-α, IL-1β, and TGF-β1 (Harris et al. 1995). In addition, the levels of these cytokines decrease as the chronic wound begins to heal, indicating a significant correlation between nonhealing wounds and an increased level of proinflammatory cytokines (Trengove et al. 2000). TGF-β plays a critical role in regulating multiple cellular responses that occur in all three phases of wound healing (Finnson et al. 2013). Of the many cytokines shown to influence the wound healing process, TGF-β has the broadest spectrum of action because it affects the behavior of a wide variety of cell types and mediates a diverse range of cellular functions (Ashcroft and Roberts 2000). The TGF-β signaling pathway is considered as a promising target for the treatment of many pathological skin conditions, including chronic nonhealing wounds (Finnson et al. 2013). Platelets are thought to be the primary source of TGF-β at the wound site; also, activation of latent TGF-β occurs immediately after wounding (Finnson et al. 2013). Keratinocytes, fibroblasts, and monocytes are among the targeted cells in the TGF-β management of the wound (Ashcroft and Roberts 2000). Monocytes, macrophages, and fibroblasts then

The Human Placenta in Wound Healing   61

contribute to autocrine-perpetuated high concentrations of TGF-β at the wound site (Ashcroft and Roberts 2000).

TGF-β exerts its cellular effects by increasing the phosphorylation of members of the receptor-activated (R-) Smad family (Smad2 and 3), although non-Smad pathways are also activated, including the ERK, JNK, and p38 mitogen-activated protein kinase pathways, the tyrosine kinase Src, and phosphatidylinositol 3'-kinase (Moustakas and Heldin 2005; Mu et al. 2012). After receptor-induced phosphorylation, R-Smads form complexes with the common-mediator Smad4, which are translocated to the nucleus (Nicolas et al. 2004; Pierreux et al. 2000) where they, in cooperation with other transcription factors, coactivators, and corepressors, regulate the transcription of certain genes (Heldin and Moustakas 2012).

The effects of TGF-β on full-thickness wound re-epithelialization have been studied in a transgenic mouse with an overexpressed TGF-β at the epidermis level that exhibits a decrease in re-epithelialization (Chan et al. 2002; Yang et al. 2001). The study in the ear mouse model suggests that TGF-β has an inhibitory effect on epithelialization when the wound involves all the layers of the skin (Tredget et al. 2005). Abolishing part of the TGF-β signaling pathway has been suggested as a solution to improve wound healing, so abolishing part of the TGF-β–stimulated Smad pathways may improve wound healing and benefit the effect of TGF-β signaling over matrix synthesis by fibroblasts, for example (Ashcroft and Roberts 2000). TGF-β causes the growth arrest of epithelial cells. The mechanisms, which differ somewhat between different cell types, involve inhibition of the expression of the transcription factor Myc and members of the Id family, and the transcriptional induction of the cell-cycle inhibitors *CDKN2B* (*p15*) and *CDKN1A* (*p21*) (Heldin et al. 2009). The downregulation of Smad3 has been suggested as a possible way of improving wound healing (Ashcroft and Roberts 2000). Notably, the ability of keratinocytes to sense TGF-β through Smad3 prevents the cell proliferation of keratinocytes and consequently prevents wound healing resolution when the levels of TGF-β are high (Ashcroft et al. 1999).

To further unravel the molecular mechanism by which AM may contribute to the epithelialization and wound border proliferation in chronic posttraumatic wounds, Alcaraz et al. (2015) studied the relationship between TGF-β signaling and AM regulation in wound healing by using keratinocytes. Strikingly, AM was able to attenuate the TGF-β–induced phosphorylation of Smad2 and Smad3 in HaCaT cells. Both the strength and duration of TGF-β signaling, expressed as continuous phosphorylation of Smads, are required to achieve proper cell responses to TGF-β; the failure to do so produces a loss of the cell-cycle arrest in response to TGF-β (Nicolas and Hill 2003).

AM attenuates TGF-β–induced Smad2 and Smad3 phosphorylation and hence attenuates *CDKN2B* (*p15*) and *CDKN1A* (*p21*) expression (Alcaraz et al. 2015), which has been related to cell-cycle regulation (Nicolas and Hill 2003). The presence of AM may therefore counteract the cell-cycle arrest induced by TGF-β on keratinocytes, releasing them from the brake imposed by TGF-β (Alcaraz et al. 2015). The effect of AM on TGF-β–regulated genes is not indiscriminate, and not all genes are affected by the presence of AM. Interestingly, genes that positively participate in wound healing such as *SNAI-2* and *PAI-1* were synergistically upregulated by the presence of AM and TGF-β (Alcaraz et al. 2015). Finally, the expression of c-Jun was maximal when both TGF-β and AM were present in either HaCaT or primary keratinocyte cells (Alcaraz et al. 2015).

It has been suggested that AM might exert its wound healing effect by accelerating keratinocyte migration from the wound edge (Lee and Tseng 1997). Growth factors and progenitor cells released by AM (Niknejad et al. 2008) might mediate the stimulatory effect on epithelialization. AM induces cell migration in a wound healing assay in a cell model (Alcaraz et al. 2015). Furthermore, inhibition of cell proliferation with mitomycin C does not affect the migrating properties of AM. In the same study, use of JNK1 inhibitors prevented AM-induced cell migration. Moreover, a closer examination of the wound margins of the scratch wound healing assays showed a high expression of c-Jun in the AM-stimulated cells engaged in the migratory front. The AM-induced high expression of c-Jun at the wound border was prevented by inhibitors SP600125 and PD98059, which is consistent with the fact that AM induces the activation of a signaling cascade that produces the phosphorylation of ERK1/2 and JNK1/2. AM also induces a local increase in c-Jun in the patient wound border that is consistent with its effect on cell migration, and indeed, in examination of patient wound borders a few days after AM application, a clear proliferation and migration were observed (Alcaraz et al. 2015). This finding correlates with the robust expression of c-Jun at the wound border that is especially strong at the *stratum basale* of the epidermis coinciding with the keratinocyte tongue, the area where the migration of keratinocyte is produced to close the wound (Alcaraz et al. 2015). The researchers showed that the application of AM is able to promote healing in chronic wounds by modifying the genetic program induced by TGF-β, stimulating keratinocyte proliferation and migration (Alcaraz et al. 2015). In addition, there might be a synergy of AM and TGF-β signaling for the resolution of chronic wounds (Alcaraz et al. 2015; Ruiz-Canada et al. submitted). At the time this chapter was written, a clinical trial was being conducted in our hospital to evaluate the effect of AM on chronic posttraumatic wounds, with excellent results.

The Smad pathway stimulated by TGF-β has also been involved in the production of fibrosis and inflammation in response to TGF-β. Thus, interfering with TGF-β signaling may be a good way of interfering with fibrosis and improving the evolution of wound healing (Ashcroft and Roberts 2000). Indeed, the application of AM is able to ameliorate fibrosis in different experimental models (Cargnoni et al. 2009, 2012, 2014; Hodge et al. 2014). It will be very attractive to explore whether the application of AM is able to reduce fibrosis and inflammation in its application on chronic wounds.

4.8 CONCLUDING REMARKS

To summarize, AM is a biological dressing that promotes proper epithelialization in chronic wounds. It has several advantages; among them, it is inexpensive, easy to obtain, and in constant supply. In addition, AM can be cryopreserved at a low temperature while maintaining all its biological functions. Finally, it can be applied as an ambulatory treatment in the outpatient clinic, which decreases costs even more. Thus, AM must be kept in mind as a consolidated treatment for chronic wounds.

The Human Placenta in Wound Healing

REFERENCES

Akle, C., I. McColl, M. Dean, et al. 1985. Transplantation of amniotic epithelial membranes in patients with mucopolysaccharidoses. *Exp Clin Immunogenet* 2 (1): 43–48.

Alcaraz, A., A. Mrowiec, C. L. Insausti, et al. 2015. Amniotic membrane modifies the genetic program induced by TGFß, stimulating keratinocytes proliferation and migration in chronic wounds. *PLoS One* 10 (8): e0135324.

Angel, P., A. Szabowski, and M. Schorpp-Kistner. 2001. Function and regulation of AP-1 subunits in skin physiology and pathology. *Oncogene* 20 (19): 2413–23. doi: 10.1038/sj.onc.1204380.

Ashcroft, G. S., and A. B. Roberts. 2000. Loss of Smad3 modulates wound healing. *Cytokine Growth Factor Rev* 11 (1–2): 125–31.

Ashcroft, G. S., X. Yang, A. B. Glick, et al. 1999. Mice lacking Smad3 show accelerated wound healing and an impaired local inflammatory response. *Nat Cell Biol* 1 (5): 260–6. doi: 10.1038/12971.

Baradaran-Rafii, A., H. R. Aghayan, B. Arjmand, and M. A. Javadi. 2007. Amniotic membrane transplantation. *Iran J Ophthalmic Res* 2 (1): 58–75.

Baskovich, B., E. M. Sampson, G. S. Schultz, and L. K. Parnell. 2008. Wound dressing components degrade proteins detrimental to wound healing. *Int Wound J* 5 (4): 543–51. doi: 10.1111/j.1742-481X.2007.00422.x.

Bello, Y. M., and T. J. Phillips. 2000. Recent advances in wound healing. *JAMA* 283 (6): 716–18.

Boukamp, P., R. T. Petrussevska, D. Breitkreutz, et al. 1988. Normal keratinization in a spontaneously immortalized aneuploid human keratinocyte cell line. *J Cell Biol* 106 (3): 761–71.

Bovill, E., P. E. Banwell, L. Teot, et al. 2008. Topical negative pressure wound therapy: A review of its role and guidelines for its use in the management of acute wounds. *Int Wound J* 5 (4): 511–29. doi: 10.1111/j.1742-481X.2008.00437.x.

Brandt, F. T., C. D. Albuquerque, and F. R. Lorenzato. 2000. Female urethral reconstruction with amnion grafts. *Int J Surg Investig* 1 (5): 409–14.

Cargnoni, A., L. Gibelli, A. Tosini, et al. 2009. Transplantation of allogeneic and xenogeneic placenta-derived cells reduces bleomycin-induced lung fibrosis. *Cell Transplant* 18 (4): 405–22. doi: 10.3727/096368909788809857.

Cargnoni, A., E. C. Piccinelli, L. Ressel, et al. 2014. Conditioned medium from amniotic membrane-derived cells prevents lung fibrosis and preserves blood gas exchanges in bleomycin-injured mice-specificity of the effects and insights into possible mechanisms. *Cytotherapy* 16 (1): 17–32. doi: 10.1016/j.jcyt.2013.07.002.

Cargnoni, A., L. Ressel, D. Rossi, et al. 2012. Conditioned medium from amniotic mesenchymal tissue cells reduces progression of bleomycin-induced lung fibrosis. *Cytotherapy* 14 (2): 153–61. doi: 10.3109/14653249.2011.613930.

Chan, T., A. Ghahary, J. Demare, et al. 2002. Development, characterization, and wound healing of the keratin 14 promoted transforming growth factor-beta 1 transgenic mouse. *Wound Repair Regen* 10 (3): 177–87.

Clark, R. A. 1989. Wound repair. *Curr Opin Cell Biol* 1 (5): 1000–8.

Clark, R. A., J. M. Lanigan, P. DellaPelle, et al. 1982. Fibronectin and fibrin provide a provisional matrix for epidermal cell migration during wound reepithelialization. *J Invest Dermatol* 79 (5): 264–69.

Clark, R. A., L. D. Nielsen, M. P. Welch, and J. M. McPherson. 1995. Collagen matrices attenuate the collagen-synthetic response of cultured fibroblasts to TGF-beta. *J Cell Sci* 108 (Pt 3): 1251–61.

Clark, R. A. F. 1996. Wound repair. Overview and general considerations. In R. A. F. Clark (Ed.), *The Molecular and Cellular Biology of Wound Repair*, pp. 3–50. New York: Plenum Press.

Colocho, G., W. P. Graham, 3rd, A. E. Greene, D. W. Matheson, and D. Lynch. 1974. Human amniotic membrane as a physiologic wound dressing. *Arch Surg* 109 (3): 370–73. doi: 10.1001/archsurg.1974.01360030022006.

Davis J. W. 1910. Skin transplantation with a review of 550 cases at the Johns Hopkins Hospital. *Johns Hopkins Med J* 15: 307–396.

de Rötth, A. 1940. Plastic repair of conjunctival defects with fetal membranes. *Arch Ophthalmol* 23 (3): 522–5. doi: 10.1001/archopht.1940.00860130586006.

Díaz-Prado, S., M. E. Rendal-Vázquez, E. Muiños-López, et al. 2010. Potential use of the human amniotic membrane as a scaffold in human articular cartilage repair. *Cell Tissue Banking* 11 (2): 183–95. doi: 10.1007/s10561-009-9144-1.

Dua, Harminder, S., J. A. P. Gomes, A. J. King, and V. Senthil Maharajan. 2004. The amniotic membrane in ophthalmology. *Survey Ophthalmol* 49 (1): 51–77. doi: 10.1016/j.survophthal.2003.10.004.

Enoch, S., and Price, P. 2004. Cellular, molecular and biochemical differences in the pathophysiology of healing between acute wounds, chronic wounds and wounds in the aged. [online] Worldwidewounds.com. Available at: http://www.worldwidewounds.com/2004/august/Enoch/Pathophysiology-Of-Healing.html

Faulk, W. P., P. J. Stevens, H. Burgos, et al. 1980. Human amnion as an adjunct in wound healing. *Lancet* 315 (8179): 1156–8. doi: 10.1016/S0140-6736(80)91617-7.

Finnson, K. W., S. McLean, G. M. Di Guglielmo, and A. Philip. 2013. Dynamics of transforming growth factor beta signaling in wound healing and scarring. *Adv Wound Care (New Rochelle)* 2 (5): 195–214. doi: 10.1089/wound.2013.0429.

Fishman, I. J., F. N. Flores, F. B. Scott, H. J. Spjut, and B. Morrow. 1987. Use of fresh placental membranes for bladder reconstruction. *J Urol* 138 (5): 1291–94.

Forbes, J., and D. E. Fetterolf. 2012. Dehydrated amniotic membrane allografts for the treatment of chronic wounds: A case series. *J Wound Care* 21 (6): 290, 292, 294–96. doi: 10.12968/jowc.2012.21.6.290.

Gajiwala, K., and A. Lobo Gajiwala. 2003. Use of banked tissue in plastic surgery. *Cell Tissue Bank* 4 (2–4): 141–6. doi: 10.1023/B:CATB.0000007023.85139.c5.

Ganatra, M. A. 2003. Amniotic membrane in surgery. *J Pak Med Assoc* 53 (1): 29–32.

Gangnuss, S., A. J. Cowin, I. S. Daehn, et al. 2004. Regulation of MAPK activation, AP-1 transcription factor expression and keratinocyte differentiation in wounded fetal skin. *J Invest Dermatol* 122 (3): 791–804. doi: 10.1111/j.0022-202X.2004.22319.x.

Georgy, M. S., and N. L. Aziz. 1996a. Vaginoplasty using amnion graft: New surgical technique using the laparoscopic transillumination light. *J Obstet Gynaecol* 16 (4): 262–64. doi: 10.3109/01443619609020728.

Gomes, J. A., A. Romano, M. S. Santos, and H. S. Dua. 2005. Amniotic membrane use in ophthalmology. *Curr Opin Ophthalmol* 16 (4): 233–40.

Greaves, N. S., S. A. Iqbal, M. Baguneid, and A. Bayat. 2013. The role of skin substitutes in the management of chronic cutaneous wounds. *Wound Repair Regen* 21 (2): 194–210. doi: 10.1111/wrr.12029.

Grinnell, F. 1994. Fibroblasts, myofibroblasts, and wound contraction. *J Cell Biol* 124 (4): 401–4.

Grinnell, F., R. E. Billingham, and L. Burgess. 1981. Distribution of fibronectin during wound healing in vivo. *J Invest Dermatol* 76 (3): 181–89.

Harris, I. R., K. C. Yee, C. E. Walters, et al. 1995. Cytokine and protease levels in healing and non-healing chronic venous leg ulcers. *Exp Dermatol* 4 (6): 342–49.

Hasegawa, T., M. Mizoguchi, K. Haruna, et al. 2007. Amnia for intractable skin ulcers with recessive dystrophic epidermolysis bullosa: Report of three cases. *J Dermatol* 34 (5): 328–32. doi: 10.1111/j.1346-8138.2007.00281.x.

Heldin, C. H., M. Landstrom, and A. Moustakas. 2009. Mechanism of TGF-beta signaling to growth arrest, apoptosis, and epithelial-mesenchymal transition. *Curr Opin Cell Biol* 21 (2): 166–76. doi: 10.1016/j.ceb.2009.01.021.

The Human Placenta in Wound Healing 65

Heldin, C. H., and A. Moustakas. 2012. Role of Smads in TGF-beta signaling. *Cell Tissue Res* 347 (1): 21–36. doi: 10.1007/s00441-011-1190-x.

Hodge, A., D. Lourensz, V. Vaghjiani, et al. 2014. Soluble factors derived from human amniotic epithelial cells suppress collagen production in human hepatic stellate cells. *Cytotherapy* 16 (8): 1132–44. doi: 10.1016/j.jcyt.2014.01.005.

Insausti, C. L., A. Alcaraz, E. M. Garcia-Vizcaino, et al. 2010a. Amniotic membrane induces epithelialization in massive posttraumatic wounds. *Wound Repair Regen* 18 (4): 368–77. doi: 10.1111/j.1524-475X.2010.00604.x.

Insausti, C. L., M. Blanquer, P. Bleda, et al. 2010b. The amniotic membrane as a source of stem cells. *Histol Histopathol* 25 (1): 91–98.

Kim, J. C., and S. C. Tseng. 1995. Transplantation of preserved human amniotic membrane for surface reconstruction in severely damaged rabbit corneas. *Cornea* 14 (5): 473–84.

Kruse, F. E., A. M. Joussen, K. Rohrschneider, et al. 2000. Cryopreserved human amniotic membrane for ocular surface reconstruction. *Graefes Arch Clin Exp Ophthalmol* 238 (1): 68–75.

Kubo, M., Y. Sonoda, R. Muramatsu, and M. Usui. 2001. Immunogenicity of human amniotic membrane in experimental xenotransplantation. *Invest Ophthalmol Vis Sci* 42 (7): 1539–46.

Kurkinen, M., A. Vaheri, P. J. Roberts, and S. Stenman. 1980. Sequential appearance of fibronectin and collagen in experimental granulation tissue. *Lab Invest* 43 (1): 47–51.

Lazarus, G. S., D. M. Cooper, D. R. Knighton, et al. 1994. Definitions and guidelines for assessment of wounds and evaluation of healing. *Arch Dermatol* 130 (4): 489–93.

Lee, S. H., and S. C. Tseng. 1997. Amniotic membrane transplantation for persistent epithelial defects with ulceration. *Am J Ophthalmol* 123 (3): 303–12.

Li, G., C. Gustafson-Brown, S. K. Hanks, et al. 2003. c-Jun is essential for organization of the epidermal leading edge. *Dev Cell* 4 (6): 865–77.

Lorenz, H. P., and M. Longaker. 2003. Wounds: Biology, pathology, and management. In Norton, J., R. R. Bollinger, A. E. Chang, S. F. Lowry, S. J. Mulvihill, H. I. Pass, R. W. Thompson, and M. Li (Eds.), *Essential Practice of Surgery*, pp. 77–88. New York: Springer.

Margolis, D. J., L. Allen-Taylor, O. Hoffstad, and J. A. Berlin. 2004. The accuracy of venous leg ulcer prognostic models in a wound care system. *Wound Repair Regen* 12 (2): 163–8. doi: 10.1111/j.1067-1927.2004.012207.x.

Marinkovich, M. P., D. R. Keene, C. S. Rimberg, and R. E. Burgeson. 1993. Cellular origin of the dermal-epidermal basement membrane. *Dev Dyn* 197 (4): 255–67. doi: 10.1002/aja.1001970404.

Martin, P. 1997. Wound healing—Aiming for perfect skin regeneration. *Science* 276 (5309): 75–81.

Martin, P., and S. J. Leibovich. 2005. Inflammatory cells during wound repair: The good, the bad and the ugly. *Trends Cell Biol* 15 (11): 599–607. doi: 10.1016/j.tcb.2005.09.002.

Mermet, I., N. Pottier, J. M. Sainthillier, et al. 2007. Use of amniotic membrane transplantation in the treatment of venous leg ulcers. *Wound Repair Regen* 15 (4): 459–64. doi: 10.1111/j.1524-475X.2007.00252.x.

Miki, T., and S. C. Strom. 2006. Amnion-derived pluripotent/multipotent stem cells. *Stem Cell Rev* 2 (2): 133–42. doi: 10.1007/s12015-006-0020-0.

Morton, K. E., and C. J. Dewhurst. 1986. Human amnion in the treatment of vaginal malformations. *Br J Obstet Gynaecol* 93 (1): 50–54.

Morykwas, M. J., L. C. Argenta, E. I. Shelton-Brown, and W. McGuirt. 1997. Vacuum-assisted closure: A new method for wound control and treatment: Animal studies and basic foundation. *Ann Plast Surg* 38 (6): 553–62.

Moustakas, A., and C. H. Heldin. 2005. Non-Smad TGF-beta signals. *J Cell Sci* 118 (Pt 16): 3573–84. doi: 10.1242/jcs.02554.

Mu, Y., S. K. Gudey, and M. Landstrom. 2012. Non-Smad signaling pathways. *Cell Tissue Res* 347 (1): 11–20. doi: 10.1007/s00441-011-1201-y.

Mulder, G. D., and J. S. Vande Berg. 2002. Cellular senescence and matrix metalloproteinase activity in chronic wounds: Relevance to debridement and new technologies. *J Am Podiatr Med Assoc* 92 (1): 34–37.

Nicolas, F. J., K. De Bosscher, B. Schmierer, and C. S. Hill. 2004. Analysis of Smad nucleo-cytoplasmic shuttling in living cells. *J Cell Sci* 117 (Pt 18): 4113–25. doi: 10.1242/jcs.01289.

Nicolas, F. J., and C. S. Hill. 2003. Attenuation of the TGF-beta-Smad signaling pathway in pancreatic tumor cells confers resistance to TGF-beta-induced growth arrest. *Oncogene* 22 (24): 3698–711. doi: 10.1038/sj.onc.1206420.

Niknejad, H., H. Peirovi, M. Jorjani, et al. 2008. Properties of the amniotic membrane for potential use in tissue engineering. *Eur Cell Mater* 15: 88–99.

Parolini, O., F. Alviano, G. P. Bagnara, et al. 2008. Concise review: Isolation and characterization of cells from human term placenta: Outcome of the first international Workshop on Placenta Derived Stem Cells. *Stem Cells* 26 (2): 300–11. doi: 10.1634/stemcells.2007-0594.

Parolini, O., and M. Soncini. 2006. Human placenta: A source of progenitor/stem cells? *J Reprod Med Endocrinol* 3 (2): 117–26.

Pierreux, C. E., F. J. Nicolas, and C. S. Hill. 2000. Transforming growth factor beta-independent shuttling of Smad4 between the cytoplasm and nucleus. *Mol Cell Biol* 20 (23): 9041–54.

Redondo, P., A. G. de Azcarate, L. Marqués, et al. 2011. Amniotic membrane as a scaffold for melanocyte transplantation in patients with stable vitiligo. *Dermatol Res Pract* 2011: 532139. doi: 10.1155/2011/532139.

Riches, D. W. H. 1996. Macrophage involvement in wound repair, remodeling, and fibrosis. In R. A. F. Clark (Ed.), *The Molecular and Cellular Biology of Wound Repair*, pp. 95–141. New York: Plenum Press.

Roberts, H. R., and A. H. Tabares. 1995. Overview of the coagulation reactions. In K. A. High and H. R. Roberts (Eds.), *Molecular Basis of Thrombosis and Hemostasis*, pp. 35–50. New York: Dekker.

Ronai, Z. 2004. JNKing revealed. *Mol Cell* 15 (6): 843–4. doi: 10.1016/j.molcel.2004.09.011.

Ruiz-Canada, C., A. Bernabé–García, G. Castellanos, et al. Submitted. Amniotic membrane stimulates cell migration by modulating TFG-β signaling. *Wound Repair Regen*.

Sabapathy, K., K. Hochedlinger, S. Y. Nam, et al. 2004. Distinct roles for JNK1 and JNK2 in regulating JNK activity and c-Jun-dependent cell proliferation. *Mol Cell* 15 (5): 713–25. doi: 10.1016/j.molcel.2004.08.028.

Sabella, N. 1913. Use of fetal membranes in skin grafting. *Med Records NY* 83: 478–80.

Sangwan, V. S., H. P. Matalia, G. K. Vemuganti, et al. 2006. Clinical outcome of autologous cultivated limbal epithelium transplantation. *Indian J Ophthalmol* 54 (1): 29–34.

Santoro, M. M., and G. Gaudino. 2005. Cellular and molecular facets of keratinocyte reepithelization during wound healing. *Exp Cell Res* 304 (1): 274–86. doi: 10.1016/j.yexcr.2004.10.033.

Serena, T. E., M. J. Carter, L. T. Le, et al. 2014. A multicenter, randomized, controlled clinical trial evaluating the use of dehydrated human amnion/chorion membrane allografts and multilayer compression therapy vs. multilayer compression therapy alone in the treatment of venous leg ulcers. *Wound Repair Regen* 22 (6): 688–93. doi: 10.1111/wrr.12227.

Shah, A. P. 2014. Using amniotic membrane allografts in the treatment of neuropathic foot ulcers. *J Am Podiatr Med Assoc* 104 (2): 198–202. doi: 10.7547/0003-0538-104.2.198.

Sheikh, E. S., E. S. Sheikh, and D. E. Fetterolf. 2014. Use of dehydrated human amniotic membrane allografts to promote healing in patients with refractory non healing wounds. *Int Wound J* 11 (6): 711–17. doi: 10.1111/iwj.12035.

Singer, A. J., and R. A. Clark. 1999. Cutaneous wound healing. *N Engl J Med* 341 (10): 738–46. doi: 10.1056/NEJM199909023411006.

Singh, R., U. S. Chouhan, S. Purohit, et al. 2004. Radiation processed amniotic membranes in the treatment of non-healing ulcers of different etiologies. *Cell Tissue Bank* 5 (2): 129–34. doi: 10.1023/B:CATB.0000034077.05000.29.

Sorsby, A., and H. M. Symons. 1946. Amniotic membrane grafts in caustic burns of the eye: (Burns of the second degree). *Br J Ophthalmol* 30 (6): 337–45.

Stern, M. 1913. The grafting of preserved amniotic membrane to burned and ulcerated surfaces, substituting skin grafts: A preliminary report. *J Am Med Assoc* 60 (13): 973–974. doi: 10.1001/jama.1913.04340130021008.

Tolhurst, D. E., and T. W. van der Helm. 1991. The treatment of vaginal atresia. *Surg Gynecol Obstet* 172 (5): 407–14.

Tonnesen, M. G., X. Feng, and R. A. Clark. 2000. Angiogenesis in wound healing. *J Investig Dermatol Symp Proc* 5 (1): 40–6. doi: 10.1046/j.1087-0024.2000.00014.x.

Tredget, E. B., J. Demare, G. Chandran, et al. 2005. Transforming growth factor-beta and its effect on reepithelialization of partial-thickness ear wounds in transgenic mice. *Wound Repair Regen* 13 (1): 61–7. doi: 10.1111/j.1067-1927.2005.130108.x.

Trelford, J. D., and M. Trelford-Sauder. 1979. The amnion in surgery, past and present. *Am J Obstet Gynecol* 134 (7): 833–45.

Trengove, N. J., H. Bielefeldt-Ohmann, and M. C. Stacey. 2000. Mitogenic activity and cytokine levels in non-healing and healing chronic leg ulcers. *Wound Repair Regen* 8 (1): 13–25.

Troensegaard-Hansen, E. 1950. Amniotic grafts in chronic skin ulceration. *Lancet* 255 (6610): 859–60. doi: 10.1016/S0140-6736(50)90693-3.

Wahl, L. M., and S. M. Wahl. 1992. Inflammation. In I. K. Cohen, R. F. Diegelmann, and W. J. Lindblad (Eds.), *Wound Healing, Biochemical and Clinical Aspects*, pp. 40–62. Elsevier Saunders, PA: Philadelphia.

Ward, D. J., and J. P. Bennett. 1984. The long-term results of the use of human amnion in the treatment of leg ulcers. *Br J Plast Surg* 37 (2): 191–3. doi: 10.1016/0007-1226(84)90009-2.

Ward, D. J., J. P. Bennett, H. Burgos, and J. Fabre. 1989. The healing of chronic venous leg ulcers with prepared human amnion. *Br J Plast Surg* 42 (4): 463–67. doi: 10.1016/0007-1226(89)90015-5.

Welch, M. P., G. F. Odland, and R. A. Clark. 1990. Temporal relationships of F-actin bundle formation, collagen and fibronectin matrix assembly, and fibronectin receptor expression to wound contraction. *J Cell Biol* 110 (1): 133–45.

Werner, S., and R. Grose. 2003. Regulation of wound healing by growth factors and cytokines. *Physiol Rev* 83 (3): 835–70. doi: 10.1152/physrev.00031.2002.

Wilshaw, S. P., J. N. Kearney, J. Fisher, and E. Ingham. 2006. Production of an acellular amniotic membrane matrix for use in tissue engineering. *Tissue Eng* 12 (8): 2117–29. doi: 10.1089/ten.2006.12.2117.

Woodley, D. T. 1996. Reepithelialization. In R. A. F. Clark (Ed.), *The Molecular and Cellular Biology of Wound Repair*, pp. 339–50. New York: Plenum Press.

Yang, L., T. Chan, J. Demare, et al. 2001. Healing of burn wounds in transgenic mice overexpressing transforming growth factor-beta 1 in the epidermis. *Am J Pathol* 159 (6): 2147–57.

Yang, L., Y. Shirakata, M. Shudou, et al. 2006. New skin-equivalent model from de-epithelialized amnion membrane. *Cell Tissue Res* 326 (1): 69–77. doi: 10.1007/s00441-006-0208-2.

Yates, S., and T. E. Rayner. 2002. Transcription factor activation in response to cutaneous injury: Role of AP-1 in reepithelialization. *Wound Repair Regen* 10 (1): 5–15.

Yeager, A. M., H. S. Singer, J. R. Buck, et al. 1985. A therapeutic trial of amniotic epithelial cell implantation in patients with lysosomal storage diseases. *Am J Med Genet* 22 (2): 347–55. doi: 10.1002/ajmg.1320220219.

Zelen, C. M. 2013. An evaluation of dehydrated human amniotic membrane allografts in patients with DFUs. *J Wound Care* 22 (7): 347–48, 350–51. doi: 10.12968/jowc.2013.22.7.347.

Zelen, C. M., A. Poka, and J. Andrews. 2013. Prospective, randomized, blinded, comparative study of injectable micronized dehydrated amniotic/chorionic membrane allograft for plantar fasciitis—A feasibility study. *Foot Ankle Int* 34 (10): 1332–9. doi: 10.1177/1071100713502179.

Zelen, C. M., T. E. Serena, and D. E. Fetterolf. 2014a. Dehydrated human amnion/chorion membrane allografts in patients with chronic diabetic foot ulcers: A long-term follow-up study. *Wound Med* 4: 1–4. doi: 10.1016/j.wndm.2013.10.008.

Zelen, C. M., T. E. Serena, and R. J. Snyder. 2014b. A prospective, randomised comparative study of weekly versus biweekly application of dehydrated human amnion/chorion membrane allograft in the management of diabetic foot ulcers. *Int Wound J* 11 (2): 122–8. doi: 10.1111/iwj.12242.

Zelen, C. M., R. J. Snyder, T. E. Serena, and W. W. Li. 2015. The use of human amnion/chorion membrane in the clinical setting for lower extremity repair: A review. *Clin Podiatr Med Surg* 32 (1): 135–46. doi: 10.1016/j.cpm.2014.09.002.

Zohar, Y., Y. P. Talmi, Y. Finkelstein, et al. 1987. Use of human amniotic membrane in otolaryngologic practice. *Laryngoscope* 97 (8 Pt 1): 978–80.

5 Cell Populations Isolated from Amnion, Chorion, and Wharton's Jelly of Human Placenta

Francesco Alviano, Roberta Costa, and Laura Bonsi

CONTENTS

Preface 69
5.1 Placenta as a Source of Intriguing Cells 70
5.2 Amniotic Membrane 70
    5.2.1 Histology of Tissue 70
    5.2.2 Cells Comprising the Amniotic Membrane 72
    5.2.3 Cells Isolated from Amniotic Membrane 72
        5.2.3.1 hAECs 72
        5.2.3.2 Human Amniotic Membrane Mesenchymal Stromal Cells 75
        5.2.3.3 Trophic Functions of Human Amniotic
                Membrane–Derived Cells 76
5.3 Chorionic Membrane 78
    5.3.1 Histology of Chorionic Membrane 79
    5.3.2 Human Chorionic MSCs 79
5.4 Wharton's Jelly 81
    5.4.1 Histology of Tissue 81
    5.4.2 Cells Isolated from the Umbilical Cord 81
    5.4.3 Human Wharton's Jelly MSCs 81
5.5 Concluding Remarks 84
References 84

PREFACE

The placenta represents an interesting source of cells with an intriguing potential, by virtue of the tissue properties acquired during embryological development and maintained during the perinatal period. Its embryological development, its metabolic role in the maternal-fetal interface, and its function in the maintenance of an "immunological sanctuary" have all raised considerable attention of researchers in placental tissue, which is usually considered as medical waste.

70   Placenta: The Tree of Life

In the attempt to find a source of adult stem cells able to overcome the limitations of research linked to the use of the embryonic stem cells (ESCs) and of some adult tissues that are not easily available, more and more research groups are looking to the human placenta and extraembryonic tissues (e.g., fetal membranes and umbilical cord) as potential sources of stem cells

5.1 PLACENTA AS A SOURCE OF INTRIGUING CELLS

The placental tissues derive from a stage of early embryonic development, lending support to the hypothesis that they may contain cells that have retained the plasticity typical of primitive cells. Furthermore, these embryonic appendages are considered young tissues, compared to adult tissues that accompany the body for the duration of life; thus, cells derived from them may be characterized by a reduced risk of damage to their genetic heritage. Last but not least, the placenta plays a key role in maintaining a state of maternal-fetal tolerance during pregnancy, preventing the mother's immune system from reacting to the allogeneic fetus. This characteristic suggests that the cells derived from the placenta can themselves maintain an immunomodulatory behavior that should allow them to control and inhibit activation of the immune system, thus making them an ideal tool for therapeutic applications based on the use of stem cells.

From the macroscopic point of view, the placenta is a discoid-shaped organ with a diameter of 15–20 cm, a thickness of 2–3 cm, and a weight of approximately 500–600 g. The placenta is composed of a chorionic plate, from which the chorionic villi depart, that is in direct contact with the uterine wall during pregnancy. At the edges of chorionic plate, the fetal membranes, composed of amnion and chorion, extend and cover an area of approximately 700–1200 cm^2 (Figure 5.1a, middle panel). The chorionic disc, directly overlooking the amniotic cavity, consists of several structures. The chorionic villi extend from the maternal side of the disc: they anchor the placenta through the maternal endometrium, whereas trophoblast cells perform different functions essential for the exchange of gases and nutrients between mother and fetus. From the fetal side, the placental disc presents the fetal membranes that are composed of the entire amnion and the chorion laeve, and the umbilical cord, which departs from the chorionic disc and contains the fetal vessels through which the mother-fetus exchanges occur.

5.2 AMNIOTIC MEMBRANE

5.2.1 HISTOLOGY OF TISSUE

Amniotic membrane (or amnion) is the inner of the two fetal membranes that form the amniotic sac that surrounds and protects the fetus. Amnion is a thin, semitransparent, semipermeable and avascular membrane attached to the chorionic membrane, and it covers the entire chorionic plate, continuing over the umbilical cord with the fetal skin. This membrane is composed of two layers: an epithelial layer in direct contact with amniotic fluid and an underling mesenchymal layer attached to the chorionic membrane. The epithelial layer, named amniotic epithelium, is composed of columnar and cuboidal epithelial cells, and it is attached to a basement membrane

Cell Populations Isolated from Human Placenta

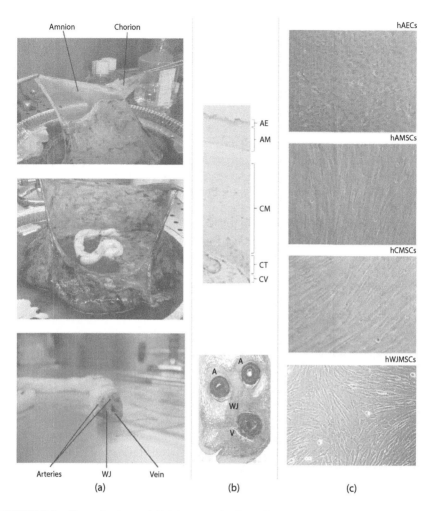

FIGURE 5.1 **(See color insert.)** (a) Macroscopic view of human term placenta demonstrating the mechanical separation of amniotic membrane from chorion, and a cord section showing the umbilical vessels surrounded by Wharton's jelly. (b) Histology of fetal membranes and umbilical blood. A section of human fetal membranes (hematoxylin and eosin stain). (c) *In vitro* morphology of cells isolated from amniotic epithelium, amniotic mesoderm, chorionic mesoderm, and Wharton's jelly. AE, amniotic epithelium; AM, amniotic mesoderm; CM, chorionic mesoderm; CT, chorionic trophoblast; CV, chorionic villi. A section of human umbilical cord (Mallory stain). A, arteries; V, vein; WJ, Wharton's jelly; hAECs, human amniotic epithelial cells; hAMSCs, human amniotic membrane mesenchymal stromal cells; hCMSCs, human chorionic membrane mesenchymal stromal cells; hWJMSCs, human Wharton's jelly mesenchymal stromal cells.

that, in turn, is in contact with the mesenchymal layer. The latter layer is compact and composed of fibronectin and collagen (type I and III), and it hosts a network of dispersed mesenchymal cells and a rare population of macrophages. Below the mesenchymal layer, a spongy layer of collagen fibers separates amnion from chorion (Evangelista et al. 2008; Diaz-Prado et al. 2011) (Figure 5.1b top panel).

5.2.2 Cells Comprising the Amniotic Membrane

The amniotic membrane hosts cell populations that show pluripotent properties, reflecting the embryological origin of this tissue and providing a rationale for the clinical application of these cells (Parolini et al. 2008; Castillo-Melendez et al. 2013). The innermost layer of the human amnion, the amniotic epithelium, is 8–12 μm thick and consists of a single layer of homogeneous epithelial cells, called human amniotic epithelial cells (hAECs) (Hebertson et al. 1986; Iwasaki et al. 2003). In contrast, dispersed in the underlying mesenchymal layer, there are two distinct cell populations: one population consists of mesenchymal cells, named human amniotic membrane mesenchymal stromal cells (hAMSCs), and the other population consists of cells with characteristics similar to monocytes and macrophages.

Different studies have shown that hAMSCs and hAECs display many characteristics of both embryonic and pluripotent stem cells, with the potential to differentiate into a range of different cell types (Ilancheran et al. 2007; Parolini et al. 2008; Castillo-Melendez et al. 2013).

5.2.3 Cells Isolated from Amniotic Membrane

Cells from the epithelial and mesenchymal layers of amniotic membrane can be easily isolated with a combination of enzymatic and mechanical treatments of the tissue. First, amniotic membrane is peeled off of the chorion and then digested with specific enzymes (Figure 5.1a, top panel).

Different protocols for isolating cells from human amniotic membrane have been published, and these methods generally entail a mechanical digestion followed by a treatment with trypsin or collagenase or with a mix of digestive enzymes in different concentrations and for different times (Diaz-Prado et al. 2011). Amniotic membrane–derived cells can be isolated from amnion at the first, second, and third trimester, but the majority of studies have been performed on cells from term amnion (Evangelista et al. 2008).

5.2.3.1 hAECs

A term amniotic membrane can yield up to 200 million epithelial cells (Ilancheran et al. 2009) that can be isolated with subsequent treatments in trypsin of 20–40 min each (Miki et al. 2005; Parolini et al. 2008). Trypsin, and other digestive enzymes, is able to cut intercellular and cell-to-basal lamina bonds and to finally release hAECs from the underlying basal membrane.

Isolated cells rapidly adhere to culture plates, forming cell clusters with cells in direct and strict contact with each other; these clusters subsequently grow out in a monolayer of cells (Bilic et al. 2008). A stable cell culture can be established *in vitro* by using a simple basal media, such as Dulbecco's modified Eagle's medium (DMEM), supplemented with 10% fetal bovine serum (FBS) and epidermal growth factor (EGF), which has a mitogenic effect on hAECs. hAECs are small-sized cells and display a typical epithelial cuboidal morphology *in vitro*, described as "cobblestone-like" (Terada et al. 2000; Parolini et al. 2008; Diaz-Prado et al. 2011).

hAECs grow easily when seeded at high density (1×10^5 cells/cm^2), whereas they do not proliferate at low density, and they can be expanded for four to five passages *in vitro* before cells show signs of senescence and cease proliferation (Miki et al. 2007; Bilic et al. 2008; Parolini et al. 2008). This condition of early replicative senescence has been addressed by investigating hAEC telomerase activity. Mosquera et al. (1999) evaluated telomerase activity of hAECs in comparison with HeLa cells, observing that telomerase activity of hAECs was significantly lower and it decreased over time (days) in culture. This reduced telomerase activity could account for early senescence of hAEC cultures (Miki and Strom 2006).

Amniotic epithelium hosts cells that express different levels of different surface cell markers, suggesting that hAECs could display heterogeneity of phenotype. A high percentage of hAECs are positive for cytokeratins, a typical epithelial marker, both immediately after isolation and during cell culture. Moreover, the immunophenotype of hAECs is characterized by the expression of surface markers normally expressed on embryonic and pluripotent stem cells, such as SSEA3, SSEA4, TRA1-60, and TRA1-81, and by the expression of other additional surface antigens, such as CD9, CD24, CD29, CD73, CD166, ATP-binding cassette transporter G2 (ABCG2/BCRP), E-cadherin, $\alpha 6/\beta 1$ integrin, and c-met (HGF receptor). Other cell markers, such as CD117 and CCR4, are generally absent or expressed at very low levels on the surface of hAECs, and other classical mesenchymal markers such as CD90 and vimentin are absent or expressed at very low levels in freshly isolated cells, but their expression increases during *in vitro* culture. Coherently with their origin, hAECs are negative for hematopoietic markers CD34 and CD45 (Miki et al. 2005; Miki and Strom 2006; Toda et al. 2007; Bilic et al. 2008). The immunological profile of hAECs shows that they have low levels of major histocompatibility complex class I (MHC-I) surface antigens and that they do not express MHC class II (Evangelista et al. 2008; Bilic et al. 2008; Parolini et al. 2008) and costimulatory molecules such as CD80, CD86, CD40, and CD40 ligand (Diaz-Prado et al. 2011). Interestingly, hAECs express the nonclassical human leukocyte antigen G (HLA-G); HLA-G has been shown to have immunomodulatory functions and is associated with tolerogenic properties (Diaz-Prado et al. 2011). This immunological profile allows us to define hAECs as immune-privileged cells that could have an active role in reducing the risk of rejection in allogeneic transplants (Kubo et al. 2001; Kamiya et al. 2005; Li et al. 2005; Ilancheran et al. 2007).

The embryological origin of amniotic epithelium, which derives from the epiblast before gastrulation, may suggest that some cells in this tissue could show pluripotency. Different groups have studied hAECs for the expression of molecular markers of pluripotency, such as OCT-4 (octamer-binding protein 4), SOX2 (SRY-related HMG-box gene 2), FGF4, REX1, and NANOG (Miki et al. 2005; Miki and Strom 2006; Parolini et al. 2008). OCT-4 is known to have a central role in maintaining pluripotency and self-renewal of stem cells, but the only expression of OCT-4 observed in most hAECs is not sufficient to confirm hAEC pluripotential activity. To assess pluripotency *in vitro*, a clonal expansion from a single hAEC should be demonstrated, but hAECs do not survive as single cells in culture, they do not keep their stem cell characteristics, and they rapidly fall into senescence. The teratoma formation assay could also be used to demonstrate *in vitro* pluripotency, but hAECs do not form teratomas when injected in immunodeficient mice (Miki et al. 2005; Ilancheran et al. 2007). Based on these data, we can affirm that pluripotency of hAECs has yet to been proven and further investigations are needed (Miki and Strom 2006).

Although pluripotency of hAECs has yet to be confirmed, several studies have produced information that suggests multipotency of hAECs. Different research groups have shown that amniotic epithelial cells are able to differentiate into cells of all three germ layers under adequate culture conditions (Miki and Strom 2006; Ilancheran et al. 2007; Diaz-Prado et al. 2010; Insausti et al. 2010).

Sakuragawa and coworkers demonstrated that naïve hAECs show characteristics of glial and neuronal progenitor cells since they express neural markers, such as glial fibrillary acid protein (GFAP), microtubule-associated protein 2 (MAP2), myelin basic protein, and neurofilament M (Sakuragawa et al. 1996; Miki et al. 2005). Upon culturing of hAECs in appropriate neural differentiating medium, they take on an elongated neuronal morphology and express or upregulate the expression of neuron-specific genes, such as nestin and glutamic acid decarboxylase. Moreover, it has been demonstrated that hAECs synthesize and secrete neurotrophic factors such as nerve growth factor, brain-derived neurotrophic factor, neurotrophin-3, and neurotransmitters (acetylcholine, dopamine, and norepinephrine) (Ilancheran et al. 2009). Various researchers independently demonstrated neural differentiation abilities and neurogenic and neuroprotective properties of hAECs *in vitro*, and these results are also supported by preclinical studies in animal models of different neural diseases (neurodegenerative disease, cerebral artery occlusion, spinal cord injury) (Evangelista et al. 2008; Ilancheran et al. 2009; Manuelpillai et al. 2011; Miki 2011).

hAECs have also been found to be able to differentiate toward cells of endodermal lineage. Different research groups (Sakuragawa et al. 2000; Takashima et al. 2004) showed that naïve hAECs express characteristic hepatocyte genes such as albumin, alpha-1 antitrypsin, and alpha-fetoprotein. Moreover, upon culture in hepatic differentiation conditions, hAECs expressed additional hepatocyte-related genes, including transthyretin and tyrosine aminotransferase, transcription factors such as hepatocyte nuclear factor 3γ and enhancer-binding protein (CEBP α and β), and several drug-metabolizing genes (cytochrome P450) (Ilancheran et al. 2009; Diaz-Prado et al. 2011).

hAECs are also able to differentiate into pancreatic cells. Wei et al. (2003) demonstrated that amniotic epithelial cells, cultured in presence of specific molecules, expressed insulin and were also able to normalize levels of serum glucose in diabetic mice for several months. Later, other research groups demonstrated that hAECs, when cultured in pancreatic differentiation medium, expressed pancreatic α and β cell markers, including transcription factors PDX1 (pancreatic duodenum homeobox 1), PAX6 (paired box homeotic gene 6), and NKX2.2 (NK2 transcription factor-related locus 2), and markers of pancreatic endocrine cells such as insulin and glucagon (Miki et al. 2005). Ilancheran et al. (2007) also investigated ultrastructural organization of differentiated hAECs, similar to exocrine beta-acinar cells.

In addition to hepatic and pancreatic commitment, hAECs have been reported to be able to differentiate into other endodermal cell types, such as lung epithelium. Moodley et al. (2010) demonstrated that hAECs differentiate into lung epithelium,

Cell Populations Isolated from Human Placenta

and in particular into type II pneumocytes after transplantation into a mouse model of lung injury.

The differentiation potential of hAECs toward the mesodermal lineage was first demonstrated by Miki et al. (2005), who showed that cardiac-specific genes (atrial and ventricular myosin light chain 2A and 2V - MLC-2A and MLC-2B) and transcription factors (GATA4 and NKX2.5) were expressed by hAECs under specific culture conditions. Moreover, Ilancheran et al. (2007) reported that hAECs differentiate into cells expressing characteristic markers of mesodermal cells, such as myocytes, adipocytes, and osteocytes.

All of these studies support the hypothesis that amniotic epithelium contains cells with progenitor and stem cell characteristics. Although these data are very promising for the use of hAECs in regenerative medicine approaches, further studies are necessary to elucidate the optimal differentiation conditions to obtain functional cell types from hAECs.

5.2.3.2 Human Amniotic Membrane Mesenchymal Stromal Cells

hAMSCs derive from extraembryonic mesoderm. They are obtained by mincing amniotic tissue into small pieces that are then treated with trypsin, to remove the amniotic epithelium, and finally with collagenase or collagenase and DNase to completely release mesenchymal stromal cells (MSCs). Different studies have shown that a whole amniotic membrane can yield up to 40 million hAMSCs (Bilic et al. 2008; Parolini et al. 2008). hAMSCs adhere to plastic and show a fibroblast-like morphology. They easily proliferate in DMEM + 10% FBS for at least 15 passages without morphological changes (Alviano et al. 2007).

The surface marker profile of hAMSCs is similar to the adult bone marrow MSC profile. Immunophenotypic characterization of hAMSCs shows the presence of classical, well-defined human MSC markers, such as CD13, CD29, CD44, CD73, CD90, and CD105, and the absence of hematopoietic (CD34 and CD45) and monocyte (CD14) markers. Moreover, hAMSCs express low levels of HLA-ABC (MHC-I) and lack HLA-DR (MHC-II) and costimulatory molecules, suggesting that these cells have an immune-privileged status. hAMSCs constitutively express HLA-G, and HLA-G has been shown to have interesting immunomodulatory functions, supporting the potential utility of hAMSCs in clinical transplantation approaches (Wolbank et al. 2007).

The expression of pluripotency markers in hAMSCs is still debated. hAMSCs have been shown to be positive for SSEA3, SSEA4, and STRO-1 expression in flow cytometry (Bilic et al. 2008), whereas immunofluorescence staining of amniotic membrane did not detect SSEA3 and SSEA4 in the mesenchymal layer (Miki et al. 2007). Moreover, only OCT-4 transcripts were detected in hAMSCs, whereas Oct-4 protein has been detected by immunocytochemistry in hAECs and not in hAMSCs. It remains unclear as to whether hAMSCs are positive for Oct-4 expression given that the presence only of gene transcripts is not predictive of stemness (Bilic et al. 2008).

hAMSCs have been also investigated for their pluripotency, and several studies demonstrate that they can be differentiated toward classical mesodermal lineages, such as osteogenic, adipogenic, and chondrogenic. Moreover, hAMSCs have been investigated for their differentiation potential toward other mesodermal cell types (cardiomyocytes, skeletal muscle, and endothelial) and toward ectodermal (neural) and endodermal (hepatic) lineages (Alviano et al. 2007; Evangelista et al. 2008; Parolini et al. 2008; Diaz-Prado et al. 2011).

Regarding differentiation into mesodermal lineages, In 't Anker et al. (2004) investigated osteogenic and adipogenic differentiation potential of hAMSCs, observing calcium deposits and the presence of lipid vacuoles. Later, Portmann-Lanz et al. (2006) showed that hAMSCs were also able to differentiate toward the chondrogenic lineage, showing the presence of abundant collagen in the extracellular matrix. This same research group also investigated myogenic differentiation potential of hAMSCs, showing the expression of mRNA for myogenic transcription factor (MyoD and myogenin) and of proteins such as desmin in cells cultured with specific differentiation medium. These results on myogenic differentiation were confirmed by Alviano et al. (2007), who in addition showed angiogenic differentiation potential of hAMSCs. In this study, amniotic mesenchymal cells reacted to the presence of Vascular endothelial growth factor (VEGF) in classical culture medium by expressing endothelial specific markers, including receptors of the vascular endothelial growth factor 1 and 2 (FLT-1 and KDR), intracellular adhesion molecule 1, CD34, and von Willebrand factor. Regarding cardiomyogenic differentiation, it has been demonstrated that hAMSCs express specific cardiac genes, such as GATA4, myosin light chain 2a and 2v (MLC-2a, -2v), cardiac troponin I (cTnI), and cardiac troponin T (cTnT), after treatment with cardiomyogenic differentiation medium (Tanaka et al. 1999; Zhao et al. 2005). hAMSC neural differentiation has been demonstrated by the expression of neural (nestin, MAP2) and glial (GFAP) markers after culture in specific induction medium (Sakuragawa et al. 2004; Portmann-Lanz et al. 2009). Similarly, Tamagawa et al. (2007) showed that naïve hAMSCs expressed hepatocytic transcripts, such as albumin, α-fetoprotein, and cytokeratin 18, and after specific hepatic induction they were capable to express glucose-6-phosphatase and glycogen storage.

In addition to *in vitro* differentiation potential, hAMSCs have been used for *in vivo* transplantation in preclinical studies. Bailo et al. (2004) demonstrated that hAMSCs can be used for xenotransplantation and that these cells successfully and persistently engraft in host tissues. Transplanted hAMSCs have been discovered in different organs after intravenous or intraperitoneal injection, suggesting that these cells can migrate. This characteristic is supported by the expression of adhesion and migration molecules on hAMSCs (L-selectin, VLA-5, CD29) as well as by the expression of cellular matrix proteinases (MMP-2 and -9) (Bailo et al. 2004).

5.2.3.3 Trophic Functions of Human Amniotic Membrane–Derived Cells

The human amniotic membrane has a long history in clinical applications, above all in surgery and ophthalmology, since it has been used for the treatment of skin wounds, burn injuries, and chronic ulcers; in head and neck surgery; and in the prevention of tissue adhesion in surgery (Parolini et al. 2009). The use of amnion in clinical approaches is justified by its ability to enhance epithelialization and wound healing, to suppress inflammation and fibrosis, and to inhibit angiogenesis, all of which are aspects that are mediated by secretion of specific molecules and cytokines (Parolini et al. 2010). In addition to these properties, the wide differentiation potential of human amniotic cells has stirred increasing interest in the potential application of these cells for regenerative medicine approaches. Human amnion–derived cells have been used in preclinical studies in different animal disease models, and they were shown to make a modest contribution in replacing damaged cells and a more significant contribution via their anti-inflammatory and anti-scarring effects.

Cell replacement therapy has been proposed as new strategy for treatment of a wide range of neurological disorders, and the differentiation potential of hAECs, particularly directed toward the neural lineage, suggested their use for treatment of brain injury. hAECs have been used in a rat model of Parkinson's disease, where they were able to prevent the death of dopaminergic neurons by the secretion of neurotrophic factors (Kakishita et al. 2000). hAEC injection into a mouse model of multiple sclerosis also demonstrated that these cells can suppress clinical symptoms and decrease central nervous system (CNS) inflammation, demyelination, and axonal degeneration. This beneficial effect seems to be attributed to the ability of hAECs to reduce infiltration and proliferation of T lymphocytes and monocytes and macrophages in the CNS and to decrease their secretion of proinflammatory cytokines (McDonald et al. 2011; Liu et al. 2012). hAECs transplanted into an animal model of spinal cord injury were able to survive without signs of inflammation, suggesting that hAECs can avoid immunological rejection in the CNS; moreover, they improved performance in locomotor tests, suggesting that hAECs also have a neuroprotective effect and can improve functionality of motor neuron tracts (Sankar et al. 2003). Ischemic stroke is a serious neurological disorder characterized by a strong inflammation that follows the initial stroke episode and that greatly contributes to secondary cell death. In this perspective, the immunomodulatory effect of human amnion–derived cells could have a beneficial effect on stroke outcome. Injection of hAECs into a rat model of stroke resulted in reduced infarct volume, improved behavioral and neurological outcome, and reduced apoptotic events (Liu et al. 2008). Similar results were obtained by injecting hAMSCs into rat model of stroke (Tao et al. 2012). Human amniotic cells could exert a positive effect on neural diseases through different mechanisms. First, the ability of amnion-derived cells to secrete neurotrophic factors promotes neuronal recovery and synaptogenesis to recreate lost neural connections. Moreover, when transplanted into the CNS, amniotic cells can secrete cytokines, hormones, growth factors, and neurotransmitters that can restore cellular function and also modulate inflammatory responses, thereby protecting neurons from immune cell-mediated damages and apoptosis. These conditions seem to have a preponderant effect on neural recovery, working in concert with the positive effects obtained through the differentiation of amniotic cells into neuronal cells that can replace dead or damaged cells (Broughton et al. 2013).

Human amniotic epithelial and MSCs have been used as a cellular treatment for lung and liver fibrosis. Different factors can cause inflammation in liver and lung that lead to collagen deposition in response to wound healing. Immune cell infiltration, the subsequent release of proinflammatory cytokines, and the activation of collagen deposition lead to an altered architecture and to a compromised tissue function. Human amniotic cells (hAECs and hAMSCs) have been injected in preclinical animal models of lung inflammation and fibrosis, and they were able to reduce immune activated cell infiltration as well as proinflammatory cytokine production, thereby leading to limited collagen production and improved tissue architecture (Cargnoni et al. 2009; Moodley et al. 2010). Amniotic membrane–derived cells are able to limit damage to the lung, but the exact mechanisms that reduce inflammation, and as a consequence fibrosis, have not been completely identified (Moodley et al. 2010). hAECs have also been used in the treatment of a model of hepatic fibrosis, and similar to the effect observed for lung fibrosis, they were able to reduce collagen production through the reduction of profibrotic cytokines (Manuelpillai et al. 2010, 2012). However, it was unclear whether the observed effect was mediated by the modulation of profibrotic macrophages or through a direct effect of hAECs on hepatic stellate cells (HSCs), which are responsible for collagen deposition. Thus, Hodge et al. (2014) analyzed the effect of the conditioned medium of hAECs on activated HSCs, and although they observed an antifibrotic effect, they did not fully identify the specific pattern of soluble factors released by hAECs that mediated this effect.

All of the beneficial effects reported above have been associated with the release of soluble factors that exert a paracrine effect that is able to promote protection, survival, and regeneration of host cells (Manuelpillai et al. 2011). Indeed, the amniotic membrane has been reported to release immunomodulatory and anti-inflammatory cytokines (interleukin [IL]-10 and IL-6) and growth factors involved in wound healing, angiogenesis (VEGF and platelet-derived growth factor), cell proliferation (EGF, keratinocyte growth factor, hepatocyte growth factor, and basic fibroblast growth factor [bFGF]), and differentiation (transforming growth factor beta [TGF-β]) (Manuelpillai et al. 2011). The exact mechanisms whereby treatments based on amniotic cells exert their positive therapeutic effect are still poorly understood, but it is increasingly accepted that the observed improvements in tissue functionality are related to paracrine and trophic effect of these cells, rather than to tissue-specific differentiation of transplanted amniotic cells themselves (Parolini et al. 2011).

5.3 CHORIONIC MEMBRANE

The most important characteristic of human embryonic development is the close relationship between the embryo and the mother by which the embryo acquires oxygen and nutrients and eliminates wastes. The fetal-maternal tissues involved in this relationship are placenta and chorion, both of which are derived from the trophoblast.

During implantation, the trophoblastic layer surrounding the blastocyst differentiates into two tissues: the cytotrophoblastic inner layer and the syncytiotrophoblastic outer layer. In human embryology, during the second week of development, the chorion is defined as the layer consisting of the cytotrophoblast, syncytiotrophoblast, and extraembryonic mesoderm. The extraembryonic mesoderm arises from the proliferation of epiblastic cells at about 12 days after fertilization and becomes the tissue supporting the epithelium of secondary villi. The secondary villi become tertiary villi when blood cells infiltrate the mesoderm core. However, the invasion of the primary villi by extraembryonic mesoderm and blood vessels occurs preferentially in the villi nearest the uterine wall named *decidua basalis*. The latter, which contains the chorionic villi, is termed the chorion frondosum and finally becomes the placenta; the remaining smooth region of the chorion is the chorion laevae (Carlson 2014).

Cell Populations Isolated from Human Placenta

5.3.1 Histology of Chorionic Membrane

As shown in the top panel of Figure 5.1b, the term fetal membranes surrounding the placenta are merged like a continuous tissue. Whereas amniotic membrane is composed of two tissues, the epithelial and stromal layers, the chorionic membrane has a mesodermal stromal layer and a trophoblastic region. Because the stromal layers of the amniochorionic membrane are weakly linked, the chorionic membrane can be separated from the amnion manually by simple mechanical traction (Figure 5.1a top panel). Macroscopically, the chorion appears as an opaque and yellowish layer due to its vascularization, and it has thick and stiff consistency.

5.3.2 Human Chorionic MSCs

Considering the origin of the chorion, two classes of primitive cells may be distinguished: human chorionic mesenchymal stem/stromal cells (hCMSCs) and human chorionic villous/placenta mesenchymal stem/stromal cells (hPMSCs).

Few studies have focused on hPMSCs in the first trimester, because of the invasive techniques required to procure them, such as chorionic villus sampling. However, no significant differences have been shown between early and late pregnancy–isolated cells, suggesting that MSCs maintain their characteristics after the first trimester (Witkowska-Zimny et al. 2011). Castrechini et al. (2010) have identified a stem cell niche within the chorionic villi whose cells are able to contribute to vessel maturation and stabilization. The hPMSC population, consisting of spindle-shaped and large cells, was positive for MSC cell surface markers and negative for hematopoietic and endothelial markers. Moreover, hPMSCs were positive for pericyte cell surface markers such as STRO-1 and 3G5 (Castrechini et al. 2010).

Experimental studies performed *in vitro* that were, in part, confirmed *in vivo* demonstrated hPMSC pluripotency through the ability of these cells to differentiate into both mesodermic and endodermic or neural tissues. Moreover, Zhang et al. (2004) have shown the ability of these cells to support and feed hematopoietic stem cells.

Therefore, MSCs (hCMSCs) can be isolated from chorionic membrane at the first, second, or third trimester, through enzymatic (collagenase or collagenase plus DNase) treatment of the tissue, although most of the studies are based on MSCs isolated from placenta at term. The trophoblastic layer can be easily removed with dispase.

Several studies carried out to evaluate the yield of isolation of MSCs from fetal membranes have shown that about 80–100 million hCMSCs can be obtained from an entire chorionic membrane, considering an average area for total membrane of about 1000 cm^2. hCMSCs adhere and proliferate easily on plastic cell culture flasks, and they can be expanded up to the 10th–15th passage in culture, with a proliferative capacity slightly higher than the counterpart isolated from the amnion. Indeed, after 10–15 passages, hCMSCs have been shown to display discrete morphological changes (Fukuchi et al. 2004).

Morphologically, hCMSCs isolated *in vitro* show a fibroblast-like shape (Figure 5.1c). Transmission electron microscopy shows a cytoplasmic organization simpler than that of hAMSCs. The ultrastructural characteristics, such as stacks of rough endoplasmic reticulum cisternae, and dispersed mitochondria and glycogen lakes, resemble those found in hematopoietic progenitors and in blue, small, round tumor cells of Ewing sarcomas; on the contrary, they lack features of higher specialization, suggesting an ancestral aspect of these cells (Pasquinelli et al. 2007).

In regard to the surface marker profile, they express typical antigens associated with mesenchymal cells isolated from bone marrow; indeed, they are positive for CD13, CD29, CD44, CD54, CD73, CD90, CD105, and CD166 and are negative for the expression of hematopoietic (CD34 and CD45), monocyte (CD14), lymphocyte (CD3), and endothelial (CD31) markers (Zhang et al. 2004). The expression of surface antigens typical of ESCs, such as SSEA3 and SSEA4, and of multi- and pluripotent stem cells, such as OCT-4, have also been reported, although they have not been confirmed by immunofluorescence analysis. A subpopulation of human chorionic stem cells expressing both OCT-4 and c-kit has also been identified in a chorionic membrane population (Resta et al. 2013). Finally, the analysis of surface markers for hCMSCs showed limited expression of HLA-ABC and a lack of expression of HLA-DR, indicating a potential immune-privileged status for this cell population (Parolini et al. 2008; Kim et al. 2011).

hCMSCs have been stimulated to differentiate into classic mesenchymal lineages (osteogenic, adipogenic, and chondrogenic) in specific culture media (In 't Anker et al. 2004; Soncini et al. 2007). In addition, differentiation potential toward cell types from all three germ layers was also studied: it has been shown that these cells can differentiate into ectodermal (neural), mesodermal (muscle cells, endothelial cells, and cardiomyocytes), and endodermal (liver) lineages (Portmann-Lanz et al. 2006). Possible angiogenic-promoting mechanisms are suggested by high secretions of VEGF and bFGF by hCMSCs with increased expressions of angiogenic genes (Fariha et al. 2013).

The role of the fetal membrane in preventing rejection of the fetus is of particular interest. Many studies have examined the immunosuppressive ability of cells isolated from fetal membranes in mixed lymphocyte cultures through the production of anti-inflammatory cytokines (e.g., IL-10, TGF-β, HLA-G5) and the reduction of proinflammatory cytokine release from alloreactive T cells (Karlsson et al. 2012). Furthermore, chorionic stromal cells were shown to possess a higher suppressive effect compared to hAECs. With respect to the immunomodulatory effects of bone marrow and adipose MSCs, using a coculture system with activated T cells, Lee et al. (2012) detected higher levels of (cytokines) IL-2, IL-4, IL-13, and granulocyte macrophage–colony-stimulating factor, supporting the use of hCMSCs in cell therapy.

hCMSCs have also been described to be a notable source of growth and neurotrophic factors, in particular of ciliary neutrophic factor that has neuroprotective effects in nerve lesion models as well as for motoneurons (Paradisi et al. 2014).

Finally, MSCs from fetal membranes have been successfully transplanted *in vivo*. As previously reported, Bailo et al. (2004) have shown that hCMSCs can also be used in xenogeneic transplants in rats and pigs, showing success and good persistence of the transplant. MSCs used for transplantation were found in many organs and tissues after intravenous or intraperitoneal infusion, highlighting the migratory capacity of these cells.

5.4 WHARTON'S JELLY

5.4.1 Histology of Tissue

The umbilical cord is an extraembryonic tissue through which nourishment and the metabolic exchange occurs between the mother and the fetus. The umbilical cord originates when the amniotic cavity expands and wraps the connecting stalk and the yolk sac, joining them together. The umbilical cord, at the end of its development, is covered externally by the amniotic membrane and contains a vein that carries oxygenated blood rich in nutrients and hormones, and two umbilical arteries that carry poorly oxygenated blood, rich in carbon dioxide and waste substances. The vessels of the umbilical cord are coated by a layer of smooth muscle cells and immersed within a connective tissue called Wharton's jelly (Figure 5.1a–b bottom panels).

Wharton's jelly (named for Thomas Wharton, an English anatomist who first described this structure in 1656) is a mucosal tissue consisting of fibroblasts surrounded by an extracellular matrix rich in glycosaminoglycans, and different types of collagen fibers, mainly composed of collagen types I and III. The structure lacks elastic fibers, and hyaluronic acid is the main representative of glycosaminoglycans. In particular, the rich concentration of hyaluronic acid makes the matrix gelatinous, with the function of protecting the vessels from trauma, crucial for the efficient exchange of nutrients between the mother and fetus.

5.4.2 Cells Isolated from the Umbilical Cord

The cellular component of Wharton's jelly is composed of fibroblast cells; these cells express vimentin, desmin, and α-smooth muscle actin (Conconi et al. 2011). The cellular composition is not the same throughout all of the Wharton's jelly; it has been demonstrated that stromal cells resident in the tissue show different degrees of differentiation, ranging from mesenchymal cells to myofibroblasts (Can et al. 2007). The most immature, highest proliferating cells are located near the surface of the amniotic layer, and this justifies the presence of MSCs also in the amniotic fluid. Moving away from the amniotic surface, stromal cells acquire characteristics of contractile cells, and close to the vascular zone the stromal cells are differentiated into myofibroblasts.

5.4.3 Human Wharton's Jelly MSCs

Cells isolated from Wharton's jelly display characteristics very similar to MSCs obtained from classic adult sources such as bone marrow. They grow in adhesion to the plastic surfaces of the culture plates, and they present two distinct morphologies: flat, wide cytoplasmatic cells and slender fibroblast-like cells (Karahuseyinoglu et al. 2007). These two cell populations differ in their cytoskeletal filament content: vimentin (mesenchymal marker) and pancytokeratin (ecto-endodermal marker). The cell population positive for both vimentin and pancytokeratin appears flattened and is localized in the perivascular region (type I cell). The cell population with a more fusiform and elongated (type II cell) cytoplasm is positive only for vimentin, and it is located in the intervascular region. Both of these cell types are able to differentiate toward different commitments according to the mesengenic potential of mesenchymal cells (Baksh et al. 2007).

The mucosal connective tissue can be processed through a mechanical treatment and digested with an enzymatic solution (collagenase I) to isolate the stromal cellular component. These fibroblast-like adherent cells are considered as umbilical cord mesenchymal stromal/stem cells and are usually called umbilical cord mesenchymal stromal cells or Wharton's jelly mesenchymal stromal cells (WJMSCs). Therefore, human Wharton's jelly mesenchymal stromal cells (hWJMSCs) includes cells derived from the perivascular space, the intravascular space, and the subamniotic region of the umbilical cord (Figure 5.1c).

Several methods have been proposed for the isolation of human MSCs from Wharton's jelly (Salehinejad et al. 2012). Isolation techniques can be distinguished based on enzymatic treatment or tissue fragment cultures. The isolation process based on enzymatic degradation is the most efficient method to release cells from the Wharton's jelly, with a recovery of approximately 1.5×10^6 cells per centimeter of umbilical cord (Weiss et al. 2006). The latter is a method characterized by the reduction of tissue into very small fragments, and the success of the isolation procedure depends on the migratory capacity of the cells (Saward et al. 1997). The enzymatic methods consist of an "enzyme cocktail" for the dissociation and release of the various cell types from the tissue. The type of enzymes used determines the efficiency of isolation in terms of yield, viability, and toxicity. Specifically, for hWJMSCs we can summarize the following isolation techniques:

- the cord, cut longitudinally, is deprived of the vessels and the remaining tissue is treated enzymatically (with collagenase and trypsin or collagenase/trypsin/hyaluronidase); cells obtained are washed and cultured (Salehinejad et al. 2012);
- after blood vessel removal, the tissue is simply treated by mechanical dissociation, and the fragments obtained are transferred to plastic plates, allowing for cell migration (Mitchell et al. 2003); and
- the umbilical cord is cut into pieces of a few centimeters, and these pieces are placed in six-well plates so that the amniotic surface is facing upward. The internal tissue is further etched to increase the extension of the surface that should be in contact with the plastic and from which cells should migrate (La Rocca et al. 2009).

According to the minimal criteria proposed by Mesenchymal and Tissue Committee of the International Society for Cellular Therapy for defining MSCs (Dominici et al. 2006), hWJMSCs derived from umbilical cord possess plastic adherence ability under standard culture conditions; have the capacity to differentiate *in vitro* along adipogenic, chondrogenic, and osteogenic commitments; and express a minimal surface antigenic pattern. Their immunophenotype does not differ significantly from that described for MSCs of adult origin such as bone marrow. In terms of flow cytometry characterization, cells derived from Wharton's jelly express CD13, CD29, CD44, CD73, CD90, CD105, and CD166 and HLA-ABC, and they lack typical hematopoietic and endothelial markers CD14, CD28, CD31, CD34, CD45 CD66, CD80, CD86, CD133, and HLA-DR (Moodley et al. 2009).

hWJMSCs exhibit MSC features, and they are usually compared to mesenchymal cells derived from different adult or other perinatal sources, such as bone marrow, adipose tissue, or fetal membranes. However, they also express characteristics attributed to pluripotent stem cells such as ESCs; specifically, they display human ESC surface markers Tra-1-6, Tra-1-81, SSEA-1, and SSEA-4 and the well-established transcriptional triumvirate that maintain stemness: OCT-4, Sox-2, and Nanog. Although these cells have lower pluripotency marker expression compared to ESCs (Fong et al. 2011), they are formed during the earliest ontogenic development period and when compared to adult mesenchymal cells such as bone marrow–derived MSCs, show higher differentiation potential, telomerase activity, colony-forming unit-fibroblastic and proliferative potential, cumulative population doubling, and lower acquisition of a senescence phenotype with a longer time maintenance of stem cell potency, Taken together, hWJMSCs appear to be a more primitive mesenchymal cell population than those derived from classic sources such as bone marrow and adipose tissue (Troyer et al. 2008).

hWJMSCs retain the expression of markers typically associated to all three primordial germ layers, which is reflected in their multipotent differentiation potential. WJMSCs can be differentiated toward several commitments, including adipocytes, osteocytes, chondrocytes, skeletal myocytes, and endothelial cells (Wang et al. 2004); cardiomyocytes and hepatocyte-like cells (Anzalone et al. 2010); insulin-producing cells (Anzalone et al. 2011); neural-like cells (Can et al. 2011); dermal fibroblasts (Han et al. 2011); and epithelial keratinocyte-like cells (Tran et al. 2011).

The embryological origin of these tissues bestows another gift to the cells derived from the umbilical cord: immunomodulatory potential. hWJMSCs are immune privileged and possess immunosuppressive abilities comparable to those of bone marrow– and adipose-derived MSCs (Girdlestone et al. 2009). Several studies have shown that hWJMSCs can suppress *in vitro* proliferation of stimulated immune cells in a xenograft or allogeneic transplant model (Weiss et al. 2008). They do not stimulate the proliferation of allogeneic or xenogeneic immune cells. These immune properties can be, in part, related to the production of the immunosuppressive isoform of the well-known tolerogenic molecule HLA-G; the absent expression of immune-related surface antigens such as CD40, CD80, and CD86; and the secretion of cytokines such as VEGF and IL-6, which are endowed with immunosuppressive activity (Djouad et al. 2007). These results show that hWJMSCs have low immunogenicity and possess immunomodulatory ability *in vitro*.

Furthermore, several studies reached the conclusion that hWJMSCs can be tolerated in allogeneic transplantation with no evidence of *in vivo* immune rejection of undifferentiated cells (Conconi et al. 2006; Troyer et al. 2008).

More recently, hWJMSCs have been demonstrated to maintain the expression of critical immunomodulatory molecules after *in vitro* differentiation commitment toward osteogenic, adipogenic, and chondrogenic lineages. This result suggests that even after the acquisition of a mature phenotype, hWJMSCs may maintain their immune privilege. Therefore, the immunosuppressive ability of hWJMSCs suggests their potential applications in graft-versus-host disease and autoimmune disorders (La Rocca et al. 2013). Altogether, these results lend support to the therapeutic utility of hWJMSCs. Considering the features of hWJMSCs, various studies explored their use in regenerative and reparative approaches, providing encouraging results in several pathological settings, including rat liver fibrosis, mouse diabetes, human type 2 diabetes, and rodent models of retinal disease and Parkinson's disease (Weiss et al. 2006; Lund et al. 2007; Tsai et al. 2009; Wang et al. 2011; Liu et al. 2014). Undifferentiated hWJMSCs express markers associated with neural progenitor cells, and under specific neurogenic conditions, they may acquire a neuron-like phenotype. hWJMSCs may also contribute through neurotrophic secretion to endogenous stimulation of CNS regeneration through reparative mechanisms (Mitchell et al. 2003; Weiss et al. 2006; Paradisi et al. 2014).

hWJMSCs display an endothelial differentiation potential and the ability to support hematopoiesis, suggesting that these cells may also ameliorate the engraftment of hematopoietic stem cells (Lu et al. 2006). Cellular therapy using these cells has also been applied in cancer research: taking advantage of mesenchymal cell homing capacities and efficient transfectability, hWJMSCs have been engineered to target cancer, resulting in tumor reduction (Rachakatla et al. 2008).

5.5 CONCLUDING REMARKS

We can summarize the advantages of perinatal tissue as a source of cells with therapeutic potential by virtue of the fact that they are derived without invasive procedures, from uncontroversial and abundant specimens that are usually discarded after birth. In addition, these cells retain high expansion potential and genetic stability, do not induce teratomas, and possess anticancer effects and important immunosuppressive properties. In the near future, perinatal cells may represent an important source for cell therapy approaches. Future investigations are therefore required to better comprehend the characteristics of these cells and to broaden their clinical applications.

REFERENCES

Alviano, F., V. Fossati, C. Marchionni, et al. 2007. Term amniotic membrane is a high throughput source for multipotent mesenchymal stem cells with the ability to differentiate into endothelial cells in vitro. *BMC Dev Biol* 7: 11.

Anzalone, R., M. Lo Iacono, S. Corrao, et al. 2010. New emerging potentials for human Wharton's jelly mesenchymal stem cells: Immunological features and hepatocyte-like differentiative capacity. *Stem Cells Dev* 19 (4): 423–38.

Anzalone, R., M. Lo Iacono, T. Loria, et al. 2011. Wharton's jelly mesenchymal stem cells as candidates for beta cells regeneration: Extending the differentiative and immunomodulatory benefits of adult mesenchymal stem cells for the treatment of type 1 diabetes. *Stem Cell Rev* 7 (2): 342–63.

Bailo, M., M. Soncini, E. Vertua, et al. 2004. Engraftment potential of human amnion and chorion cells derived from term placenta. *Transplantation* 78 (10): 1439–48.

Baksh, D., R. Yao, and R. S. Tuan. 2007. Comparison of proliferative and multilineage differentiation potential of human mesenchymal stem cells derived from umbilical cord and bone marrow. *Stem Cells* 25 (6): 1384–92.

Bilic, G., S. M. Zeisberger, A. S. Mallik, R. Zimmermann, and A. H. Zisch. 2008. Comparative characterization of cultured human term amnion epithelial and mesenchymal stromal cells for application in cell therapy. *Cell Transplant* 17 (8): 955–68.

Cell Populations Isolated from Human Placenta

Broughton, B. R., R. Lim, T. V. Arumugam, et al. 2013. Post-stroke inflammation and the potential efficacy of novel stem cell therapies: Focus on amnion epithelial cells. *Front Cell Neurosci* 6: 66.

Can, A., and D. Balci. 2011. Isolation, culture, and characterization of human umbilical cord stroma-derived mesenchymal stem cells. *Methods Mol Biol* 698: 51–62.

Can, A., and S. Karahuseyinoglu. 2007. Concise review: Human umbilical cord stroma with regard to the source of fetus-derived stem cells. *Stem Cells* 25 (11): 2886–95.

Cargnoni, A., M. Di Marcello, M. Campagnol, et al. 2009. Amniotic membrane patching promotes ischemic rat heart repair. *Cell Transplant* 18 (10): 1147–59.

Carlson, B. M. 2014. *Human Embryology and Developmental Biology*. Elsevier Saunders, PA: Philadelphia.

Castillo-Melendez, M., T. Yawno, G. Jenkin, and S. L. Miller. 2013. Stem cell therapy to protect and repair the developing brain: A review of mechanisms of action of cord blood and amnion epithelial derived cells. *Front Neurosci* 7: 194.

Castrechini, N. M., P. Murthi, N. M. Gude, et al. 2010. Mesenchymal stem cells in human placental chorionic villi reside in a vascular Niche. *Placenta* 31 (3): 203–12.

Conconi, M. T., P. Burra, R. Di Liddo, et al. 2006. CD105(+) cells from Wharton's jelly show in vitro and in vivo myogenic differentiative potential. *Int J Mol Med* 18 (6): 1089–96.

Conconi, M. T., R. D. Liddo, M. Tommasin, C. Calore, and P. P. Parnigotto. 2011. Phenotype and differentiation potential of stromal populations obtained from various zones of human umbilical cord: An overview. *Open Tissue Eng Regen Med J* 4 (1): 6–20.

Diaz-Prado, S., E. Muiños-López, T. Hermida-Gómez, et al. 2010. Multilineage differentiation potential of cells isolated from the human amniotic membrane. *J Cell Biochem* 111 (4): 846–57.

Diaz-Prado, S., E. Muiños-López, T. Hermida-Gómez, et al. 2011. Human amniotic membrane as an alternative source of stem cells for regenerative medicine. *Differentiation* 81 (3): 162–71.

Djouad, F., L. M. Charbonnier, C. Bouffi, et al. 2007. Mesenchymal stem cells inhibit the differentiation of dendritic cells through an interleukin-6- dependent mechanism. *Stem Cells* 25 (8): 2025–32.

Dominici, M., K. Le Blanc, I. Mueller, et al. 2006. Minimal criteria for defining multipotent mesenchymal stromal cells. The International Society for Cellular Therapy position statement. *Cytotherapy* 8 (4): 315–31.

Evangelista, M., M. Soncini, and O. Parolini. 2008. Placenta-derived stem cells: New hope for cell therapy? *Cytotechnology* 58 (1): 33–42.

Fariha, M. M., K. H. Chua, G. C. Tan, Y. H. Lim, and A. R. Hayati. 2013. Pro-angiogenic potential of human chorion-derived stem cells: In vitro and in vivo evaluation. *J Cell Mol Med* 17 (5): 681–92.

Fong, C. Y., L. L. Chak, A. Biswas, et al. 2011. Human Wharton's jelly stem cells have unique transcriptome profiles compared to human embryonic stem cells and other mesenchymal stem cells. *Stem Cell Rev Rep* 7 (1): 1–16.

Fukuchi, Y., H. Nakajima, D. Sugiyama, et al. 2004. Human placenta-derived cells have mesenchymal stem/progenitor cell potential. *Stem Cells* 22 (5): 649–58.

Girdlestone, J., V. A. Limbani, A. J. Cutler, et al. 2009. Efficient expansion of mesenchymal stromal cells from umbilical cord under low serum conditions. *Cytotherapy* 11 (6): 738–48.

Han, Y., J. Chai, T. Sun, D. Li, and R. Tao. 2011. Differentiation of human umbilical cord mesenchymal stem cells into dermal fibroblasts in vitro. *Biochem Biophys Res Commun* 413 (4): 561–5.

Hebertson, R. M., M. E. Hammond, and M. J. Bryson. 1986. Amniotic epithelial ultrastructure in normal, polyhydramnic, and oligohydramnic pregnancies. *Obstet Gynecol* 68 (1): 74–9.

Hodge, A., D. Lourensz, V. Vaghjiani, et al. 2014. Soluble factors derived from human amniotic epithelial cells suppress collagen production in human hepatic stellate cells. *Cytotherapy* 16 (8): 1132–44.

Ilancheran, S., A. Michalska, G. Peh, et al. 2007. Stem cells derived from human fetal membranes display multi-lineage differentiation potential. *Biol Reprod* 77 (3): 577–88.

Ilancheran, S., Y. Moodley, and U. Manuelpillai. 2009. Human fetal membranes: A source of stem cells for tissue regeneration and repair? *Placenta* 30 (1): 2–10.

Insausti, C. L., A. Alcaraz, E. M. García-Vizcaíno, et al. 2010. Amniotic membrane induces epithelialization in massive posttraumatic wounds. *Wound Repair Regen* 18 (4): 368–77.

In 't Anker, P. S., S. A. Scherjon, C. Kleijburg-van der Keur, et al. 2004. Isolation of mesenchymal stem cells of fetal or maternal origin from human placenta. *Stem Cells* 22 (7): 1338–45.

Iwasaki, R., S. Matsubara, T. Takizawa, et al. 2003. Human amniotic epithelial cells are morphologically homogeneous: Enzyme histochemical, tracer, and freeze-substitution fixation study. *Eur J Histochem* 47 (3): 223–32.

Kakishita, K., M. A. Elwan, N. Nakao, T. Itakura, and N. Sakuragawa. 2000. Human amniotic epithelial cells produce dopamine and survive after implantation into the striatum of a rat model of Parkinson's disease: A potential source of donor for transplantation therapy. *Exp Neurol* 165 (1): 27–34.

Kamiya, K., M. Wang, S. Uchida, et al. 2005. Topical application of culture supernatant from human amniotic epithelial cells suppresses inflammatory reactions in cornea. *Exp Eye Res* 80 (5): 671–79.

Karahuseyinoglu, S., O. Cinar, E. Kilic, et al. 2007. Biology of stem cells in human umbilical cord stroma: In situ and in vitro surveys. *Stem Cells* 25 (2): 319–31.

Karlsson, H., T. Erkers, S. Nava, et al. 2012. Stromal cells from term fetal membrane are highly suppressive in allogeneic settings in vitro. *Clin Exp Immunol* 167 (3): 543–55.

Kim, M. J., K. S. Shin, J. H. Jeon, et al. 2011. Human chorionic-plate-derived mesenchymal stem cells and Wharton's jelly-derived mesenchymal stem cells: a comparative analysis of their potential as placenta-derived stem cells. *Cell Tissue Res* 346 (1): 53–64.

Kubo, M., Y. Sonoda, R. Muramatsu, and M. Usui. 2001. Immunogenicity of human amniotic membrane in experimental xenotransplantation. *Invest Ophthalmol Vis Sci* 42 (7): 1539–46.

La Rocca, G., R. Anzalone, S. Corrao, et al. 2009. Isolation and characterization of Oct-4+/HLA-G+ mesenchymal stem cells from human umbilical cord matrix: Differentiation potential and detection of new markers. *Histochem Cell Biol* 131 (2): 267–82.

La Rocca, G., M. Lo Iacono, T. Corsello, et al. 2013. Human Wharton's jelly mesenchymal stem cells maintain the expression of key immunomodulatory molecules when subjected to osteogenic, adipogenic and chondrogenic differentiation in vitro: New perspectives for cellular therapy. *Curr Stem Cell Res Ther* 8 (1): 100–13.

Lee, J. M., J. Jung, H. J. Lee, et al. 2012. Comparison of immunomodulatory effects of placenta mesenchymal stem cells with bone marrow and adipose mesenchymal stem cells. *Int Immunopharmacol* 13 (2): 219–24.

Li, H., J. Y. Niederkorn, S. Neelam, et al. 2005. Immunosuppressive factors secreted by human amniotic epithelial cells. *Invest Ophthalmol Vis Sci* 46 (3): 900–7.

Liu, T., J. Wu, Q. Huang, et al. 2008. Human amniotic epithelial cells ameliorate behavioral dysfunction and reduce infarct size in the rat middle cerebral artery occlusion model. *Shock* 29 (5): 603–11.

Liu, X., P. Zheng, X. Wang, et al. 2014. A preliminary evaluation of efficacy and safety of Wharton's jelly mesenchymal stem cell transplantation in patients with type 2 diabetes mellitus. *Stem Cell Res Ther* 5 (2): 57.

Liu, Y. H., V. Vaghjiani, J. Y. Tee, et al. 2012. Amniotic epithelial cells from the human placenta potently suppress a mouse model of multiple sclerosis. *PLoS One* 7 (4): e35758.

Lu, L. L., Y. J. Liu, S. G. Yang, et al. 2006. Isolation and characterization of human umbilical cord mesenchymal stem cells with hematopoiesis-supportive function and other potentials. *Haematologica* 91 (8): 1017–26.

Lund, R. D., S. Wang, B. Lu, et al. 2007. Cells isolated from umbilical cord tissue rescue photoreceptors and visual functions in a rodent model of retinal disease. *Stem Cells* 25 (3): 602–11.

Manuelpillai, U., D. Lourensz, V. Vaghjiani, et al. 2012. Human amniotic epithelial cell transplantation induces markers of alternative macrophage activation and reduces established hepatic fibrosis. *PLoS One* 7 (6): e38631.

Manuelpillai, U., Y. Moodley, C. V. Borlongan, and O. Parolini. 2011. Amniotic membrane and amniotic cells: Potential therapeutic tools to combat tissue inflammation and fibrosis? *Placenta* 32 (Suppl 4): S320–25.

Manuelpillai, U., J. Tchongue, D. Lourensz, et al. 2010. Transplantation of human amnion epithelial cells reduces hepatic fibrosis in immunocompetent CCl(4)-treated mice. *Cell Transplant* 19 (9): 1157–68.

McDonald, C., C. Siatskas, and C. C. A. Bernard. 2011. The emergence of amnion epithelial stem cells for the treatment of multiple sclerosis. *Inflamm Regen* 31: 256–71.

Miki, T. 2011. Amnion-derived stem cells: In quest of clinical applications. *Stem Cell Res Ther* 2: 25.

Miki, T., T. Lehmann, H. Cai, et al. 2005. Stem cell characteristics of amniotic epithelial cells. *Stem Cells* 23 (10): 1549–59.

Miki, T., K. Mitamura, M.A. Ross, D. B. Stolz, and S. C. Strom. 2007. Identification of stem cell marker-positive cells by immunofluorescence in term human amnion. *J Reprod Immunol* 75 (2): 91–96.

Miki, T., and S. Strom. 2006. Amnion-derived pluripotent/multipotent stem cells. *Stem Cell Rev Rep* 2 (2): 133–42.

Mitchell, K. E., M. L. Weiss, B. M. Mitchell, et al. 2003. Matrix cells from Wharton's jelly form neurons and glia. *Stem Cells* 21 (1): 50–60.

Moodley, Y., D. Atienza, U. Manuelpillai, et al. 2009. Human umbilical cord mesenchymal stem cells reduce fibrosis of bleomycin-induced lung injury. *Am J Pathol* 175 (1): 303–13.

Moodley, Y., S. Ilancheran, C. Samuel, et al. 2010. Human amnion epithelial stem cell transplantation abrogates lung fibrosis and augments repair. *Am J Respir Crit Care Med* 182 (5): 643–51.

Mosquera, A., J. L. Fernández, A. Campos, et al. 1999. Simultaneous decrease of telomere length and telomerase activity with ageing of human amniotic fluid cells. *J Med Genet* 36 (6): 494–96.

Paradisi, M., F. Alviano, S. Pirondi, et al. 2014. Human mesenchymal stem cells produce bioactive neurotrophic factors: Source, individual variability and differentiation issues. *Int J Immunopathol Pharmacol* 27 (3): 391–402.

Parolini, O., F. Alviano, G. P. Bagnara, et al. 2008. Concise review: Isolation and characterization of cells from human term placenta: Outcome of the First International Workshop on placenta derived stem cells. *Stem Cells* 26 (2): 300–11.

Parolini, O., F. Alviano, I. Bergwerf, et al. 2010. Toward cell therapy using placenta-derived cells: Disease mechanisms, cell biology, preclinical studies, and regulatory aspects at the round table. *Stem Cells Dev* 19 (2):143–54.

Parolini, O., F. Alviano, A. G. Betz, et al. 2011. Meeting report of the first conference of the International Placenta Stem Cell Society (IPLASS). *Placenta* 32 (Suppl 4): S285–90.

Parolini, O., M. Soncini, M. Evangelista, and D. Schmidt. 2009. Amniotic membrane and amniotic fluid-derived cells: Potential tools for regenerative medicine? *Regen Med* 4 (2): 275–91.

Pasquinelli, G., P. Tazzari, R. Ricci, et al. 2007. Ultrastructural characteristics of human mesenchymal stromal (stem) cells derived from bone marrow and term placenta. *Ultrastruct Pathol* 31 (1): 23–31.

Portmann-Lanz, C., A. Schoeberlein, R. Portmann, et al. 2009. Turning placenta into brain: Placental mesenchymal stem cells differentiate into neurons and oligodendrocytes. *Am J Obstet Gynecol* 202 (3): 294.e1–294.e11.

Portmann-Lanz, C. B., A. Schoeberlein, A. Huber, et al. 2006. Placental mesenchymal stem cells as potential autologous graft for pre and perinatal neuroregeneration. *Am J Obstet Gynecol* 194 (3): 664–73.

Rachakatla, R. S., M. M. Pyle, R. Ayuzawa, et al. 2008. Combination treatment of human umbilical cord matrix stem cell-based interferon-beta gene therapy and 5-fluorouracil significantly reduces growth of metastatic human breast cancer in SCID mouse lungs. *Cancer Invest* 26 (7): 662–70.

Resta, E., M. Zavatti, L. Bertoni, et al. 2013. Enrichment in c-kit enhance mesodermal and neural differentiation of human chorionic placental cells. *Placenta* 34 (7): 526–35.

Sakuragawa, N., S. Enosawa, T. Ishii, et al. 2000. Human amniotic epithelial cells are promising transgene carriers for allogeneic cell transplantation into liver. *J Hum Genet* 45 (3): 171–76.

Sakuragawa, N., K. Kakinuma, A. Kikuchi, et al. 2004. Human amnion mesenchyme cells express phenotypes of neuroglial progenitor cells. *J Neurosci Res* 78 (2): 208–14.

Sakuragawa, N., R. Thangavel, M. Mizuguchi, M. Hirasawa, and I. Kamo. 1996. Expression of markers for both neuronal and glial cells in human amniotic epithelial cells. *Neurosci Lett* 209 (1): 9–12.

Salehinejad, P., N. B. Alitheen, A. M. Ali, et al. 2012. Comparison of different methods for the isolation of mesenchymal stem cells from human umbilical cord Wharton's jelly. *In vitro Cell Dev Biol Anim* 48 (2): 75–83.

Sankar, V., and R. Muthusamy. 2003. Role of human amniotic epithelial cell transplantation in spinal cord injury repair research. *Neuroscience* 118 (1): 11–17.

Saward, L., and P. Zahradka. 1997. Coronary artery smooth muscle in culture: Migration of heterogeneous cell populations from vessel wall. *Mol Cell Biochem* 176 (1–2): 53–9.

Soncini, M., E. Vertua, L. Gibelli, et al. 2007. Isolation and characterization of mesenchymal cells from human fetal membranes. *J Tissue Eng Regen Med* 1 (4): 296–305.

Takashima, S., H. Ise, P. Zhao, T. Akaike, and T. Nikaido. 2004. Human amniotic epithelial cells possess hepatocyte-like characteristics and functions. *Cell Struct Funct* 29 (3): 73–84.

Tamagawa, T., S. Oi, I. Ishiwata, H. Ishikawa, and Y. Nakamura. 2007. Differentiation of mesenchymal cells derived from human amniotic membranes into hepatocyte-like cells in vitro. *Human Cell* 20 (3): 77–84.

Tanaka, M., Z. Chen, S. Bartunkova, N. Yamasaki, and S. Izumo. 1999. The cardiac homeobox gene Csx/Nkx2.5 lies genetically upstream of multiple genes essential for heart development. *Development* 126 (6): 1269–80.

Tao, J., F. Ji, B. Liu, et al. 2012. Improvement of deficits by transplantation of lentiviral vector-modified human amniotic mesenchymal cells after cerebral ischemia in rats. *Brain Res* 1448: 1–10.

Terada, S., K. Matsuura, S. Enosawa, et al. 2000. Inducing proliferation of human amniotic epithelial (HAE) cells for cell therapy. *Cell Transplant* 9 (5): 701–4.

Toda, A., M. Okabe, T. Yoshida, and T. Nikaido. 2007. The potential of amniotic membrane/amnion-derived cells for regeneration of various tissues. *J Pharmacol Sci* 105 (3): 215–28.

Tran, C. T., D. T. Huynh, C. Gargiulo, et al. 2011. In vitro culture of Keratinocytes from human umbilical cord blood mesenchymal stem cells: The Saigonese culture. *Cell Tissue Bank* 12 (2): 125–33.

Troyer, D. L., and M. L. Weiss. 2008. Wharton's jelly-derived cells are a primitive stromal cell population. *Stem Cells* 26 (3): 591–99.

Tsai, P. C., T. W. Fu, Y. M. Chen, et al. 2009. The therapeutic potential of human umbilical mesenchymal stem cells from Wharton's jelly in the treatment of rat liver fibrosis. *Liver Transplant* 15 (5): 484–95.

Wang, H. S., S. C. Hung, S. T. Peng, et al. 2004. Mesenchymal stem cells in the Wharton's jelly of the human umbilical cord. *Stem Cells* 22 (7): 1330–37.

Wang, H. S., J. F. Shyu, W. S. Shen, et al. 2011. Transplantation of insulin producing cells derived from umbilical cord stromal mesenchymal stem cells to treat NOD mice. *Cell Transplant* 20 (3): 455–66.

Wei J. P., T. S. Zhang, S. Kawa, et al. 2003. Human amnion-isolated cells normalize blood glucose in streptozotocin-induced diabetic mice. *Cell Transplant* 2 (5): 545–52.

Weiss, M. L., C. Anderson, S. Medicetty, et al. 2008. Immune properties of human umbilical cord Wharton's jelly-derived cells. *Stem Cells* 26 (11): 2865–74.

Weiss, M. L., S. Medicetty, A. R. Bledsoe, et al. 2006. Human umbilical cord matrix stem cells: Preliminary characterization and effect of transplantation in a rodent model of Parkinson's disease. *Stem Cells* 24 (3): 781–92.

Witkowska-Zimny, M., and E. Wrobel. 2011. Perinatal sources of mesenchymal stem cells: Wharton's jelly, amnion and chorion. *Cell Mol Biol Lett* 16 (3): 493–514.

Wolbank, S., A. Peterbauer, M. Fahrner, et al. 2007. Dose-dependent immunomodulatory effect of human stem cells from amniotic membrane: A comparison with human mesenchymal stem cells from adipose tissue. *Tissue Eng* 13 (6): 1173–83.

Zhang, Y., C. Li, X. Jiang, et al. 2004. Human placenta-derived mesenchymal progenitor expansion of long-term culture-initiating cells from cord blood CD34+cells. *Exp Hematol* 32 (7): 657–65.

Zhao, P., H. Ise, M. Hongo, et al. 2005.Human amniotic mesenchymal cells have some characteristics of cardiomyocytes. *Transplantation* 79 (5): 528–35.

6 The Immunomodulatory Features of Mesenchymal Stromal Cells Derived from Wharton's Jelly, Amniotic Membrane, and Chorionic Villi
In Vitro *and* In Vivo *Data*

Marta Magatti, Mohamed H. Abumaree,
Antonietta R. Silini, Rita Anzalone,
Salvatore Saieva, Eleonora Russo,
Maria Elena Trapani, Giampiero La Rocca,
and Ornella Parolini

CONTENTS

Preface...92
6.1　Immunogenicity of MSCs Derived from WJ, AM, and Chorionic Villi........92
　　6.1.1　Expression of HLA Molecules 93
　　6.1.2　Expression of B7 Costimulatory Family Molecules...........................96
　　6.1.3　Expression of TNF/TNFR Superfamily Molecules 97
　　6.1.4　Expression of Hematopoietic Markers...98
　　6.1.5　*In Vitro* Studies 98
　　6.1.6　*In Vivo* Studies..99
6.2　Immunomodulatory Properties of MSCs Derived from WJ, AM,
　　and Chorionic Villi 99
　　6.2.1　Effects on T Cells 100
　　6.2.2　Effects on B Cells 104
　　6.2.3　Effects on NK Cells...104
　　6.2.4　Effects on APCs (DCs)...105
　　6.2.5　Effects on APCs (Macrophages) 107
　　6.2.6　Effects on Neutrophils 110

91

6.3	Mechanisms of Immunosuppression by MSCs Derived from WJ, AM, and Chorionic Villi	111
	6.3.1 Interleukin 10	111
	6.3.2 Transforming Growth Factor Beta	112
	6.3.3 Hepatocyte Growth Factor	112
	6.3.4 Prostaglandin E2	113
	6.3.5 Human Leukocyte Antigen G	113
	6.3.6 Indoleamine 2,3-Dioxygenase	114
	6.3.7 Inflammatory Cytokines (INF-γ and IL-1β)	115
	6.3.8 Other Secreted Factors	115
References		116

PREFACE

This chapter focuses on the immunomodulatory properties of placental mesenchymal stromal cells (MSCs) derived from the amniotic membrane, umbilical cord, and chorionic villi. Within the amniotic membrane (AM), we discuss the immunomodulatory properties of the two main cell populations that can be isolated from AM: human amniotic mesenchymal stromal cells (hAMSCs) and human amniotic epithelial cells (hAECs). Within the umbilical cord, several compartments have been described, including the amniotic compartment, the Wharton's jelly (WJ) compartment, and the vascular and perivascular compartment, but herein attention is focused on the properties of human WJ MSCs (hWJMSCs). Since different isolation protocols have been used, umbilical cord matrix stem cells, umbilical cord stromal cells, and human umbilical cord MSCs (herein hUCMSCs) were also included. This chapter also discusses the properties of human chorionic villi/placenta MSCs (hCVMSCs/hPLMSCs).

The nomenclature of placental MSCs used in this chapter was chosen on the basis of consensus from the First International Workshop on Placenta-Derived Stem Cells for hAECs and hAMSCs (Parolini et al. 2008). Since different nomenclature is reported for the other placental MSCs, the following nomenclature is used: hWJMSCs/hUCMSCs for MSCs from WJ (and from umbilical cord where described isolation protocols include WJ) and hCVMSCs/hPLMSCs for MSCs from chorionic villi and the placenta/trophoblast compartment.

6.1 IMMUNOGENICITY OF MSCs DERIVED FROM WJ, AM, AND CHORIONIC VILLI

Immunogenicity is referred to as the ability to induce a humoral immune response, a cell-mediated immune response, or both. The response of T cells to their cognate antigens is principally governed by two distinct molecular signals. The first signal results from ligation of the T cell receptor by antigens associated with human leukocyte antigen (HLA) molecules. The second signal (costimulatory signal) occurs through costimulatory molecules found on the surface of antigen-presenting cells (APCs) and their receptors on T cells (immunoglobulin superfamily, such as CD28/B7 pathway and inducible costimulator of T cells [ICOS]/B7h pathway, tumor necrosis

The Immunomodulatory Features of Mesenchymal Stromal Cells 93

factor [TNF]/TNF receptor (TNFR) superfamily, and T cell immunoglobulin [Ig] domain and mucin domain family) (Bretscher 1999; Li et al. 2009).

The immunological profile described for cells isolated from the AM (hAM-SCs and hAECs), WJ (hWJMSCs), and chorionic villi (hCVMSCs) has led to the hypothesis that these cells are poorly immunogenic. This feature is generally associated to the absence of HLA class II (HLA-DR) and the major costimulatory molecules responsible for T cell activation, specifically B7-1 (CD80) and B7-2 (CD86) (Table 6.1). In addition, hWJMSCs, hAMSCs, hAECs, and hCVMSCs express molecules known to have immunoregulatory properties, such as HLA-G (Hunt et al. 2005), PD-L1 and PD-L2 (Murakami and Riella 2014), and B7-H3 (Leitner et al. 2009) (Table 6.1). All these features render these cells poor stimulators of T cells *in vitro*, and consequently, they are able to survive for prolonged periods after xenogeneic and allogeneic transplantation into immunocompetent animals without the use of immunosuppressants. However, it remains to be clarified whether these cells truly lack immunogenicity (due to the absence of HLA-DR and B7-1 and B7-2 costimulatory molecules), possess immunotolerance (mainly due to HLA-G expression), or are simply exerting immunomodulation (placental cells interact with and modulate functions of immune cells, as discussed in Section 6.2).

6.1.1 EXPRESSION OF HLA MOLECULES

hWJMSCs, hAMSCs, hAECs, and hCVMSCs have been shown to express the HLA class I (HLA-ABC), but not HLA class II (HLA-DR) (Abumaree et al. 2013b; Banas et al. 2008; Bilic et al. 2008; Donders et al. 2015; Gu et al. 2013; Igura et al. 2004; Kang et al. 2012; Kronsteiner et al. 2011b; La Rocca et al. 2013b; Magatti et al. 2015; Moodley et al. 2010; Murphy et al. 2010; Portmann-Lanz et al. 2006; Prasanna et al. 2010; Pratama et al. 2011; Raicevic et al. 2011; Roelen et al. 2009; Tipnis et al. 2010; Tsuji et al. 2010; Weiss et al. 2006; Wolbank et al. 2009b; Zhang et al. 2006; Zhou et al. 2011). Interestingly, quantitative reverse transcription-polymerase chain reaction (qRT-PCR) studies have shown that hWJMSCs produce an immunosuppressive isoform of HLA-I (Wang et al. 2014). Furthermore, hWJMSCs and hAECs do not express HLA class II -DP and -DQ (Chen et al. 2012; Moodley et al. 2010; Murphy et al. 2010; Pratama et al. 2011).

One of the most interesting characteristics of hWJMSCs, hAMSCs, hAECs, and hCVMSCs is that they express HLA-G, a nonclassical HLA class Ib molecule (Anam et al. 2013; Banas et al. 2008; Chang et al. 2006; Donders et al. 2015; Kronsteiner et al. 2011b; La Rocca et al. 2013b; Lefebvre et al. 2000; Pratama et al. 2011; Roelen et al. 2009), known to have immune regulatory properties (Hunt et al. 2005). Furthermore, the treatment of hAMSCs and hAECs with inflammatory cytokine interferon (IFN)-γ increases the expression of HLA-G (Anam et al. 2013; Banas et al. 2008; Kronsteiner et al. 2011b; Lefebvre et al. 2000). HLA-G interacts with Ig-like transcript (ILT) receptors (ILT-2, ILT-3, and ILT-4), which are expressed by T and B lymphocytes, as well as natural killer (NK) cells and mononuclear phagocytes (Allan et al. 2000), but not by hAECs (Banas et al. 2008). Through the interaction with inhibitory receptors on leukocytes, HLA-G displays relevant immune functions and appears to be one of the factors contributing to fetal escape from maternal immune

TABLE 6.1
Immunological Phenotype of hWJMSCs/hUCMSCs, hAECs, hAMSCs, and hCVMSCs/hPLMSCs

		hWJMSCs/hUCMSCs		hAECs		hAMSCs		hCVMSCs/hPLMSCs	
		Not Stimulation	INF-γ Stimulation	Not Stimulation	INF-γ Stimulation	Not Stimulation	INF-γ Stimulation	Not Stimulation	INF-γ Stimulation
HLA class I	HLA-ABC	+[1–5]	+ ↑[1]	+[6–12]	+ ↑[9]	+[7,11–16]	+[14]	+[7,17–19]	NA
Nonclassical HLA class I	HLA-G	+[2,20]	NA	−[9], +[10,21,22]	+ ↑[9,21,22]	−[23], +[14]	+ ↑[14]	+[24]	NA
	HLA-E	+[2,20]	NA	NA	NA	NA	NA	NA	NA
	HLA-F	+[2]	NA	NA	NA	NA	NA	NA	NA
Nonclassical HLA class I receptors	ILT-2	NA	NA	−[9]	−[9]	NA	NA	NA	NA
	ILT-3	NA	NA	−[9]	−[9]	NA	NA	NA	NA
	ILT-4	NA	NA	−[9]	−[9]	NA	NA	NA	NA
HLA class II	HLA-DP	−[25]	NA	−[6,10]	NA	NA	NA	NA	NA
	HLA-DQ	−[25]	NA	−[6,10], +[8]	NA	NA	NA	NA	NA
	HLA-DR	−[1–5,20,26]	−[1,4], +[3]	−[6–12]	−[9]	−[7,11–16,23,27]	+[14]	−[17–19,24,28–30]	−[24]
B7 costimulatory molecules	B7-1 (CD80)	−[1–4,20,31]	−[1,3,4]	−[9,10]	−[9]	−[14,23]	−[14]	−[19,29,30]	NA
	B7-2 (CD86)	−[1–4,20,31]	−[1,3,4]	−[9,10]	−[9]	−[23]; −/+[14]	−/+[14]	−[19,29,30]	NA
	B7-H2 (CD275)	NA	NA	−[32]	NA	−[14]	−[14]	−[19]	NA
	B7-H3 (CD276)	+[2]	NA	+[32]	NA	NA	NA	+[19]	NA
	PD-1 (CD279)	NA	NA	−[9]	−[9]	−[33]	NA	−[19,30]	NA
	PD-L1 (CD274)	−[2]; +[3]	+ ↑[3]	−[9,32]	+[9]	+[14,33]	+ ↑[14]	−[24]; +[19,30]	+[24]
	PD-L2 (CD273)	−[3]	−[3]	−[9,32]	+[9]	+[14]	+ ↑[14]	−[30]; +[19]	NA

(Continued)

TABLE 6.1 (Continued)
Immunological Phenotype of hWJMSCs/hUCMSCs, hAECs, hAMSCs, and hCVMSCs/hPLMSCs

		hWJMSCs/hUCMSCs		hAECs		hAMSCs		hCVMSCs/hPLMSCs	
		Not Stimulation	INF-γ Stimulation	Not Stimulation	INF-γ Stimulation	Not Stimulation	INF-γ Stimulation	Not Stimulation	INF-γ Stimulation
TNF/TNF receptor costimulatory molecules	CD40	−(3,4,20,31)	+[a](4)	−(9), +(10)	+(9)	−(14,23)	+(14)	−(19,29)	NA
	CD40L (CD154)	−(31)	NA	−(9)	−(9)	NA	NA	NA	NA
	Fas (CD95)	NA	NA	+(9)	+(9)	NA	NA	−(30)	NA
	FasL (CD178)	NA	NA	−(9)	−(9)	NA	NA	+(30)	NA
	GITR-L	NA	NA	NA	NA	−(14)	−(14)	NA	NA
	CD271	+(34)	NA	−(35)	NA	+(35)	NA	NA	NA
Hematopoietic markers	CD34	−(1,20,26,36)	NA	−(6,7,9,12,37–39)	NA	−(7,12–16,23,27,37–41)	−(14)	−(17,18,28,30,36,42)	NA
	CD14	−(20,26,36)	NA	−(7,9,11,39)	NA	−(7,11,15,16,23,27,39–41)	NA	−(7,29,36,42)	NA
	CD45	−(1,3,20,25,26,36)	NA	−(6,7,9,11,12,37–39)	NA	−(7,11–16,23,27,37–41)	−(14)	−(7,17–19,28–30,36,42)	NA

FasL, Fas ligand; GITR, glucocorticoid-induced tumor necrosis factor receptor; hAECs, human amniotic epithelial cells; hAMSCs, human amniotic membrane mesenchymal stromal cells; hCVMSCs/hPLMSCs, human chorionic villous/placenta mesenchymal stem/stromal cells; hWJMSCs/hUCMSCs, human Wharton's jelly mesenchymal stromal cells/human umbilical cord mesenchymal stromal cells; HLA, human leukocyte antigen; IFN, interferon; ILT, immunoglobulin-like transcript; NA, not available; TNF, tumor necrosis factor. −, negative; −/+, weak expression; +, positive; ↑, increase expression compared to not stimulated cells.

[1] Prasanna et al. 2010; [2] La Rocca et al. 2013b; [3] Tipnis et al. 2010; [4] Raicevic et al. 2011; [5] Zhou et al. 2011; [6] Moodley et al. 2010; [7] Portmann-Lanz et al. 2006; [8] Murphy et al. 2010; [9] Banas et al. 2008; [10] Pratama et al. 2011; [11] Magatti et al. 2015; [12] Bilic et al. 2008; [13] Kang et al. 2012; [14] Kronsteiner et al. 2011b; [15] Wolbank et al. 2009b; [16] Jaramillo-Ferrada et al. 2012; [17] Igura et al. 2004; [18] Zhang et al. 2006; [19] Abumaree et al. 2013b; [20] Donders et al. 2015; [21] Lefèbvre et al. 2000; [22] Anam et al. 2013; [23] Roelen et al. 2009; [24] Chang et al. 2006; [25] Chen et al. 2012; [26] Weiss et al. 2006; [27] Tsuji et al. 2010; [28] Mathews et al. 2015; [29] Abomaray et al. 2015; [30] Gu et al. 2013; [31] Wang et al. 2014; [32] Petroff and Perchellet 2010; [33] Wu et al. 2014; [34] Margossian et al. 2012; [35] Soncini et al. 2007; [36] Li et al. 2014; [37] Diaz-Prado et al. 2010; [38] Roda et al. 2009; [39] Stadler et al. 2008; [40] Wegmeyer et al. 2013; [41] Alviano et al. 2007; [42] Poloni et al. 2008.

[a] Inflammatory cocktail containing IL-1β, IFN-γ, TNF-α, and INF-α.

surveillance (Petroff 2005). Furthermore, hWJMSCs have been shown to express not only HLA-G but also HLA-E and -F. Interestingly, the expression of class Ib major histocompatibility complex (MHC) molecules by hWJMSCs has been shown to be consistent even after differentiation *in vitro* (Donders et al. 2015; La Rocca et al. 2013a, 2013b). Similar to HLA-G, HLA-E and HLA-F are known to be involved in the tolerance process between the mother and fetus (Ishitani et al. 2003), playing a key role through selective binding to NK cells (Goodridge et al. 2013; Lee et al. 1998).

Of note, several differences in HLA expression have been described, mainly due to culture conditions such as cell number, passage, and culture medium. For example, some reports have shown a low-to-moderate expression of HLA-ABC on hAECs (Moodley et al. 2010; Portmann-Lanz et al. 2006), whereas others have reported high expression HLA-ABC (Banas et al. 2008; Bilic et al. 2008; Magatti et al. 2015; Murphy et al. 2010; Pratama et al. 2011). Moreover, one study has shown that in hAECs from passages 0 to 5, the expression of HLA-DQ increases and that of HLA-ABC decreases (Murphy et al. 2010). Furthermore, HLA-G expression on hAECs has been reported to vary with cell expansion and culture medium. Specifically, whereas passage (P)0 hAECs were HLA-G positive, the number of HLA-G–producing cells decreased significantly at P5 when expansion was made in Dulbecco's modified Eagle's medium/ F12 + 10% fetal calf serum, and HLA-G was absent in cells expanded in serum-free EpiLife medium (Pratama et al. 2011). Finally, when activated with IFN-γ, hAMSCs express HLA-DR (Kronsteiner et al. 2011b; Tipnis et al. 2010), whereas hWJMSCs have been reported to have either minimal (Prasanna et al. 2010) or high (Tipnis et al. 2010) HLA-DR expression after IFN-γ activation.

6.1.2 Expression of B7 Costimulatory Family Molecules

hWJMSCs, hAMSCs, hAECs, and hCVMSCs do not express B7-1 (CD80) and B7-2 (CD86), the major costimulatory molecules belonging to the B7 family and responsible for T cell activation (Abomaray et al. 2015; Abumaree et al. 2013b; Banas et al. 2008; Donders et al. 2015; Gu et al. 2013; Kronsteiner et al. 2011b; La Rocca et al. 2013b; Prasanna et al. 2010; Pratama et al. 2011; Raicevic et al. 2011; Roelen et al. 2009; Tipnis et al. 2010; Wang et al. 2014). Other molecules belonging to the B7 family are programmed death ligands (PD-Ls) (PD-L1, also known as B7-H1 and CD274) and PD-L2 (B7-DC and CD273). hAMSCs (Kronsteiner et al. 2011b; Wu et al. 2014) and hCVMSCs/hPLMSCs (Abumaree et al. 2013b; Gu et al. 2013) express both PD-L1 and PD-L2, and IFN-γ treatment has been shown to increase PD-L1 and PD-L2 expression in hAMSCs (Kronsteiner et al. 2011b), and also on hAECs, which do not constitutively express these costimulatory molecules (Banas et al. 2008; Petroff and Perchellet 2010). In contrast, IFN-γ treatment has been shown to increase PD-L1, but not PD-L2, on hWJMSCs (Tipnis et al. 2010).

PD-L1 and PD-L2 possess both costimulatory and co-inhibitory actions on T cells. Although their stimulatory function is not well understood, many studies have focused on their inhibitory function, which occurs by signaling through PD-1 (CD279) receptor (Petroff and Perchellet 2010). hAECs, hAMSCs, and hCVMSCs/ hPLMSCs do not express PD-1 (Abumaree et al. 2013b; Banas et al. 2008; Gu et al. 2013; Wu et al. 2014), but it is expressed on activated T cells, B cells, and myeloid

The Immunomodulatory Features of Mesenchymal Stromal Cells 97

cells; and, given its suppressive role in controlling reactive T cells, it seems to play critical roles in maintaining immunologic tolerance. Indeed, the PD-1:PD-L1/PD-L2 pathway inhibits T cell receptor (TCR) signaling, T cell survival, cytokine production, and induces T regulatory (Treg) cells (Murakami and Riella 2014).

Another costimulatory molecule that belongs to the B7 family is the B7-H3 (CD276) protein. Flow cytometry and immunohistochemical analyses have shown that it is expressed by hWJMSCs (La Rocca et al. 2013b), hAECs (Petroff and Perchellet 2010), and hCVMSCs (Abumaree et al. 2013b). In addition, the molecule has been shown to be expressed not only in naive hWJMSCs but also in their *in vitro*–differentiated (toward osteogenic, adipogenic, and chondrogenic lineages) counterparts, a characteristic described above also for HLA-G, -E, and -F (La Rocca et al. 2013b). B7-H3 is an important T cell–immunosuppressive molecule that has been reported to consistently downregulate human T cell cytokine production and proliferation (Leitner et al. 2009). Finally, in contrast to PD-L1, PD-L2, and B7-H3, B7-H2 (CD275 or ICOS-L), the ligand for ICOS and provider of positive costimulatory effects promoting T cell activation, differentiation, and effector responses (Coyle et al. 2000; Petroff and Perchellet 2010; Yoshinaga et al. 1999), is not expressed by hAECs (Petroff and Perchellet 2010), hAMSCs (Kronsteiner et al. 2011b), and hCVMSCs (Abumaree et al. 2013b).

6.1.3 Expression of TNF/TNFR Superfamily Molecules

Among the costimulatory molecules belong to the TNF/TNFR superfamily, several studies have investigated the presence of glucocorticoid-induced TNFR (GITR)-related protein ligand (GITR-L), Fas (CD95) and Fas ligand (FasL) (CD95L, CD178), CD40 and CD40 ligand (CD40L, CD154), and CD271.

Upon T cell activation, GITR-L transmits the signal through GITR expressed on T cells. GITR abrogates the suppressive activity of CD4+CD25+ Treg cells, while exhibiting a costimulatory function on CD4+CD25− and CD8+ T cells (Nocentini and Riccardi 2005; Shimizu et al. 2002). Overall, GITR-L is mainly associated to overstimulation of the immune system, and it is not expressed by hAMSCs (Kronsteiner et al. 2011b).

Fas and FasL trigger apoptotic cell death, essentially by FasL-induced cell death of Fas-positive cells (Strasser et al. 2009). Therefore, FasL-Fas signaling plays critical role in the control of the immune system. hAECs have been shown to express Fas, but not FasL (Banas et al. 2008), indicating that hAECs could be the target of the apoptotic process, but not the trigger. On the contrary, a role for FasL expression in the induction of T cell apoptosis has been suggested for hPLMSCs, which have been shown to express FasL, but not Fas (Gu et al. 2013).

Furthermore, there are several studies that investigate the expression of CD40 and CD40L, able to activate T cells (Li et al. 2009; van Kooten and Banchereau 2000), on placental cells (Li et al. 2009; van Kooten and Banchereau 2000). hCVMSCs and hWJMSCs do not express CD40 (Abumaree et al. 2013b; Donders et al. 2015; Raicevic et al. 2011; Tipnis et al. 2010; Wang et al. 2014), and the latter of which lack also CD40L (Wang et al. 2014). hAMSCs have been shown to lack or marginally express CD40 (Kronsteiner et al. 2011b; Roelen et al. 2009), which can be induced after INF-γ stimulation (Kronsteiner et al. 2011b). hAECs do not express the CD40L

(Banas et al. 2008) and discrepancies have been reported regarding the expression of CD40. For example, Banas et al. (2008) suggested that hAECs are negative for CD40 and that the expression increases after INF-γ stimulation, whereas Pratama et al. (2011) found a moderate-to-high constitutive expression of CD40 on hAECs.

Finally, CD271, another member of the TNFR subfamily and a receptor for neurotrophins, such as nerve growth factor (Rogers et al. 2008), has been reported to define a subset of MSCs with immunomodulatory functions (Kuci et al. 2010), and to be expressed on hAMSCs (Soncini et al. 2007) and hWJMSCs (Margossian et al. 2012).

6.1.4 Expression of Hematopoietic Markers

hWJMSCs, hAMSCs, hAECs, and hCVMSCs are usually described to be negative for CD45, CD34, or CD14, a trait that distinguishes them from hematopoietic cells (Abomaray et al. 2015; Abumaree et al. 2013b; Alviano et al. 2007; Bilic et al. 2008; Chen et al. 2012; Diaz-Prado et al. 2010; Donders et al. 2015; Gu et al. 2013; Igura et al. 2004; Jaramillo-Ferrada et al. 2012; Kang et al. 2012; Kronsteiner et al. 2011b; Li et al. 2014; Magatti et al. 2015; Mathews et al. 2015; Poloni et al. 2008; Portmann-Lanz et al. 2006; Prasanna et al. 2010; Roda et al. 2009; Roelen et al. 2009; Stadler et al. 2008; Tipnis et al. 2010; Tsuji et al. 2010; Wegmeyer et al. 2013; Weiss et al. 2006; Wolbank et al. 2009b; Zhang et al. 2006). However, in freshly isolated (P0) hAMSC preparations, there is subpopulation of cells (range 5–15%) that have been shown to express the monocyte/macrophage markers CD45, CD14, CD11b, and HLA-DR that could account for their stimulatory properties (Magatti et al. 2008). Of note, the percentage of these cells drastically decreases (range 0.5–2%) after culture passages (Magatti et al. 2008, 2015). In addition, hWJMSCs express CD68 (La Rocca et al. 2009), classically known as a macrophage-specific antigen (Micklem et al. 1989), even though its expression has been demonstrated in cell types other than myeloid cells (Gottfried et al. 2008).

6.1.5 In Vitro Studies

In vitro studies have supported the low immunogenicity of hWJMSCs, hAMSCs, and hAECs, previously attributed to cell immunophenotype. Indeed, different groups have shown that these cells fail to induce the proliferation of unstimulated allogenic peripheral blood mononuclear cells (PBMCs) (Bailo et al. 2004; Banas et al. 2008; Magatti et al. 2008; Prasanna et al. 2010; Weiss et al. 2008; Wolbank et al. 2007). The inability of hUCMSCs to induce T cell proliferation remained unaltered even after exposing the cells to IFN-γ (Tipnis et al. 2010). However, when hWJMSCs were primed with TNF-α, only minimal PBMC proliferation was observed (Prasanna et al. 2010). Moreover, low cell concentrations of hWJMSCs, hAECs, hAMSCs, and hCVMSCs have been shown to stimulate PBMC proliferation (Donders et al. 2015; Karlsson et al. 2012; Wolbank et al. 2007), and hAMSCs at P0 can induce the proliferation of purified T cells cultured with anti-CD3 (Magatti et al. 2008). Since stimulation with anti-CD3 is unable to induce proliferation of T cells unless APCs are also present, this suggests that hAMSCs are capable of providing costimulatory signals, and could act as APC and activate immune responses.

The Immunomodulatory Features of Mesenchymal Stromal Cells

6.1.6 *In Vivo* Studies

Implantation of hWJMSCs, hAMSCs, and hAECs into immunocompetent animals has been used to verify their *in vivo* immune status. Long-term engraftment has been observed after intravenous and intraperitoneal injection of human amniotic cells into newborn swine and rats, with human DNA microchimerism detected in several organs (Bailo et al. 2004). In addition, hAMSCs and hAECs have been shown to survive for prolonged periods after xenogeneic transplantation into immunocompetent animals without the use of immunosuppressants, including rabbits (Avila et al. 2001), mice (Manuelpillai et al. 2010; Moodley et al. 2010), rats (Kubo et al. 2001; Tsuji et al. 2010; Wei et al. 2009; Zhao et al. 2005), guinea pigs (Yuge et al. 2004), and bonnet monkeys (Sankar and Muthusamy 2003). Interestingly, several clinical trials in humans have proven that allogeneic transplantation of the AM (Sakuragawa et al. 1992; Scaggiante et al. 1987; Tylki-Szymanska et al. 1985) or hAECs (Akle et al. 1981; Scaggiante et al. 1987; Yeager et al. 1985) does not induce acute immune rejection in the absence of immunosuppressive treatment. Moreover, the AM has been used in clinical studies for treatment of skin wounds, burn injuries, and chronic leg ulcers; prevention of tissue adhesion in surgical procedures; and in ocular surface reconstruction to promote the development of normal corneal or conjunctival epithelium, all in absence of immunosuppressive treatment and without acute rejection (Colocho et al. 1974; Faulk et al. 1980; Gomes et al. 2005; Gruss et al. 1978; Subrahmanyam 1995; Ward and Bennett 1984).

Even hWJMSCs were not able to elicit an immune response under xenotransplantation settings and in the absence of immune suppression. Indeed, human hWJMSCs were able to survive for 16 weeks post-transplantation in an immunocompetent rat model of spinal cord injury and in the absence of immunosuppressive drugs (Vawda and Fehlings 2013; Yang et al. 2008). Moreover, hWJMSCs were able to survive in the liver and pancreas when transplanted into diabetic rats (Wang et al. 2014). Similarly, long-term survival has also been observed for porcine umbilical-cord MSCs (pUCMSCs), for up to 4 weeks in the brains of a rat model of Parkinson's disease. In addition, no immune infiltrate was observed at the site of transplantation (Medicetty et al. 2004). In line with these results, another study has shown that pUCMSCs did not induce a detectable immune response across a full MHC barrier (Cho et al. 2008). However, under certain circumstances—specifically when injected in a site of inflammation, injected repeatedly in the same region, or stimulated with IFN-γ prior to injection—pUCMSCs were shown to be immunogenic, stimulating alloantibody production and accelerating skin graft rejection (Cho et al. 2008).

6.2 IMMUNOMODULATORY PROPERTIES OF MSCs DERIVED FROM WJ, AM, AND CHORIONIC VILLI

Different *in vitro* and *in vivo* studies have demonstrated that hWJMSCs/hUCMSCs, hAECs, hAMSCs, and hCVMSCs/hPLMSCs exhibit immunomodulatory activities on different immune cell subpopulations of the innate (macrophages, dendritic cells [DCs], neutrophils, and NK cells) and adaptive (T and B cells) immunity.

Indeed, they were shown to suppress T and B lymphocyte proliferation and to modulate APCs (monocytes, DCs, and macrophages), neutrophils, and the functions of NK cells.

6.2.1 EFFECTS ON T CELLS

T and B cells are the two major cellular subsets that coordinate adaptive immune responses. The recognition of antigenic peptides presented by MHC molecules through the TCR, combined with costimulatory signals on APCs, provide effective activation of naïve T cells. Activated T cells rapidly proliferate, orchestrate the lysis of infected cells (cytotoxic CD8+ T cells), and produce inflammatory cytokines (CD4+ T helper [Th] cells) that can be toxic to the target cells, stimulate B cell antibody production, and also mobilize others inflammatory cells (Janeway 2001; Medzhitov 2007; von Boehmer 1994). On the basis of phenotype and cytokine profile, activated CD4+ T cells can be subdivided into different subsets. Th1 cells are generated in the presence of interleukin (IL)-12 are potent inflammatory cells able to destroy intracellular pathogens and produce IFN-γ and IL-2. Th2 cells are generated by IL-4 priming of naïve T cells and secrete IL-4, IL-5, and IL-13. IL-4 inhibits Th1 and Th17 cells and induces B cells to produce IgE antibodies, whereas IL-5 activates eosinophils. Th17 cells can be induced by the combination of transforming growth factor beta (TGF-β) and IL-6, or by IL-1β, IL-21, and IL-23. Th17 cells produce IL-17 and IL-22 and are important for immune responses against fungi and extracellular bacteria, and they are involved in several chronic inflammation and autoimmune diseases. Th9 and Th22 cells produce IL-9 and IL-22, respectively, but not IFN-γ, IL-4, or IL-17 (Geginat et al. 2013, 2014). Treg cells are immunosuppressive cells able to suppress the activation, proliferation, and effector functions of a different immune cells, such as CD4+ and CD8+ T cells, NK and NK T (NKT) cells, B cells, and APCs. They are central to the prevention of unwanted immune reactions (autoimmune disease, immunopathology, and allergy) and for the maintenance of tolerance. Treg cells can be divided into different subsets based on the expression of CD25 and FoxP3 and/or the production of IL-10, TGF-β, and IL-35 (Sakaguchi et al. 2010). CD8+ T cells are usually recognized as cytotoxic effector cells that produce IFN-γ and kill transformed or infected target cells. Similar to CD4+ T cells, different subsets of CD8+ T cells have been described, such as Tc2 cells that secrete IL-4 (Le Gros and Erard 1994; Maggi et al. 1995) and Tc17 cells that produce IL-17(Maggi et al. 2010). Moreover, certain CD8+CD28− T cells have also been described as adaptive regulatory cells (Elrefaei et al. 2007; Geginat et al. 2013; Siegmund et al. 2009).

Many *in vitro* studies have shown that different placental cell populations are able to inhibit T lymphocyte proliferation; these populations include hWJMSCs (Li et al. 2014; Najar et al. 2010b; Prasanna et al. 2010; Raicevic et al. 2011; Wang et al. 2014; Weiss et al. 2008; Zhou et al. 2011), the entire cell population obtained from the AM (Bailo et al. 2004; Ueta et al. 2002), as well as both hAECs (Banas et al. 2008; Pratama et al. 2011; Wolbank et al. 2007) and hAMSCs (Kang et al. 2012; Kronsteiner et al. 2011b; Magatti et al. 2008; Roelen et al. 2009; Wolbank et al. 2009b), and hCVMSCs/hPLMSCs (Chang et al. 2006; Gu et al. 2013; Li et al. 2007, 2014).

The Immunomodulatory Features of Mesenchymal Stromal Cells **101**

Moreover, dose-dependent inhibition was observed when T cell proliferation was induced by a variety of stimuli, such as by allogenic PBMCs in mixed lymphocyte cultures (MLCs) (Bailo et al. 2004; Banas et al. 2008; Chang et al. 2006; Chen et al. 2010; Karlsson et al. 2012; Kronsteiner et al. 2011b; Li et al. 2007; Magatti et al. 2008; Manochantr et al. 2013; Prasanna et al. 2010; Raicevic et al. 2011; Roelen et al. 2009; Ueta et al. 2002; Weiss et al. 2008; Wolbank et al. 2007), allogeneic DCs (Chang et al. 2006; Tipnis et al. 2010), T cell receptor crosslinking (anti-CD3/anti-CD28) (Kronsteiner et al. 2011b; Magatti et al. 2008), mitogens such as concavillin A (Banas et al. 2008; Donders et al. 2015; Kang et al. 2012; Pratama et al. 2011; Yamahara et al. 2014), phytohemagglutinin (PHA) (Amari et al. 2015; Chang et al. 2006; Chen et al. 2010; Gu et al. 2013; Kronsteiner et al. 2011b; Li et al. 2007; Prasanna et al. 2010; Roelen et al. 2009; Wang et al. 2014; Wolbank et al. 2007, 2009b; Zhou et al. 2011), or PHA/IL-2 (Chang et al. 2006; Li et al. 2014; Najar et al. 2012), or by cytomegalovirus (CMV) recall antigen (Banas et al. 2008). Interestingly, AM cells and hWJMSCs can also suppress proliferation of PBMCs isolated from patients with rheumatoid arthritis (Parolini et al. 2014) or immune thrombocytopenia (Ma et al. 2012), respectively. Furthermore, INF-γ enhances antiproliferative properties of hWJMSCs (Donders et al. 2015; Prasanna et al. 2010), hAMSCs (Kronsteiner et al. 2011b), and placenta MSCs/hCVMSCs (Chang et al. 2006) on stimulated PBMCs.

Although several groups have reported that the antiproliferative effect is dependent on cell-cell contact (Banas et al. 2008; Wolbank et al. 2007), other groups have shown that hWJMSCs, hAECs, hAMSCs, and hCVMSCs/hPLMSCs are able to suppress lymphocyte proliferation also when cultured in a transwell system (Chang et al. 2006; Gu et al. 2013; Kronsteiner et al. 2011b; Li et al. 2007, 2014; Magatti et al. 2008; Roelen et al. 2009), demonstrating that inhibition occurs regardless of cell-cell contact. Moreover, the conditioned medium (CM) generated from the culture of hWJMSCs (Zhou et al. 2011), hAMSCs (Kang et al. 2012; Magatti et al. 2008; Rossi et al. 2012), and hAECs (Li et al. 2005) has been shown to possess antiproliferative effects on lymphocytes, similar to their cellular counterpart, thus supporting the fundamental role of paracrine bioactive factors secreted by these cells in their immunosuppressive activity.

The inhibition induced by hAMSCs on lymphocyte proliferation seems to be transient and reversible. Indeed, responder cells previously exposed to hAMSCs are able to proliferate in new MLCs against either the original or new allogeneic PBMC stimulator cells (Magatti et al. 2008). Culture passages (Wolbank et al. 2007), cell immortalization (obtained inducing human telomerase reverse transcriptase overexpression in hAMSCs by using a retroviral transfection system) (Wolbank et al. 2009b), or hWJMSC lentiviral transduction (Amari et al. 2015), do not alter the suppressive properties. Instead, a significant reduction in the immunomodulatory potential of hAMSCs after cryopreservation was found (Wolbank et al. 2007). Interestingly, T cell activation status and differentiation were reported to be a critical element for inhibition. Indeed, Banas et al. (2008) reported that hAECs do not inhibit IL-2–dependent T cell blast proliferation. Stimulated naïve or memory T cells have been shown to be more prone to inhibition by hAECs, whereas activated T cells are less affected (Banas et al. 2008).

hWJMSCs/hUCMSCs (Donders et al. 2015; Najar et al. 2010b; Wang et al. 2010), hCVMSCs/hPLMSCs (Chang et al. 2006; Li et al. 2007), and CM from hAMSCs (Pianta et al. 2015) have been shown to suppress the proliferation of both CD4 and CD8 cytotoxic T lymphocytes stimulated by allogeneic PBMCs, TCR, or PHA. Furthermore, hWJMSCs/hUCMSCs, hAECs, hAMSCs, and hCVMSCs/hPLMSCs were also able to decrease proinflammatory IFN-γ in activated PBMCs (Anam et al. 2013; Chen et al. 2010; Donders et al. 2015; Gu et al. 2013; Kang et al. 2012; Li et al. 2007; Parolini et al. 2014; Pianta et al. 2015; Raicevic et al. 2011; Wang et al. 2014; Zhou et al. 2011). Moreover, hAECs and hAMSCs were able to decrease specific subset-related cytokines, such as Th1- (IFN-γ, TNF-α, IL-1β, and IL-12p70), Th2- (IL-5, IL-6, and IL-13), Th9- (IL-9), and Th17 (IL-17A, IL-22)-associated cytokines (Anam et al. 2013; Kang et al. 2012; Karlsson et al. 2012; Parolini et al. 2014; Pianta et al. 2015). Interestingly, CM-hAMSCs were able to effectively reduce the expression of markers associated to the Th1 (T-bet+CD119+) and Th17 (RORγt+CD161+), while having no effect on the Th2 population (GATA3+CD193+/GATA3+CD294+cells) (Pianta et al. 2015). Moreover, cytokines such as IL-21, IL-12/IL-23p40, RANTES, IP-10, MIG, MIP-1alpha, MIP-1beta, MCP-1, and the soluble FasL and sCD40L were also suppressed by hAMSCs in PBMCs activated in MLC (Kronsteiner et al. 2011b). Although amniotic cells and their CM decreased IL-17 secretion (Kang et al. 2012; Karlsson et al. 2012; Parolini et al. 2014; Pianta et al. 2015), hWJMSCs and hCVMSCs were shown to induce the production of IL-17 in MLC (Karlsson et al. 2012). Interestingly, all placental stromal cells (hWJMSCs, amniotic-derived cells, and hCVMSCs) have been shown to promote the production of low levels of IL-17 from unstimulated PBMCs (Karlsson et al. 2012). Whether IL-17/Th17 suppression or promotion is involved in the immunosuppressive capacity of these cells requires further investigation.

Cells that could significantly contribute to suppressive activities are regulatory T cells. These cells are found to be increased in activated PBMCs cultured with hWJMSCs (Amari et al. 2015; Najar et al. 2010a), amniotic-derived cells (Parolini et al. 2014), CM from hAMSC culture (Pianta et al. 2015), hAECs (Anam et al. 2013), and hCVMSCs/hPLMSCs (Chang et al. 2006).

Interestingly, in hWJMSCs, promotion and expansion of regulatory T cells were independent of the hWJMSC/T cell ratio (Najar et al. 2010a). Whether hWJMSCs, amniotic-derived cells, and hCVMSCs/hPLMSCs act directly on T cells, inducing formation of Treg cells, or whether they act indirectly on other immune cells (e.g., NK cells, B cells, and monocytes) present within PBMCs, is still unknown. Interestingly, hAMSCs have shown significantly higher suppressive capacity versus stimulated PBMCs, rather than versus T cells (Kronsteiner et al. 2011b). Since it is known that hWJMSCs, amniotic-derived cells, and hCVMSCs/hPLMSCs could influence other immune cells present within PBMCs (see next sections), it is very probable that other immune cells could, in turn, contribute to enhance the immunosuppressive capacity observed with PBMCs.

The immunosuppressive activity of hWJMSCs, hAECs, hAMSCs, and hCVMSCs/hPLMSCs observed *in vitro* have been largely confirmed in *in vivo* studies in which these cells have been applied in animal models of diseases where inflammation plays a critical role.

The Immunomodulatory Features of Mesenchymal Stromal Cells

For example, hWJMSCs (Anzalone et al. 2010), hAECs (Manuelpillai et al. 2010), AM (Sant'Anna et al. 2011), and chorionic-derived MSCs (Jung et al. 2013; Lee et al. 2010) have all been shown to have beneficial effects in models of chronic liver disease. Specifically, hAEC infusion in CCl_4-treated mice was able to decrease hepatic inflammation, leading to a decrease in T cell infiltration and lower protein levels of inflammatory cytokines TNF-α and IL-6 in the liver and an increase in the protein level of inhibitory cytokine IL-10 (Manuelpillai et al. 2010, 2012). Furthermore, hAECs and hAMSCs were reported to reduce fibrosis and chronic inflammation in a mouse model of lung disease. The reduced inflammation was associated to decreased levels of proinflammatory MCP-1, TNF-α, IL-1, INF-γ, and IL-6 and profibrotic TGF-β, PDGF-α, and PDGF-β cytokines and growth factors in the mouse lungs (Moodley et al. 2010, 2013; Murphy et al. 2011; Vosdoganes et al. 2013). Inflammatory cytokines TNF-α, IL-1β, and IL-6 were also reduced by hAECs in inflamed lungs of fetal sheep exposed to intrauterine lipopolysaccharide (LPS) (Vosdoganes et al. 2011).

Therapeutic actions of amniotic-derived cells have also been observed in a mouse model of collagen-induced arthritis (Parolini et al. 2014). Treatment with amniotic-derived cells was shown to significantly decrease the production of Th1 (IFN-γ) and Th17 (IL-17) cytokines and highly increased the levels of IL-10. Moreover, treated mice showed induced peripheral generation of antigen-specific Treg cells with suppressive functions, able to prevent arthritis progression when transferred to mice with collagen-induced arthritis (Parolini et al. 2014). Furthermore, amniotic-derived cells have been shown to ameliorate other inflammatory or autoimmune disorders, such experimental sepsis, inflammatory bowel disease, and experimental autoimmune encephalomyelitis (EAE), an animal model for multiple sclerosis (Parolini et al. 2014). Moreover, EAE mice treated with hAECs were shown to lack or have minimal monocyte/macrophage and T cell infiltration (Liu et al. 2012b), and splenocytes from hAEC-treated mice produced less inflammatory Th1- (IFN-γ) and Th17 (IL-17)-related cytokines, and increased the number of Th2 (IL-5) cells, peripheral Treg cells, and naïve CD4+ T cells (Liu et al. 2012b; McDonald et al. 2015). In contrast, in a rat EAE model, the therapeutic potential of hWJMSCs has been reported to be related to their ability to modulate T cell (splenocytes) proliferation, but without affecting Th1, Th2, Treg, and Th17 polarization (Donders et al. 2015). Finally, hWJMSCs have been shown to modulate the inflammatory response and prevent neuronal damage after administration in a rat model of cerebral ischemia (Hirko et al. 2008; Jomura et al. 2007; Lin et al. 2011). In addition, hyperglycemic improvement, related to the immunomodulatory effects exerted by hWJMSCs, has been reported after hWJMSC transplantation into diabetic animal models (Tsai et al. 2015; Wang et al. 2014). Indeed, these studies demonstrated reduced systemic and pancreatic levels of cell populations, such as Th1 and Th17, involved in the pathogenesis of type I diabetes in nonobese diabetic (NOD) mice. Moreover, a shift toward the Th2 profile and an increase in Treg cells was found in hWJMSC-treated mice (Tsai et al. 2015).

An important aspect of hWJMSCs, hAECs, hAMSCs, and hCVMSCs/hPLMSCs is their ability to engraft in immune competent animals without the use of immunosuppressants. What remains to be fully understood is whether the long-term engraftment of fetal placental cells in xenogeneic models is due to immune cell evasion or to active host immune cell suppression. Interestingly, Tsuji et al. (2010) reported that hAMSCs transdifferentiated into cardiomyocytes and survived more than 4 weeks in the infarcted rat hearts, suggesting that hAMSCs were tolerated in the immunocompetent host. They found that FOXP3-positive Treg cells were constantly detected adjacent to the surviving hAMSC-derived cardiomyocytes, suggesting that they could be involved in maintenance of tolerance (Tsuji et al. 2010). Moreover, co-infusion of a subset of hAECs (termed human amnion–derived multipotent progenitor cells), with limited numbers of donor-unfractionated bone marrow cells, was shown to facilitate multilineage hematopoietic chimerism induction and long-term graft tolerance in a mouse skin transplantation model, through deletion of donor-reactive T cells and expansion of Treg cells (Anam et al. 2013).

6.2.2 Effects on B Cells

Together with T cells, B cells are central in adaptive immune responses. Once a B cell encounters its antigen through the B cell receptor (BCR), and receives additional signals from Th (predominately Th2) cells, it differentiates into a plasma cell, an effector cell able to produce antibodies. The antibodies bind to antigens, making them easier targets for phagocytes, and they trigger the complement cascade (Levine et al. 2000; Shlomchik and Weisel 2012). Like T cells, regulatory B (Breg) cells have been recognized as immunosuppressive cells able to dampen inflammation and favor immunological tolerance (Rosser and Mauri 2015).

Cells and CM derived from fetal placental tissues have been described to also target B cells. In fact, CM from hAEC culture was shown to induce murine B cell apoptosis and inhibit B cell proliferative responses to LPS (Li et al. 2005). Moreover, hWJMSCs or their extracts (CM and cell lysates) were reported to suppress the proliferation of Burkitt's lymphoma cell lines through the induction of oxidative stress pathways leading to cell death (Lin et al. 2014). Furthermore, hWJMSCs were shown to dramatically suppress the proliferation of autoreactive B lymphocytes in PBMCs from immune thrombocytopenia patients (Ma et al. 2012).

6.2.3 Effects on NK Cells

NK cells are important cytolytic effector cells that play a relevant role in innate immunity, participating in the first line of defense against viruses and pathogens, and also play a role in immune surveillance against tumors (Vivier et al. 2011). NK cells primarily reside in peripheral blood, bone marrow, and the spleen, but they may also reside, and even undergo differentiation, in different tissues (lymph nodes, gut, liver, uterus, and thymus), where they are greatly influenced by the microenvironment (Freud et al. 2014; Sojka et al. 2014). NK cell activation is mediated by activating receptors that recognize ligands expressed on target cells that have lost (or express low levels of) HLA class I molecules, as it occurs in tumors or virus-infected cells. Recognition of MHC class I molecules through surface receptors (KIR and NKG2A) can inhibit NK cell function, and it appears to be the most secure mechanism to prevent the NK-mediated attack against self-normal cells. After priming with various

The Immunomodulatory Features of Mesenchymal Stromal Cells

factors (such as IL-15, IL-12, and IL-18), activated NK cells rapidly release proinflammatory cytokines (such as IFN-γ and TNF-α) and chemokines involved in early inflammatory responses, which, in turn, affect DC, macrophage, and neutrophil activation and may ultimately influence the quality and strength of downstream T and B cell responses (Cichocki et al. 2014; Montaldo et al. 2014; Moretta et al. 2014; Vivier et al. 2011).

hWJMSCs/hUCMSCs, hAECs, and hAMSCs were shown to suppress the cytotoxicity and activation status of NK cells. Indeed, hUCMSCs (Chatterjee et al. 2014a) and amnion-derived cells (hAECs and hAMSCs) (Li et al. 2015) have been shown to inhibit the cytotoxicity of NK cells against K562 cells in a dose-dependent manner. The inhibition seemed irreversible since the reduced NK cytotoxicity was recovered by continuous culturing without amnion-derived cells (Li et al. 2015). Moreover, inhibition of NK cytotoxic activity was shown to correlate with down-modulation of NK-activated receptors (Chatterjee et al. 2014a; Li et al. 2015). hUCMSCs and amnion-derived cells have also been shown to suppress NK activation, significantly inhibiting IFN-γ (Chatterjee et al. 2014b; Li et al. 2015; Noone et al. 2013), TNF-α (Ribeiro et al. 2013), and perforin (Ribeiro et al. 2013) production by activated NK cells. Interestingly, CM from hUCMSCs was also able to significantly inhibit IFN-γ production (Chatterjee et al. 2014b). Moreover, prestimulation of hUCMSCs with IFN-γ enhanced NK inhibition and rendered hUCMSCs less susceptible to NK killing (Noone et al. 2013). In this case, protection was associated with elevated expression of HLA-ABC induced by IFN-γ on hUCMSCs (Noone et al. 2013). Different secreted factors produced by hWJMSCs, hAECs, and hAMSCs have been described to be involved in their immunosuppressive activity toward NK cells, such as IL-10 (Li et al. 2015), indoleamine 2,3-dioxygenase expression (IDO) (Noone et al. 2013), Activin-A (Chatterjee et al. 2014b), and prostaglandin E2 (PGE2) (Chatterjee et al. 2014a; Li et al. 2015; Noone et al. 2013). Indeed, addition of blocking antibody to these factors significantly abrogated their suppressive abilities.

6.2.4 EFFECTS ON APCS (DCS)

DCs are the most efficient APC crucial for innate and adaptive immune response and for maintenance of immune tolerance (Banchereau and Steinman 1998). Myeloid immature DCs, distributed ubiquitously in the human body, are highly potent in capturing, processing, and transporting antigens to areas rich in T cells, such as the lymphoid tissues, where they undergo complete maturation. Mature DCs are characterized by loss of phagocytic receptors and increased expression of MHC class II and costimulatory proteins (Banchereau et al. 2000; Mellman et al. 2001). The maturation process of immature DCs (iDCs) constitutes an important checkpoint in the stimulation of the immune response, since iDCs are unable to efficiently stimulate T cells and can induce immune tolerance (Dhodapkar et al. 2001; Fu et al. 1996; Hackstein et al. 2001; Jonuleit et al. 2000; Lu et al. 1995). In addition, DCs can modify the functions of both B cells and NK cells (Dubois et al. 1999; Gerosa et al. 2002). Thus, DCs are essential for the initiation of a primary immune response, and they are key targets for immunosuppressive therapy aimed at inhibiting the rejection of allografts (Josien et al. 1998; Sallusto and Lanzavecchia 1999; Waldmann 1999).

In vitro, iDCs differentiate from CD34 bone marrow stem cells or peripheral blood CD14+ monocytes (Jacobs et al. 2008). iDCs generated from monocytes with IL-4 and granulocyte macrophage–colony-stimulating factor (GM-CSF) express the DC marker CD1a, whereas the monocytic marker CD14 is downregulated (Chapuis et al. 1997; Romani et al. 1994; Sallusto and Lanzavecchia 1994). Cocktails containing different stimulatory molecules can be used to obtain fully mature DCs (mDCs). These cocktails contain LPS, TNF-α, GM-CSF, IL-4, IL-1β, IL-6, TNF-α, and PGE2 (Jacobs et al. 2008; Langenkamp et al. 2000), and they can induce the expression of MHC class II (HLA-DR), T cell costimulatory molecules (CD80, CD83, and CD86), and chemotactic signals (CCR7) (Banchereau et al. 2000).

hWJMSCs/hUCMSCs, hAECs, hAMSCs, and hCVMSCs/hPLMSCs have been found to prevent the differentiation and maturation of monocytes toward DCs when cocultured in direct contact (Abomaray et al. 2015; Banas et al. 2014; Magatti et al. 2015; Tipnis et al. 2010) and in noncontact settings (transwell) with monocytes (Abomaray et al. 2015; Donders et al. 2015; Kronsteiner et al. 2011a; Magatti et al. 2009, 2015; Saeidi et al. 2013; Tipnis et al. 2010), demonstrating the involvement of secreted inhibitory factors. Moreover, the CM generated from hAMSC and hCVMSC cultures have been shown to prevent DC development (Abomaray et al. 2015; Liu et al. 2014; Magatti et al. 2015), suggesting that the production of inhibitory secreted factors does not require paracrine signals from monocytes. However, hAEC- and hCVMSC-secreted factors seem to have a weaker effect on the differentiation of DCs compared to when they are in direct contact with monocytes (Abomaray et al. 2015; Banas et al. 2014; Magatti et al. 2015; Tipnis et al. 2010). Overall, hWJMSCs/hUCMSCs, hAECs, hAMSCs, and hCVMSCs have been shown to completely (hWJMSCs/hUCMSCs and hAMSCs) or partially (hAECs and hCVMSCs) suppress the expression of the DC marker CD1a and to preserve the monocytic marker CD14. Moreover, they have been shown to reduce the expression of costimulatory molecules CD40, CD80, CD86, and CD83, as well as of HLA-DR, on DCs (Abomaray et al. 2015; Banas et al. 2014; Deng et al. 2014; Donders et al. 2015; Kronsteiner et al. 2011a; Magatti et al. 2009, 2015; Saeidi et al. 2013; Tipnis et al. 2010). Generally, CD40, CD80, and CD86 expressed on DCs bind to CD28 and CD154 expressed on T cells, leading to the activation of T cells (Sharpe and Freeman 2002; van Kooten and Banchereau 2000). Moreover, increased expression of immunosuppressive molecules, such as HLA-G or B7-H3/B7-H4, were reported in monocytes differentiated toward DCs in the presence of hAECs (Banas et al. 2014) or hCVMSCs (Abomaray et al. 2015), respectively. Contrasting data have been reported regarding the expression of other immunomodulating molecules, such as PD-L2 (CD273) and PD-L1(CD274). For example, one group observed that hCVMSCs increased the expression of both molecules (Abomaray et al. 2015), whereas another group reported that hWJMSCs decreased the expression of PD-L2 (Tipnis et al. 2010). As discussed in the previous section, HLA-G, B7-H3, B7-H4, PD-L2, and PD-L1 are regulatory molecules that inhibit the proliferation of T cells, and they are involved in the protection of allogeneic transplants (Carter et al. 2002; Latchman et al. 2001; Ou et al. 2006; Ueno et al. 2012).

Moreover, compared to fully mature control DCs, monocytes differentiated in the presence of hAECs, hAMSCs, and hCVMSCs produce lower levels of IL-12p70,

The Immunomodulatory Features of Mesenchymal Stromal Cells 107

IL-23, TNF-α, CXCL10, MIG/CXCL9, CCL5, and MIP-1α and higher levels of IL-6, MIC-1, MCP-1/CCL2, IL-1-β, IL-10, HLA-G, and PGE2 (Abomaray et al. 2015; Banas et al. 2014; Kronsteiner et al. 2011a; Magatti et al. 2009, 2015). In addition, hCVMSCs are able to induce the expression of immunosuppressive IDO enzyme in iDCs and mDCs (Abomaray et al. 2015). As a consequence, monocyte-derived DCs differentiated in the presence of hWJMSCs, hAECs, hAMSCs, and hCVMSCs were shown to have impaired ability to stimulate allogeneic T cell proliferation (Abomaray et al. 2015; Deng et al. 2014; Magatti et al. 2009, 2015), while acquiring the ability to induce Treg cells and Th2 responses (Deng et al. 2014). Furthermore, T cells cultured with DCs in the presence of hCVMSCs secrete low levels of IL-12 and IFN-γ and high levels of IL-10 (Abomaray et al. 2015). Overall, the significant inhibition of costimulatory molecules (CD40, CD80, and CD86) and inflammatory cytokines (IL-12, IL-23, and TNF-α) and the increase of immunosuppressive molecules (HLA-G, B7-H3, and B7-H4) and factors (IL-10, PGE2, HLA-G, and IDO) are indicative of the ability of hWJMSCs, hAECs, hAMSCs, and hCVMSCs to induce monocytes with an anti-inflammatory phenotype; this could modulate T cells to acquire regulatory functions, making them able to inhibit the immune response and, ultimately, to trigger immune tolerance. Moreover, DC phagocytic activity has been shown to increase when cultured with either hWJMSCs (Donders et al. 2015), hAMSCs (Kronsteiner et al. 2011a), or hCVMSCs (Abomaray et al. 2015), indicating a more iDC phenotype, with the potential to remove cellular debris that may have accumulated in the site of inflammation as a result of cell death or tissue damage. Interestingly, as a result, DCs induced phenotypic changes in hCVMSCs that may influence their immunomodulatory functions. Indeed, iDCs were able to induce the expression of anti-inflammatory (IL-6, IL-10, B7-H3, B7-H4, CD273, CD274, and IDO) proteins in hCVMSCs, which could ultimately modulate the functions of T cells (Abomaray et al. 2015).

In accordance with the *in vitro* data, some *in vivo* studies clearly support that the immunomodulatory effects of hWJMSCs/hUCMSCs during the course of cellular therapy can be attributed, at least in part, to the disturbance of DC differentiation and maturation processes. In fact, administration of hWJMSCs in an *in vivo* model of type I diabetes in NOD mice has been shown to not only reduce Th1 and Th17 cells and increase Th2 and Treg cells but also to decrease the number of CD11c+/CD86+ cells in both spleen and pancreatic lymph nodes. Thus, infusion of hWJM-SCs suppressed the activation of T cells by decreasing the level of antigen-presenting DCs (Tsai et al. 2015).

6.2.5 EFFECTS ON APCS (MACROPHAGES)

Macrophages include a heterogeneous population of cells that are broadly distributed in many different tissues, are potent immune regulators, and have a central role in the resolution of inflammation and induction of tissue repair in many human diseases (Murray et al. 2011). Polarized macrophages have been classically referred to as M1 and M2. M1 macrophages are proinflammatory cells, with strong antimicrobial activities that stimulate Th1 responses and participate in many inflammatory diseases (Murray et al. 2011). M1 macrophages develop from monocytes in response to different inflammatory molecules, such as GM-CSF, IFN-γ, and TNF-α, and they are characterized *in vitro* by high levels of chemokine receptor CCR7 (CD197), costimulatory molecules (CD80 and CD86), and proinflammatory cytokines (IL-12, IL-23, TNF, and CXCL10), resulting in efficient antigen presentation capacity (Mantovani et al. 2004). On the contrary, M2 macrophages produce anti-inflammatory cytokines such as IL-10 and TGF-β1; can promote Th2 functions; have high phagocytic functions; and play a major role in the resolution of inflammation, tissue remodeling, and wound repair (Mantovani et al. 2013). M2 macrophage differentiation is stimulated by different molecules, such as IL-1 receptor antagonist; IL-4, IL-6, IL-10, and IL-13; glucocorticoids; macrophage–colony-stimulating factor (M-CSF); VEGF; TGF-β1; leukemia inhibitory factor (LIF); and PGE2 and human M2 macrophages express *in vitro* CD14; CD23 (Fcϵ-RII); the scavenger receptors CD163 and CD204; the mannose receptor CD206; CD209 (DC-SIGN); and B7-H4 (Buechler et al. 2000; Heusinkveld et al. 2011; Jeannin et al. 2011; Kryczek et al. 2006; Rickard and Young 2009; Svensson et al. 2011).

As discussed in Section 6.2.4, hWJMSCs/hUCMSCs, hAECs, hAMSCs, and hCVMSCs are able to prevent the differentiation of monocytes toward DCs *in vitro*. Monocytes can differentiate toward both DCs and macrophages in a reversible process that depends on the differentiation signals in their microenvironments (Palucka et al. 1998). Interestingly, hAMSCs have been shown to not only simply block DC differentiation at the monocytic level but also to skew their differentiation toward cells with macrophage features, specifically to M2 macrophages. Indeed, these cells express the M2 markers CD14, CD163, and CD23, whereas they lack or express low levels of the M1 markers CD197 and CD80 (Magatti et al. 2015). Accordingly, monocyte-derived cells from hAMSC cocultures show elevated endocytosis levels (Kronsteiner et al. 2011a), another feature that could be indicative of M2 differentiation. Moreover, hAMSCs can shift the differentiation of monocytes toward M2 not only when monocytes are differentiated into DCs *in vitro* but also into M1, or even in the absence of any differentiation or polarization stimuli (Magatti et al., unpublished data).

In addition, hCVMSCs can stimulate anti-inflammatory phenotypes in human macrophages *in vitro*. The addition of hCVMSCs to the culture of monocytes growing in GM-CSF growth medium has been shown to result in the differentiation of monocytes into cells expressing higher levels of M2-macrophage cell surface markers, such as CD14, CD36, CD163, CD204, CD206, and CD11b (Abumaree et al. 2013a). Furthermore, hCVMSCs can decrease the expression of costimulatory molecules (CD40, CD80, and CD86) and proinflammatory cytokines (IL-1β, IL-12p70, and MIP-1α) and can increase the expression of co-inhibitory molecules (B7-H4, PD-L1, and PD-L2) and anti-inflammatory cytokine IL-10, indicative of induction of an anti-inflammatory or immunosuppressive M2 phenotype. Moreover, hCVMSCs can increase the phagocytic activity of these macrophages. Remarkably, macrophages can also induce phenotype changes in hCVMSC, such as the increase expression of the immunosuppressive proteins (IL-6, IL-10, IDO, and B7-H4), that, in turn, can enhance their immunoregulatory properties (Abumaree et al. 2013a).

Murine macrophages have also been shown to be affected by placental cells. Indeed, CM from hAEC culture has been shown to inhibit the migration of murine macrophages, probably due to the presence of migration inhibitory factor in the CM

(Li et al. 2005; Tan et al. 2014). Moreover, the phenotype of murine macrophages was shifted toward M2, as demonstrated by an increase in M2 markers and in phagocytic ability (Tan et al. 2014). Furthermore, hWJMSCs (Li et al. 2013) and hAMSCs (Onishi et al. 2015) were found to modulate the activation status, cytokine expression, and phenotype of mouse RAW 264.7 macrophage cells through the inhibition of IL-1β and IL-6 expression, and by enhancing that of arginase-1 and IL-10 when cocultured with hUCMSCs (Li et al. 2013). Moreover, these cells exhibited a significant increase in the percentage of CD206+ cells among CD11b+ macrophages, indicative of an M2 phenotype induction (Li et al. 2013).

Interestingly, hAMSCs have been described to modulate the activation of human microglia, the resident macrophages of the brain and spinal cord (Gehrmann et al. 1995). Indeed, the production of inflammatory cytokine TNF-α and the proliferation of microglia isolated from the brain tissue of aborted fetuses were suppressed when cocultured with hAMSCs (Wu et al. 2014).

Importantly, hWJMSCs/hUCMSCs, hAECs, hAMSCs, and hCVMSCs were able to shift macrophages toward M2 phenotype in a noncontact (transwell) setting (Abumaree et al. 2013a; Li et al. 2013; Magatti et al. 2015), even when CM by unstimulated cells was used (Abumaree et al. 2013a; Li et al. 2005; Magatti et al. 2015; Tan et al. 2014), suggesting that a crosstalk between placental cells and immune cells is unnecessary. The exact mechanism underlining the immunoregulatory effect on macrophage differentiation is not fully understood, but recent findings support a role of soluble factors acting partially via glucocorticoid and progesterone receptors (Abumaree et al. 2013a). Interestingly, hWJMSCs, hAMSCs, and hCVMSCs secrete different prostanoids, such as PGD2, PGF2a, and PGE2, and several molecules, including IL-1R antagonist, IL-6, IL-10, M-CSF, LIF, VEGF, TGFβ-1, and B7-H4 (Abumaree et al. 2013a, 2013b; Amari et al. 2015; Rossi et al. 2012), which could contribute to the differentiation of M2 macrophages (Buechler et al. 2000; Heusinkveld et al. 2011; Holland et al. 2012; Jeannin et al. 2011; Rickard and Young 2009; Svensson et al. 2011), or could ultimately foster an immunosuppressive environment, as is the case of B7-H4 (Kryczek et al. 2006). Moreover, the PD-L1–PD1 signaling pathway has been described to be involved in the inhibitory effects of hAMSCs on microglial activation (Wu et al. 2014).

The *in vitro* immunomodulatory actions of hWJMSCs/hUCMSCs, amniotic cells, and hCVMSCs on macrophages have been confirmed in some *in vivo* studies that used these cells either to treat inflammatory and fibrotic diseases or for tissue repair.

For example, mice with CCl4-induced hepatic fibrosis transplanted with hAECs showed a reduction of established hepatic fibrosis, induced by a wound healing, M2-macrophage phenotype (Manuelpillai et al. 2012). The same group has also shown that transplantation of hAECs in a mouse model of multiple sclerosis reduced not only T cells but also monocyte/macrophage infiltration in the central nervous system, the latter of which has been correlated with progression of clinical disease (Liu et al. 2012b). A reduction of inflammatory microglia/macrophage cells has also been observed after hAEC infusion in fetal sheep brains after injury induced by LPS (Yawno et al. 2013). Others have reported improvement in locomotion of spinal cord–injured rats upon transplantation of hWJMSCs, accompanied by fewer microglia and reactive astrocytes in the spinal cord (Yang et al. 2008).

In mouse models of bleomycin-induced lung injury, administration of hAECs and hAMSCs has been shown to significantly reduce the injury primarily through modulation of the host immune response (Moodley et al. 2010, 2013; Murphy et al. 2011; Vosdoganes et al. 2013). hAECs were able to significantly reduce macrophage infiltration into the lungs (Murphy et al. 2012; Tan et al. 2014). Moreover, in saline-treated control animals, the infiltrating macrophages were predominantly M1, whereas in hAEC-treated animals the predominant macrophage phenotype was M2 (Moodley et al. 2010; Tan et al. 2014).

hUCMSCs have also been shown to reduce the infiltration of macrophages in injured tissues, while increasing the proportion of anti-inflammatory M2 macrophages, when used to treat renal repair after ischemia-reperfusion injury. Indeed, they increased in the percentage of CD206+ and CD11b+ macrophages, suppressed the expression of IL-1β and IL-6, and induced that of IL-10 at the injury sites (Li et al. 2013).

Interestingly, *in vivo* evidence corroborates the importance of the crosstalk between monocytes/macrophages and hWJMSCs or amniotic-derived cells to exert immunomodulation. For example, hWJMSCs were found in the proximity of or in contact with lung and spleen macrophages after having been administered in a chronic EAE rat model for multiple sclerosis. The interaction with macrophages was suggested to contribute to hWJMSC-mediated suppression of inflammation and to enhance tissue repair (Donders et al. 2015). Accordingly, hAECs were not able to exert reparative effects when administrated in the fibrotic lungs of surfactant protein C–deficient mice, which have impairment of macrophage function (Murphy et al. 2012). Again, it was observed that macrophage abolishment during the early renal injury phase promotes the therapeutic effects of hUCMSCs, whereas macrophage depletion during the late repair phase led to their loss (Li et al. 2013).

Overall, these findings provide evidence that macrophages are key players for the downstream immunosuppressive effects of hWJMSCs/amniotic-derived cells and, ultimately, for the improvement of the therapeutic outcome.

6.2.6 EFFECTS ON NEUTROPHILS

Neutrophils are the professional phagocytosing cells present in both blood and tissues that have the capacity to recognize, engage, phagocytose, and kill their targets by multiple cytotoxic mechanisms (Koenderman et al. 2014). Several studies have described different neutrophil phenotypes with different functions, indicating that neutrophils are both potent phagocytes and immunoregulatory cells. Indeed, neutrophils can differentiate into hybrid populations with APC function (Matsushima et al. 2013), produce a large array of cytokines (Tecchio et al. 2014), and inhibit T cell response (Pillay et al. 2012, 2013).

CM from hAECs has been shown to inhibit the migration of murine neutrophils *in vitro* (Li et al. 2005), whereas CM from placental MSCs induced human neutrophil chemotaxis (Chen et al. 2014). Moreover, CM from placental MSCs was able to reduce the production of reactive oxygen species by neutrophils and inhibit LPS-induced phagocytosis (Chen et al. 2014). Again, CM from placental MSCs rescued neutrophils from apoptosis (Chen et al. 2014), whereas CM from AM accelerated apoptosis of neutrophils (Zhou et al. 2003).

The Immunomodulatory Features of Mesenchymal Stromal Cells

In vivo studies have shown that administration of hAECs in mice with bleomycin-induced lung injury improved lung function while decreasing macrophage and neutrophil infiltration (Murphy et al. 2012). A reduction of macrophage and neutrophil infiltration was also observed after treatment with placental MSCs in ischemia-induced hind limb injury (Zhang et al. 2014). Interestingly, clinical studies have shown that AM transplantation can reduce the degree of inflammation of ocular chemical burns (Ivekovic et al. 2005; Tandon et al. 2011) and that AM transplanted in patients with acute ocular chemical burns can trap macrophages and neutrophils and even induce their apoptosis (Liu et al. 2012a). These data suggest that placental cells could also act on neutrophils by influencing their migration and functions, but further detailed studies should be performed to clarify this hypothesis.

6.3 MECHANISMS OF IMMUNOSUPPRESSION BY MSCs DERIVED FROM WJ, AM, AND CHORIONIC VILLI

The mechanisms underlying the immunosuppressive properties of hWJMSCs, hAMSCs, hAECs, and hCVMSCs are not completely understood, but many different theories have been proposed. Both cell-cell contact with immune cells and release of inhibitory secreted factors by placental cells have been described to be involved. In turn, different suppressive cell types (such as Treg cells and M2 macrophages), generated in the presence of hWJMSCs, hAMSCs, hAECs, and hCVMSCs could, in turn, modulate the microenvironment, altogether enhancing these effects; and placental cell activation with inflammatory cytokines, such as INF-γ and IL-β, can also potentiate immunosuppressive actions. Thus, it is evident that a complex crosstalk between different elements contributes to the immune modulation exerted by hWJMSCs, hAMSCs, hAECs, and hCVMSCs. In the following sections, some of the suppressive factors proposed to be involved in the mechanisms underlying the inhibitory actions of placental cells are discussed.

6.3.1 INTERLEUKIN 10

IL-10 is an inhibitory cytokine that serves to dampen inflammation (Moore et al. 2001). It has been shown to inhibit the secretion of inflammatory cytokines, including IFN-γ, TNF, IL-1, IL-2, and GM-CSF, as well as several chemokines (de Waal Malefyt et al. 1991; Macatonia et al. 1993). Moreover, the action of IL-10 can result in the downregulation of MHC class II proteins and costimulatory molecules CD80 and CD86, on the surface of target APCs (Bogdan et al. 1991; Ding et al. 1993). IL-10 has also been shown to suppress the production of reactive oxygen and nitrogen intermediates in activated macrophages (Gazzinelli et al. 1992).

The possible participation of IL-10 in the immunomodulatory effect exerted by placental cells has been suggested by the fact that hAMSCs and hCVMSCs secrete IL-10 per se (Abomaray et al. 2015; Kronsteiner et al. 2011a; Magatti et al. 2015; Rossi et al. 2012). In turn, hWJMSCs, amniotic-derived cells, and hCVMSCs/hPLMSCs can also stimulate PBMCs to increase the secretion of anti-inflammatory IL-10 (Donders et al. 2015; Gu et al. 2013; Kang et al. 2012; Karlsson et al. 2012;

Li et al. 2007; Parolini et al. 2014; Roelen et al. 2009; Rossi et al. 2012), and coculture with placental cells has been shown to upregulate IL-10 expression in NK cells (Li et al. 2015), DCs (Abomaray et al. 2015; Banas et al. 2014; Kronsteiner et al. 2011a; Magatti et al. 2015), and macrophages (Abumaree et al. 2013a; Li et al. 2013). Interestingly, the addition of anti-IL-10 can abrogate the inhibitory effect observed on activated PBMCs (Parolini et al. 2014; Roelen et al. 2009), and it tends to increase NK cytotoxicity (Li et al. 2015).

In contrast, other groups found only minimal reversion after the addition of neutralizing antibodies against IL-10 (Chang et al. 2006; Rossi et al. 2012), even when used in combination with neutralizing antibodies against other putative inhibitory secreted factors such as IL-6, hepatocyte growth factor (HGF), and TGF-β (Rossi et al. 2012). Moreover, a decrease in IL-10 production was observed when hAMSCs or total amniotic cells were cocultured with PBMCs activated in MLC (Kronsteiner et al. 2011b; Ueta et al. 2002), suggesting only a partial involvement of IL-10 in the immunosuppressive mechanisms of placental cells.

6.3.2 Transforming Growth Factor Beta

Similar to IL-10, TGF-β is a potent anti-inflammatory cytokine, able to induce both Treg (Hadaschik and Enk 2015) and Breg (Rosser and Mauri 2015) cells. Indeed TGF-β has been suggested to play a role in the modulatory activity of placental cells. For example, hWJMSCs (Zhou et al. 2011), hAMSCs (Pianta et al. 2015; Rossi et al. 2012), and hAECs (Liu et al. 2012b) can secrete TGF-β, and a significant increase in TGF-β production has been demonstrated after coculture of activated PBMCs with hAMSCs (Kang et al. 2012), CM from hAMSCs (Pianta et al. 2015), and hCVMSCs (Chang et al. 2006).

However, different results have been reported from experiments with TGF-β neutralizing antibody. For example, the presence of TGF-β neutralizing antibody has been shown to significantly abrogate the inhibition of splenocyte proliferation induced by hAECs (Liu et al. 2012b), as well as to partially reverse the inhibitory effects of placental MSCs on both CD4 and CD8 lymphocyte proliferation (Chang et al. 2006), or even to not be able to reverse the suppression of lymphocyte proliferation induced by CM from hAMSCs (Rossi et al. 2012). Taken together, these data suggest that TGF-β is very likely not the most important mediator of the immunosuppressive properties attributed to placental cells.

6.3.3 Hepatocyte Growth Factor

HGF has been shown to mediate the protection of many types of inflammatory and autoimmune diseases (Molnarfi et al. 2015). It is a factor with strong neuroprotective properties (Ebens et al. 1996), and it promotes the development of tolerogenic DCs (Molnarfi et al. 2014), modulates the immune functions of monocytes/macrophages (Galimi et al. 2001) and B cells (Gordin et al. 2010), and limits both the generation and activity of CD8+ cytotoxic T cells (Benkhoucha et al. 2013).

HGF was found constitutively expressed by hAMSCs (Kronsteiner et al. 2011a; Yamahara et al. 2014) and hWJMSCs (Najar et al. 2010b; Raicevic et al. 2011),

The Immunomodulatory Features of Mesenchymal Stromal Cells   113

and a significant increase of HGF has been demonstrated after coculture with PBMCs (Kang et al. 2012; Raicevic et al. 2011) and monocyte-derived DCs (Kronsteiner et al. 2011a). However, neutralization assays revealed that HGF is not involved in the inhibitory mechanism of activated PBMCs induced by hAMSCs (Rossi et al. 2012).

6.3.4 Prostaglandin E2

PGE2 is an immunosuppressive molecule synthesized from arachidonic acid by cyclooxygenase (COX)-1 and COX-2 enzymes. It can suppress effector functions of macrophages, neutrophils, and Th1-, CTL-, and NK cell–mediated type 1 immunity. It inhibits the attraction of proinflammatory cells, while promoting Th2, Th17, and Treg cell responses (Kalinski 2012). PGE2 has been shown to be constitutively produced by hWJMSCs/hUCMSCs (Chen et al. 2010; Najar et al. 2010b; Raicevic et al. 2011), hAMSCs (Kang et al. 2012; Kronsteiner et al. 2011a; Rossi et al. 2012; Whittle et al. 2000; Yamahara et al. 2014), hAEC (Liu et al. 2012b; Whittle et al. 2000), and hCVMSCs/PDMSCs (Abomaray et al. 2015; Liu et al. 2014). Moreover, a significant increase of PGE2 has been demonstrated after coculture of hWJMSCs/hUCMSCs, hAECs, hAMSCs, and hCVMSCs with stimulated PBMCs (Castro-Manrreza et al. 2014; Kang et al. 2012; Najar et al. 2010b; Yamahara et al. 2014), NK (Chatterjee et al. 2014a; Li et al. 2015; Noone et al. 2013) and DCs (Abomaray et al. 2015; Banas et al. 2014; Kronsteiner et al. 2011a). Interestingly, blockage of PGE2 production through of the inhibition of COX activities was able to almost completely reverse the suppressive properties of hWJMSCs/hUCMSCs (Chen et al. 2010; Najar et al. 2010b) and amniotic-derived cells (Liu et al. 2012b; Parolini et al. 2014; Rossi et al. 2012) on stimulated lymphocytes. The addition of PGE2 inhibitor also tended to increase NK cytotoxicity (Chatterjee et al. 2014a; Li et al. 2015; Noone et al. 2013) and to partially reverse the inhibitory effect on DC differentiation and maturation (Liu et al. 2014). Overall, PGE2 seems to be a major player in the immunomodulatory actions exerted by placental cells.

6.3.5 Human Leukocyte Antigen G

HLA-G is a nonclassical HLA class Ib molecule known to have immune regulatory properties mediated by the interaction with ILT receptors (ILT-2, ILT-3, and ILT-4) (Allan et al. 2000; Hunt et al. 2005). HLA-G displays relevant immune functions through the interaction with these inhibitory receptors, which are expressed by T cells, B cells, NK cells, and mononuclear phagocytes. Indeed, HLA-G modulates cytokine release from human allogeneic PBMCs (Kanai et al. 2001), prevents the proliferation of CD4-T cells, and directs them toward an immunosuppressive phenotype (LeMaoult et al. 2004). In addition, HLA-G induces CD8-T cell apoptosis (Contini et al. 2003; Fournel et al. 2000), or it can even reduce the production of CD8, a major coreceptor/activator cell surface molecule (Contini et al. 2003; Shiroishi et al. 2003). Other reports have shown that HLA-G inhibits NK cell toxicity (Khalil-Daher et al. 1999) and can lead mononuclear phagocytes into suppressive modes characterized by high production of anti-inflammatory cytokines (McIntire et al. 2004).

Moreover, HLA-G appears to play a role in the immune tolerance during pregnancy by evading a maternal immune response against the fetus and by inducing the expansion of Treg cells, which would contribute to the suppression of effector responses to alloantigens (Gregori et al. 2015; Petroff 2005).

It has already been mentioned above that hWJMSCs, hAMSCs, hAECs, and hCVMSCs/hPLMSCs express HLA-G and that the treatment of hAMSCs and hAECs with IFN-γ increases HLA-G expression (Anam et al. 2013; Banas et al. 2008; Kronsteiner et al. 2011b; Lefebvre et al. 2000). hAEC and hCVMSCs have also been shown to induce HLA-G expression in monocyte-derived DCs (Abomaray et al. 2015; Banas et al. 2014), together suggesting that some of the inhibitory effects of placental cells could be exerted by HLA-G. Interestingly, several studies have associated the presence of HLA-G with tolerance induction after allogeneic organ transplantation (Le Rond et al. 2004, 2006; Lila et al. 2000). Specifically, hAMSCs were found to be tolerated long term in the hearts of immunocompetent rats, where pretreatment of hAMSCs with IL-10 or progesterone was observed to markedly increase hAMSCs survival *in vivo*, and pretreatment with IL-10 increased the level of HLA-G expressed by hAMSCs (Tsuji et al. 2010). Thus, the authors speculated that HLA-G might play a role in the initial process of tolerance (Tsuji et al. 2010).

6.3.6 INDOLEAMINE 2,3-DIOXYGENASE

IDO is an enzyme with immunosuppressive functions that induces the degradation of tryptophan within the microenvironment. Tryptophan is an amino acid that is necessary for the proliferation of T cells; consequently, the induced activity of IDO can inhibit the proliferation of T cells (Munn et al. 2005). Induction of IDO activity is usually associated with immunosuppressive effects, such as tolerance and resolution of inflammation (Huang et al. 2010). The exposure of naive T cells to low concentrations of tryptophan *in vitro* has been shown to induce their differentiation into Treg cells, whereas the addition of DC to culture can induce the immunosuppression activities (Belladonna et al. 2007; Brenk et al. 2009). Accordingly, induction of IDO activity *in vivo* has been reported to inhibit the rejection of allogeneic grafts (Grohmann et al. 2002; Munn et al. 1998).

hWJMSCs (Donders et al. 2015), hAECs (Anam et al. 2013), hAMSCs (Rossi et al. 2012), and hPLMSCs (Chang et al. 2006) possess IDO, and its activity has been shown to increase when hAMSCs are cocultured with stimulated PBMCs (Kang et al. 2012; Rossi et al. 2012). Moreover, hCVMSCs are able to induce the expression of IDO in iDCs and mDCs (Abomaray et al. 2015). It has also been suggested that IDO can participate in the immune modulation of hWJMSCs. Indeed, inhibition of IDO counteracted the capacity of hWJMSCs to suppress T cell proliferation (Donders et al. 2015). Moreover, the addition of IDO antibody was shown to significantly abrogate the suppressive activity of hWJMSCs/hUCMSCs toward NK cells (Noone et al. 2013). In contrast, the inhibitory effect of hAMSCs on lymphocyte proliferation did not decrease in presence of IDO inhibitor (Rossi et al. 2012), suggesting that different mechanisms may be involved.

The Immunomodulatory Features of Mesenchymal Stromal Cells 115

6.3.7 Inflammatory Cytokines (INF-γ and IL-1β)

INF-γ plays an important role in bringing about acute inflammation, providing protection against intracellular pathogens. It stimulates the activation and differentiation of several lymphocyte populations, particularly the generation of CD4+ Th1 cells, which produce large amounts of INF-γ that may contribute to the recruitment and activation of immune cells, such as monocytes and granulocytes. Activated Th1, NK, and CD8+ T cytotoxic cells are the most potent sources of IFN-γ, but γδ T cells, NKT cells, macrophages, DCs, naive CD4+ T cells, and even B cells can also produce IFN-γ (de Araujo-Souza et al. 2015; Farrar et al. 1993; Frucht et al. 2001). Interestingly, several studies have reported that INF-γ enhances the immunomodulatory functions of hWJMSCs/hUCMSCs, hAECs, hAMSCs, and hCVMSCs/PDMSCs. Indeed, the treatment of these cells with IFN-γ enhanced their antiproliferative effects on stimulated PBMCs (Chang et al. 2006; Donders et al. 2015; Kronsteiner et al. 2011b; Prasanna et al. 2010) and increased NK inhibition (Noone et al. 2013). Furthermore, INF-γ was shown to increase the expression of immunosuppressive molecules, such as HLA-G (Banas et al. 2008; Kronsteiner et al. 2011b; Lefebvre et al. 2000), PD-L1 and PD-L2 (Banas et al. 2008; Kronsteiner et al. 2011b; Petroff and Perchellet 2010; Tipnis et al. 2010), and also PGE2 production (Chen et al. 2010), which could further enhance their immunomodulatory activity.

IL-1β is another inflammatory cytokine that has been shown to activate the immunomodulatory properties of hWJMSCs/hUCMSCs, hAECs, hAMSCs, and hCVMSCs/PDMSCs. IL-1β is a potent inflammatory and immunostimulatory cytokine involved in the body's natural responses to invasion, and it is a major pathogenic mediator of autoinflammatory, autoimmune, infectious, and degenerative diseases (Dinarello 1998; Garlanda et al. 2013).

As observed with INF-γ, IL-1β is able to induce the production of PGE2 in hUCMSCs (Chen et al. 2010), amniotic cells (Mitchell et al. 1993), and villous and chorion trophoblast cells (Pomini et al. 1999). Interestingly, IL-1β priming of hUCMSCs seems to be fundamental for their immunosuppressive activity toward NK cells (Chatterjee et al. 2014a). In particular, IL-1β production by NK cells has been reported to require cell-cell contact with hUCMSCs, and these hUCMSCs, in turn, are induced to secrete PGE2. It has also been demonstrated that inhibition of gamma-secretase activity, which acts on IL-1R, significantly alleviates immunosuppression (Chatterjee et al. 2014a).

6.3.8 Other Secreted Factors

In addition to the molecules discussed above, hWJMSCs/hUCMSCs, hAMSCs, hAECs, and hCVMSCs/hPLMSCs have also been reported to express mRNA, or to secrete, various factors, including angiogenin, bFGF, IL-1α, IL-1β, IL-6, IL-8, IL-11, IL-15, IL-24, IL-27, IL-32, EGF, IGF-1, granulocyte–colony-stimulating factor, GM-CFS, GRO, IFN-γ, IGF-1, leptin, LIF, MCP-1 (CCL-2), MIP-1a, MIP-1b, OSM, PGD2, PGF2a, RANTES, sCD54, Serpin-E1, TIMP-1/2, TNF-α, and VEGF (Abumaree et al. 2013b; Kronsteiner et al. 2011a, 2011b; Magatti et al. 2009, 2015;

Najar et al. 2010a; Pratama et al. 2011; Rossi et al. 2012; Wang et al. 2012; Wegmeyer et al. 2013; Weiss et al. 2008; Whittle et al. 2000; Wolbank et al. 2009a; Yamahara et al. 2014).

Some of these factors could account for their immunosuppressive properties. For example, hWJMSCs, hAMSCs, hAECs, and hCVMSCs could use IL-15 to induce Treg cells (Kagimoto et al. 2008) or IL-27 and IL-31 to induce the inhibition of T cell functions (Artis et al. 2004; Perrigoue et al. 2009). Alternatively, IL-32 has been shown to antagonize human immunodeficiency virus infection (Nold et al. 2008), and IL-24 to reduce inflammation and kill cancerous cells (Andoh et al. 2009; Su et al. 2005). Furthermore, IL-6 has been described to be involved in inhibition of DC differentiation (Djouad et al. 2007), whereas it did not contribute to immunosuppressive activity of hUCMSCs on T cells (Wang et al. 2012). Furthermore, LIF, a member of the IL-6 cytokine family, was shown to boost Treg formation (Janssens et al. 2015) and inhibit T cell proliferation by hWJMSCs (Najar et al. 2010a).

REFERENCES

Abomaray, F. M., M. A. Al Jumah, B. Kalionis, et al. 2015. Human chorionic villous mesenchymal stem cells modify the functions of human dendritic cells, and induce an anti-inflammatory phenotype in CD1+ dendritic cells. *Stem Cell Rev* 11 (3): 423–41.

Abumaree, M. H., M. A. Al Jumah, B. Kalionis, et al. 2013a. Human placental mesenchymal stem cells (pMSCs) play a role as immune suppressive cells by shifting macrophage differentiation from inflammatory M1 to anti-inflammatory M2 macrophages. *Stem Cell Rev* 9: 620–41.

Abumaree, M. H., M. A. Al Jumah, B. Kalionis, et al. 2013b. Phenotypic and functional characterization of mesenchymal stem cells from chorionic villi of human term placenta. *Stem Cell Rev* 9: 16–31.

Akle, C. A., M. Adinolfi, K. I. Welsh, S. Leibowitz, and I. McColl. 1981. Immunogenicity of human amniotic epithelial cells after transplantation into volunteers. *Lancet* 2: 1003–5.

Allan, D. S., A. J. McMichael, V. M. and Braud. 2000. The ILT family of leukocyte receptors. *Immunobiology* 202: 34–41.

Alviano, F., V. Fossati, C. Marchionni, et al. 2007. Term Amniotic membrane is a high throughput source for multipotent Mesenchymal Stem Cells with the ability to differentiate into endothelial cells in vitro. *BMC Dev Biol* 7: 11.

Amari, A., M. Ebtekar, S. M. Moazzeni, et al. 2015. Investigation of immunomodulatory properties of human Wharton's Jelly-derived mesenchymal stem cells after lentiviral transduction. *Cell Immunol* 293: 59–66.

Anam, K., Y. Lazdun, P. M. Davis, et al. 2013. Amnion-derived multipotent progenitor cells support allograft tolerance induction. *Am J Transplant* 13: 1416–28.

Andoh, A., M. Shioya, A. Nishida, et al. 2009. Expression of IL-24, an activator of the JAK1/STAT3/SOCS3 cascade, is enhanced in inflammatory bowel disease. *J Immunol* 183: 687–95.

Anzalone, R., M. Lo Iacono, S. Corrao, et al. 2010. New emerging potentials for human Wharton's jelly mesenchymal stem cells: Immunological features and hepatocyte-like differentiative capacity. *Stem Cells Dev* 19: 423–38.

Artis, D., A. Villarino, M. Silverman, et al. 2004. The IL-27 receptor (WSX-1) is an inhibitor of innate and adaptive elements of type 2 immunity. *J Immunol* 173: 5626–34.

Avila, M., M. Espana, C. Moreno, and C. Pena. 2001. Reconstruction of ocular surface with heterologous limbal epithelium and amniotic membrane in a rabbit model. *Cornea* 20: 414–20.

The Immunomodulatory Features of Mesenchymal Stromal Cells   **117**

Bailo, M., M. Soncini, E. Vertua, et al. 2004. Engraftment potential of human amnion and chorion cells derived from term placenta. *Transplantation* 78: 1439–48.

Banas, R., C. Miller, L. Guzik, and A. Zeevi. 2014. Amnion-derived multipotent progenitor cells inhibit blood monocyte differentiation into mature dendritic cells. *Cell Transplant* 23: 1111–25.

Banas, R. A., C. Trumpower, C. Bentlejewski, et al. 2008. Immunogenicity and immuno-modulatory effects of amnion-derived multipotent progenitor cells. *Hum Immunol* 69: 321–8.

Banchereau, J., F. Briere, C. Caux, et al. 2000. Immunobiology of dendritic cells. *Annu Rev Immunol* 18: 767–811.

Banchereau, J., and R. M. Steinman. 1998. Dendritic cells and the control of immunity. *Nature* 392: 245–52.

Belladonna, M. L., P. Puccetti, C. Orabona, et al. 2007. Immunosuppression via tryptophan catabolism: The role of kynurenine pathway enzymes. *Transplantation* 84: S17–20.

Benkhoucha, M., N. Molnarfi, G. Schneiter, P. R. Walker, and P. H. Lalive. 2013. The neu-rotrophic hepatocyte growth factor attenuates CD8+ cytotoxic T-lymphocyte activity. *J Neuroinflammation* 10: 154.

Bilic, G., S. M. Zeisberger, A. S. Mallik, R. Zimmermann, and A. H. Zisch. 2008. Comparative characterization of cultured human term amnion epithelial and mesenchymal stromal cells for application in cell therapy. *Cell Transplant* 17: 955–68.

Bogdan, C., Y. Vodovotz, and C. Nathan. 1991. Macrophage deactivation by interleukin 10. *J Exp Med* 174: 1549–55.

Brenk, M., M. Scheler, S. Koch, et al. 2009. Tryptophan deprivation induces inhibitory recep-tors ILT3 and ILT4 on dendritic cells favoring the induction of human CD4+CD25+ Foxp3+ T regulatory cells. *J Immunol* 183: 145–54.

Bretscher, P. A. 1999. A two-step, two-signal model for the primary activation of precursor helper T cells. *Proc Natl Acad Sci U S A* 96: 185–90.

Buechler, C., M. Ritter, E. Orso, et al. 2000. Regulation of scavenger receptor CD163 expres-sion in human monocytes and macrophages by pro- and antiinflammatory stimuli. *J Leukoc Biol* 67: 97–103.

Carter, L., L. A. Fouser, J. Jussif, et al. 2002. PD-1:PD-L inhibitory pathway affects both CD4(+) and CD8(+) T cells and is overcome by IL-2. *Eur J Immunol* 32: 634–43.

Castro-Manrreza, M. E., H. Mayani, A. Monroy-Garcia, et al. 2014. Human mesenchymal stromal cells from adult and neonatal sources: A comparative in vitro analysis of their immunosuppressive properties against T cells. *Stem Cells Dev* 23: 1217–32.

Chang, C. J., M. L. Yen, Y. C. Chen, et al. 2006. Placenta-derived multipotent cells exhibit immunosuppressive properties that are enhanced in the presence of interferon-gamma. *Stem Cells* 24: 2466–77.

Chapuis, F., M. Rosenzwajg, M. Yagello, et al. 1997. Differentiation of human dendritic cells from monocytes in vitro. *Eur J Immunol* 27: 431–41.

Chatterjee, D., N. Marquardt, D. M. Tufa, et al. 2014a. Role of gamma-secretase in human umbilical-cord derived mesenchymal stem cell mediated suppression of NK cell cyto-toxicity. *Cell Commun Signal* 12: 63.

Chatterjee, D., N. Marquardt, D. M. Tufa, et al. 2014b. Human umbilical cord-derived mesen-chymal stem cells utilize activin-A to suppress interferon-gamma production by natu-ral killer cells. *Front Immunol* 5: 662.

Chen, C. P., Y. Y. Chen, J. P. Huang, and Y. H. Wu. 2014. The effect of conditioned medium derived from human placental multipotent mesenchymal stromal cells on neutrophils: Possible implications for placental infection. *Mol Hum Reprod* 20: 1117–25.

Chen, H., N. Zhang, T. Li, et al. 2012. Human umbilical cord Wharton's jelly stem cells: Immune property genes assay and effect of transplantation on the immune cells of heart failure patients. *Cell Immunol* 276: 83–90.

Chen, K., D. Wang, W. T. Du, et al. 2010. Human umbilical cord mesenchymal stem cells hUC-MSCs exert immunosuppressive activities through a PGE2-dependent mechanism. *Clin Immunol* 135: 448–58.

Cho, P. S., D. J. Messina, E. L. Hirsh, et al. 2008. Immunogenicity of umbilical cord tissue derived cells. *Blood* 111: 430–38.

Cichocki, F., E. Sitnicka, and Y. T. Bryceson. 2014. NK cell development and function— Plasticity and redundancy unleashed. *Semin Immunol* 26: 114–26.

Colocho, G., W. P. Graham, 3rd, A. E. Greene, D. W. Matheson, and D. Lynch. 1974. Human amniotic membrane as a physiologic wound dressing. *Arch Surg* 109: 370–73.

Contini, P., M. Ghio, A. Poggi, et al. 2003. Soluble HLA-A,-B,-C and -G molecules induce apoptosis in T and NK CD8+ cells and inhibit cytotoxic T cell activity through CD8 ligation. *Eur J Immunol* 33: 125–34.

Coyle, A. J., S. Lehar, C. Lloyd, et al. 2000. The CD28-related molecule ICOS is required for effective T cell-dependent immune responses. *Immunity* 13: 95–105.

de Araujo-Souza, P. S., S. C. Hanschke, and J. P. Viola. 2015. Epigenetic control of interferon-gamma expression in CD8 T cells. *J Immunol Res* 2015: 849573.

de Waal Malefyt, R., J. Abrams, B. Bennett, C. G. Figdor, and J. E. de Vries. 1991. Interleukin 10(IL-10) inhibits cytokine synthesis by human monocytes: An autoregulatory role of IL-10 produced by monocytes. *J Exp Med* 174: 1209–20.

Deng, Y., S. Yi, G. Wang, et al. 2014. Umbilical cord-derived mesenchymal stem cells instruct dendritic cells to acquire tolerogenic phenotypes through the IL-6-mediated upregulation of SOCS1. *Stem Cells Dev* 23: 2080–92.

Dhodapkar, M. V., R. M. Steinman, J. Krasovsky, C. Munz, and N. Bhardwaj. 2001. Antigen-specific inhibition of effector T cell function in humans after injection of immature dendritic cells. *J Exp Med* 193: 233–38.

Diaz-Prado, S., E. Muinos-Lopez, T. Hermida-Gomez, et al. 2010. Multilineage differentiation potential of cells isolated from the human amniotic membrane. *J Cell Biochem* 111: 846–57.

Dinarello, C. A. 1998. Interleukin-1, interleukin-1 receptors and interleukin-1 receptor antagonist. *Int Rev Immunol* 16: 457–99.

Ding, L., P. S. Linsley, L. Y. Huang, R. N. Germain, and E. M. Shevach. 1993. IL-10 inhibits macrophage costimulatory activity by selectively inhibiting the up-regulation of B7 expression. *J Immunol* 151: 1224–34.

Djouad, F., L. M. Charbonnier, C. Bouffi, et al. 2007. Mesenchymal stem cells inhibit the differentiation of dendritic cells through an interleukin-6-dependent mechanism. *Stem Cells* 25: 2025–32.

Donders, R., M. Vanheusden, J. F. Bogie, et al. 2015. Human Wharton's jelly-derived stem cells display immunomodulatory properties and transiently improve rat experimental autoimmune encephalomyelitis. *Cell Transplant* 24: 2077–98.

Dubois, B., J. M. Bridon, J. Fayette, et al. 1999. Dendritic cells directly modulate B cell growth and differentiation. *J Leukoc Biol* 66: 224–30.

Ebens, A., K. Brose, E. D. Leonardo, et al. 1996. Hepatocyte growth factor/scatter factor is an axonal chemoattractant and a neurotrophic factor for spinal motor neurons. *Neuron* 17: 1157–72.

Elrefaei, M., F. L. Ventura, C. A. Baker, et al. 2007. HIV-specific IL-10-positive CD8+ T cells suppress cytolysis and IL-2 production by CD8+ T cells. *J Immunol* 178: 3265–71.

Farrar, M. A., and R. D. Schreiber. 1993. The molecular cell biology of interferon-gamma and its receptor. *Annu Rev Immunol* 11: 571–611.

Faulk, W. P., R. Matthews, P. J. Stevens, et al. 1980. Human amnion as an adjunct in wound healing. *Lancet* 1: 1156–58.

Fournel, S., M. Aguerre-Girr, X. Huc, et al. 2000. Cutting edge: Soluble HLA-G1 triggers CD95/CD95 ligand-mediated apoptosis in activated CD8+ cells by interacting with CD8. *J Immunol* 164: 6100–104.

The Immunomodulatory Features of Mesenchymal Stromal Cells 119

Freud, A. G., J. Yu, and M. A. Caligiuri. 2014. Human natural killer cell development in secondary lymphoid tissues. *Semin Immunol* 26: 132–37.

Frucht, D. M., T. Fukao, C. Bogdan, et al. 2001. IFN-gamma production by antigen-presenting cells: Mechanisms emerge. *Trends Immunol* 22: 556–60.

Fu, F., Y. Li, S. Qian, et al. 1996. Costimulatory molecule-deficient dendritic cell progenitors (MHC class II+, CD80dim, CD86-) prolong cardiac allograft survival in nonimmunosuppressed recipients. *Transplantation* 62: 659–65.

Galimi, F., E. Cottone, E. Vigna, et al. 2001. Hepatocyte growth factor is a regulator of monocyte-macrophage function. *J Immunol* 166: 1241–47.

Garlanda, C., C. A. Dinarello, and A. Mantovani. 2013. The interleukin-1 family: Back to the future. *Immunity* 39: 1003–18.

Gazzinelli, R. T., I. P. Oswald, S. L. James, and A. Sher. 1992. IL-10 inhibits parasite killing and nitrogen oxide production by IFN-gamma-activated macrophages. *J Immunol* 148: 1792–96.

Geginat, J., M. Paroni, F. Facciotti, et al. 2013. The CD4-centered universe of human T cell subsets. *Semin Immunol* 25: 252–62.

Geginat, J., M. Paroni, S. Maglie, et al. 2014. Plasticity of human CD4 T cell subsets. *Front Immunol* 5: 630.

Gehrmann, J., Y. Matsumoto, and G. W. Kreutzberg. 1995. Microglia: Intrinsic immuneffector cell of the brain. *Brain Res Brain Res Rev* 20: 269–87.

Gerosa, F., B. Baldani-Guerra, C. Nisii, et al. 2002. Reciprocal activating interaction between natural killer cells and dendritic cells. *J Exp Med* 195: 327–33.

Gomes, J. A., A. Romano, M. S. Santos, and H. S. Dua. 2005. Amniotic membrane use in ophthalmology. *Curr Opin Ophthalmol* 16: 233–40.

Goodridge, J. P., A. Burian, N. Lee, and D. E. Geraghty. 2013. HLA-F and MHC class I open conformers are ligands for NK cell Ig-like receptors. *J Immunol* 191: 3553–62.

Gordin, M., M. Tesio, S. Cohen, et al. 2010. c-Met and its ligand hepatocyte growth factor/scatter factor regulate mature B cell survival in a pathway induced by CD74. *J Immunol* 185: 2020–31.

Gottfried, E., L. A. Kunz-Schughart, A. Weber, et al. 2008. Expression of CD68 in non-myeloid cell types. *Scand J Immunol* 67: 453–63.

Gregori, S., G. Amodio, F. Quattrone, and P. Panina-Bordignon. 2015. HLA-G orchestrates the early interaction of human trophoblasts with the maternal niche. *Front Immunol* 6: 128.

Grohmann, U., C. Orabona, F. Fallarino, et al. 2002. CTLA-4-Ig regulates tryptophan catabolism in vivo. *Nat Immunol* 3: 1097–101.

Gruss, J. S., and D. W. Jirsch. 1978. Human amniotic membrane: A versatile wound dressing. *Can Med Assoc J* 118: 1237–46.

Gu, Y. Z., Q. Xue, Y. J. Chen, et al. 2013. Different roles of PD-L1 and FasL in immunomodulation mediated by human placenta-derived mesenchymal stem cells. *Hum Immunol* 74: 267–76.

Hackstein, H., A. E. Morelli, and A. W. Thomson. 2001. Designer dendritic cells for tolerance induction: Guided not misguided missiles. *Trends Immunol* 22: 437–42.

Hadaschik, E. N., and A. H. Enk. 2015. TGF-beta-1-induced regulatory T cells. *Hum Immunol* 76: 561–64.

Heusinkveld, M., P. J. de Vos van Steenwijk, R. Goedemans, et al. 2011. M2 macrophages induced by prostaglandin E2 and IL-6 from cervical carcinoma are switched to activated M1 macrophages by CD4+ Th1 cells. *J Immunol* 187: 1157–65.

Hirko, A. C., R. Dallasen, S. Jomura, and Y. Xu. 2008. Modulation of inflammatory responses after global ischemia by transplanted umbilical cord matrix stem cells. *Stem Cells* 26: 2893–901.

Holland, O., N. Medvedeva, M. McDowell-Hook, M. Abumaree, and L. Chamley. 2012. Syncytial nuclear aggregates, carriers of fetal alloantigens. *J Reprod Immunol* 94: 118–18.

Huang, L., B. Baban, B. A. Johnson, 3rd, and A. L. Mellor. 2010. Dendritic cells, indoleamine 2,3 dioxygenase and acquired immune privilege. *Int Rev Immunol* 29: 133–55.

Hunt, J. S., M. G. Petroff, R. H. McIntire, and C. Ober. 2005. HLA-G and immune tolerance in pregnancy. *FASEB J* 19: 681–93.

Igura, K., X. Zhang, K. Takahashi, et al. 2004. Isolation and characterization of mesenchymal progenitor cells from chorionic villi of human placenta. *Cytotherapy* 6: 543–53.

Ishitani, A., N. Sageshima, N. Lee, et al. 2003. Protein expression and peptide binding suggest unique and interacting functional roles for HLA-E, F, and G in maternal-placental immune recognition. *J Immunol* 171: 1376–84.

Ivekovic, R., E. Tedeschi-Reiner, K. Novak-Laus, et al. 2005. Limbal graft and/or amniotic membrane transplantation in the treatment of ocular burns. *Ophthalmologica* 219: 297–302.

Jacobs, B., M. Wuttke, C. Papewalis, J. Seissler, and M. Schott. 2008. Dendritic cell subtypes and in vitro generation of dendritic cells. *Horm Metab Res* 40: 99–107.

Janeway, C. A., Jr. 2001. How the immune system protects the host from infection. *Microbes Infect* 3: 1167–71.

Janssens, K., C. Van den Haute, V. Baekelandt, et al. 2015. Leukemia inhibitory factor tips the immune balance towards regulatory T cells in multiple sclerosis. *Brain Behav Immun* 45: 180–88.

Jaramillo-Ferrada, P. A., E. J. Wolvetang, and J. J. Cooper-White. 2012. Differential mesengenic potential and expression of stem cell-fate modulators in mesenchymal stromal cells from human-term placenta and bone marrow. *J Cell Physiol* 227: 3234–42.

Jeannin, P., D. Duluc, and Y. Delneste. 2011. IL-6 and leukemia-inhibitory factor are involved in the generation of tumor-associated macrophage: Regulation by IFN-gamma. *Immunotherapy* 3: 23–26.

Jomura, S., M. Uy, K. Mitchell, et al. 2007. Potential treatment of cerebral global ischemia with Oct-4+ umbilical cord matrix cells. *Stem Cells* 25: 98–106.

Jonuleit, H., E. Schmitt, G. Schuler, J. Knop, and A. H. Enk. 2000. Induction of interleukin 10-producing, nonproliferating CD4(+) T cells with regulatory properties by repetitive stimulation with allogeneic immature human dendritic cells. *J Exp Med* 192: 1213–22.

Josien, R., M. Heslan, S. Brouard, J. P. Soulillou, and M. C. Cuturi. 1998. Critical requirement for graft passenger leukocytes in allograft tolerance induced by donor blood transfusion. *Blood* 92: 4539–44.

Jung, J., J. H. Choi, Y. Lee, et al. 2013. Human placenta-derived mesenchymal stem cells promote hepatic regeneration in CCl4 -injured rat liver model via increased autophagic mechanism. *Stem Cells* 31: 1584–96.

Kagimoto, Y., H. Yamada, T. Ishikawa, et al. 2008. A regulatory role of interleukin 15 in wound healing and mucosal infection in mice. *J Leukoc Biol* 83: 165–72.

Kalinski, P. 2012. Regulation of immune responses by prostaglandin E2. *J Immunol* 188: 21–28.

Kanai, T., T. Fujii, S. Kozuma, et al. 2001. Soluble HLA-G influences the release of cytokines from allogeneic peripheral blood mononuclear cells in culture. *Mol Hum Reprod* 7: 195–200.

Kang, J. W., H. C. Koo, S. Y. Hwang, et al. 2012. Immunomodulatory effects of human amniotic membrane-derived mesenchymal stem cells. *J Vet Sci* 13: 23–31.

Karlsson, H., T. Erkers, S. Nava, et al. 2012. Stromal cells from term fetal membrane are highly suppressive in allogeneic settings in vitro. *Clin Exp Immunol* 167: 543–55.

Khalil-Daher, I., B. Riteau, C. Menier, et al. 1999. Role of HLA-G versus HLA-E on NK function: HLA-G is able to inhibit NK cytolysis by itself. *J Reprod Immunol* 43: 175–82.

Koenderman, L., W. Buurman, and M. R. Daha. 2014. The innate immune response. *Immunol Lett* 162: 95–102.

Kronsteiner, B., A. Peterbauer-Scherb, R. Grillari-Voglauer, et al. 2011a. Human mesenchymal stem cells and renal tubular epithelial cells differentially influence monocyte-derived dendritic cell differentiation and maturation. *Cell Immunol* 267: 30–38.

Kronsteiner, B., S. Wolbank, A. Peterbauer, et al. 2011b. Human mesenchymal stem cells from adipose tissue and amnion influence T-cells depending on stimulation method and presence of other immune cells. *Stem Cells Dev* 20: 2115–26.

Kryczek, I., L. Zou, P. Rodriguez, et al. 2006. B7-H4 expression identifies a novel suppressive macrophage population in human ovarian carcinoma. *J Exp Med* 203: 871–81.

Kubo, M., Y. Sonoda, R. Muramatsu, and M. Usui. 2001. Immunogenicity of human amniotic membrane in experimental xenotransplantation. *Invest Ophthalmol Vis Sci* 42: 1539–46.

Kuci, S., Z. Kuci, H. Kreyenberg, et al. 2010. CD271 antigen defines a subset of multipotent stromal cells with immunosuppressive and lymphohematopoietic engraftment-promoting properties. *Haematologica* 95: 651–59.

Langenkamp, A., M. Messi, A. Lanzavecchia, and F. Sallusto. 2000. Kinetics of dendritic cell activation: Impact on priming of TH1, TH2 and nonpolarized T cells. *Nat Immunol* 1: 311–16.

La Rocca, G., and R. Anzalone. 2013a. Perinatal stem cells revisited: Directions and indications at the crossroads between tissue regeneration and repair. *Curr Stem Cell Res Ther* 8: 2–5.

La Rocca, G., R. Anzalone, and F. Farina. 2009. The expression of CD68 in human umbilical cord mesenchymal stem cells: New evidences of presence in non-myeloid cell types. *Scand J Immunol* 70: 161–62.

La Rocca, G., M. Lo Iacono, T. Corsello, et al. 2013b. Human Wharton's jelly mesenchymal stem cells maintain the expression of key immunomodulatory molecules when subjected to osteogenic, adipogenic and chondrogenic differentiation in vitro: New perspectives for cellular therapy. *Curr Stem Cell Res Ther* 8: 100–13.

Latchman, Y., C. R. Wood, T. Chernova, et al. 2001. PD-L2 is a second ligand for PD-1 and inhibits T cell activation. *Nat Immunol* 2: 261–68.

Le Gros, G., and F. Erard. 1994. Non-cytotoxic, IL-4, IL-5, IL-10 producing CD8+ T cells: Their activation and effector functions. *Curr Opin Immunol* 6: 453–57.

Le Rond, S., C. Azema, I. Krawice-Radanne, et al. 2006. Evidence to support the role of HLA-G5 in allograft acceptance through induction of immunosuppressive/ regulatory T cells. *J Immunol* 176: 3266–76.

Le Rond, S., J. Le Maoult, C. Creput, et al. 2004. Alloreactive CD4+ and CD8+ T cells express the immunotolerant HLA-G molecule in mixed lymphocyte reactions: In vivo implications in transplanted patients. *Eur J Immunol* 34: 649–60.

Lee, M. J., J. Jung, K. H. Na, et al. 2010. Anti-fibrotic effect of chorionic plate-derived mesenchymal stem cells isolated from human placenta in a rat model of CCl(4)-injured liver: Potential application to the treatment of hepatic diseases. *J Cell Biochem* 111: 1453–63.

Lee, N., M. Llano, M. Carretero, et al. 1998. HLA-E is a major ligand for the natural killer inhibitory receptor CD94/NKG2A. *Proc Natl Acad Sci U S A* 95: 5199–204.

Lefebvre, S., F. Adrian, P. Moreau, et al. 2000. Modulation of HLA-G expression in human thymic and amniotic epithelial cells. *Hum Immunol* 61: 1095–101.

Leitner, J., C. Klauser, W. F. Pickl, et al. 2009. B7-H3 is a potent inhibitor of human T-cell activation: No evidence for B7-H3 and TREML2 interaction. *Eur J Immunol* 39: 1754–64.

LeMaoult, J., I. Krawice-Radanne, J. Dausset, and E. D. Carosella. 2004. HLA-G1-expressing antigen-presenting cells induce immunosuppressive CD4+ T cells. *Proc Natl Acad Sci U S A* 101: 7064–69.

Levine, M. H., A. M. Haberman, D. B. Sant'Angelo, et al. 2000. A B-cell receptor-specific selection step governs immature to mature B cell differentiation. *Proc Natl Acad Sci U S A* 97: 2743–48.

Li, C., W. Zhang, X. Jiang, and N. Mao. 2007. Human-placenta-derived mesenchymal stem cells inhibit proliferation and function of allogeneic immune cells. *Cell Tissue Res* 330: 437–46.

Li, H., J. Y. Niederkorn, S. Neelam, et al. 2005. Immunosuppressive factors secreted by human amniotic epithelial cells. *Invest Ophthalmol Vis Sci* 46: 900–7.

Li, J., C. Koike-Soko, J. Sugimoto, et al. 2015. Human amnion-derived stem cells have immunosuppressive properties on NK cells and monocytes. *Cell Transplant* 24: 2065–76.

Li, W., Q. Zhang, M. Wang, et al. 2013. Macrophages are involved in the protective role of human umbilical cord-derived stromal cells in renal ischemia-reperfusion injury. *Stem Cell Res* 10: 405–16.

Li, X., J. Bai, X. Ji, et al. 2014. Comprehensive characterization of four different populations of human mesenchymal stem cells as regards their immune properties, proliferation and differentiation. *Int J Mol Med* 34: 695–704.

Li, X. C., D. M. Rothstein, and M. H. Sayegh. 2009. Costimulatory pathways in transplantation: Challenges and new developments. *Immunol Rev* 229: 271–93.

Lila, N., A. Carpentier, C. Amrein, et al. 2000. Implication of HLA-G molecule in heart-graft acceptance. *Lancet* 355: 2138.

Lin, H. D., C. Y. Fong, A. Biswas, M. Choolani, and A. Bongso. 2014. Human Wharton's jelly stem cells, its conditioned medium and cell-free lysate inhibit the growth of human lymphoma cells. *Stem Cell Rev* 10: 573–86.

Lin, Y. C., T. L. Ko, Y. H. Shih, et al. 2011. Human umbilical mesenchymal stem cells promote recovery after ischemic stroke. *Stroke* 42: 2045–53.

Liu, T., H. Zhai, Y. Xu, et al. 2012a. Amniotic membrane traps and induces apoptosis of inflammatory cells in ocular surface chemical burn. *Mol Vis* 18: 2137–46.

Liu, W., A. Morschauser, X. Zhang, et al. 2014. Human placenta-derived adherent cells induce tolerogenic immune responses. *Clin Transl Immunology* 3: e14.

Liu, Y. H., V. Vaghjiani, J. Y. Tee, et al. 2012b. Amniotic epithelial cells from the human placenta potently suppress a mouse model of multiple sclerosis. *PLoS One* 7: e35758.

Lu, L., D. McCaslin, T. E. Starzl, and A. W. Thomson. 1995. Bone marrow-derived dendritic cell progenitors (NLDC 145+, MHC class II+, B7-1dim, B7-2-) induce alloantigen-specific hyporesponsiveness in murine T lymphocytes. *Transplantation* 60: 1539–45.

Ma, L., Z. Zhou, D. Zhang, et al. 2012. Immunosuppressive function of mesenchymal stem cells from human umbilical cord matrix in immune thrombocytopenia patients. *Thromb Haemost* 107: 937–50.

Macatonia, S. E., T. M. Doherty, S. C. Knight, and A. O'Garra. 1993. Differential effect of IL-10 on dendritic cell-induced T cell proliferation and IFN-gamma production. *J Immunol* 150: 3755–65.

Magatti, M., M. Caruso, S. De Munari, et al. 2015. Human amniotic membrane-derived mesenchymal and epithelial cells exert different effects on monocyte-derived dendritic cell differentiation and function. *Cell Transplant* 24: 1733–52.

Magatti, M., S. De Munari, E. Vertua, et al. 2008. Human amnion mesenchyme harbors cells with allogeneic T-cell suppression and stimulation capabilities. *Stem Cells* 26: 182–92.

Magatti, M., S. De Munari, E. Vertua, et al. 2009. Amniotic mesenchymal tissue cells inhibit dendritic cell differentiation of peripheral blood and amnion resident monocytes. *Cell Transplant* 18: 899–914.

Maggi, E., R. Manetti, F. Annunziato, and S. Romagnani. 1995. CD8+ T lymphocytes producing Th2-type cytokines (Tc2) in HIV-infected individuals. *J Biol Regul Homeost Agents* 9: 78–81.

Maggi, L., V. Santarlasci, M. Capone, et al. 2010. CD161 is a marker of all human IL-17-producing T-cell subsets and is induced by RORC. *Eur J Immunol* 40: 2174–81.

Manochantr, S., Y. U-pratya, P. Kheolamai, et al. 2013. Immunosuppressive properties of mesenchymal stromal cells derived from amnion, placenta, Wharton's jelly and umbilical cord. *Intern Med J* 43: 430–39.

The Immunomodulatory Features of Mesenchymal Stromal Cells 123

Mantovani, A., S. K. Biswas, M. R. Galdiero, A. Sica, and M. Locati. 2013. Macrophage plasticity and polarization in tissue repair and remodelling. *J Pathol* 229: 176–85.

Mantovani, A., A. Sica, S. Sozzani, et al. 2004. The chemokine system in diverse forms of macrophage activation and polarization. *Trends Immunol* 25: 677–86.

Manuelpillai, U., D. Lourensz, V. Vaghjiani, et al. 2012. Human amniotic epithelial cell transplantation induces markers of alternative macrophage activation and reduces established hepatic fibrosis. *PLoS One* 7: e38631.

Manuelpillai, U., J. Tchongue, D. Lourensz, et al. 2010. Transplantation of human amnion epithelial cells reduces hepatic fibrosis in immunocompetent CCl(4)-treated mice. *Cell Transplant* 19: 1157–68.

Margossian, T., L. Reppel, N. Makdissy, et al. 2012. Mesenchymal stem cells derived from Wharton's jelly: Comparative phenotype analysis between tissue and in vitro expansion. *Biomed Mater Eng* 22: 243–54.

Mathews, S., K. Lakshmi Rao, K. Suma Prasad, et al. 2015. Propagation of pure fetal and maternal mesenchymal stromal cells from terminal chorionic villi of human term placenta. *Sci Rep* 5: 10054.

Matsushima, H., S. Geng, R. Lu, et al. 2013. Neutrophil differentiation into a unique hybrid population exhibiting dual phenotype and functionality of neutrophils and dendritic cells. *Blood* 121: 1677–89.

McDonald, C. A., N. L. Payne, G. Sun, et al. 2015. Immunosuppressive potential of human amnion epithelial cells in the treatment of experimental autoimmune encephalomyelitis. *J Neuroinflammation* 12: 112.

McIntire, R. H., P. J. Morales, M. G. Petroff, M. Colonna, and J. S. Hunt. 2004. Recombinant HLA-G5 and -G6 drive U937 myelomonocytic cell production of TGF-beta 1. *J Leukoc Biol* 76: 1220–28.

Medicetty, S., A. R. Bledsoe, C. B. Fahrenholtz, D. Troyer, and M. L. Weiss. 2004. Transplantation of pig stem cells into rat brain: Proliferation during the first 8 weeks. *Exp Neurol* 190: 32–41.

Medzhitov, R. 2007. Recognition of microorganisms and activation of the immune response. *Nature* 449: 819–26.

Mellman, I., and R. M. Steinman. 2001. Dendritic cells: Specialized and regulated antigen processing machines. *Cell* 106: 255–58.

Micklem, K., E. Rigney, J. Cordell, et al. 1989. A human macrophage-associated antigen (CD68) detected by six different monoclonal antibodies. *Br J Haematol* 73: 6–11.

Mitchell, M. D., S. S. Edwin, S. Lundin-Schiller, et al. 1993. Mechanism of interleukin-1 beta stimulation of human amnion prostaglandin biosynthesis: Mediation via a novel inducible cyclooxygenase. *Placenta* 14: 615–25.

Molnarfi, N., M. Benkhoucha, H. Funakoshi, T. Nakamura, and P. H. Lalive. 2015. Hepatocyte growth factor: A regulator of inflammation and autoimmunity. *Autoimmun Rev* 14: 293–303.

Molnarfi, N., M. Benkhoucha, C. Juillard, K. Bjarnadottir, and P. H. Lalive. 2014. The neurotrophic hepatocyte growth factor induces protolerogenic human dendritic cells. *J Neuroimmunol* 267: 105–10.

Montaldo, E., P. Vacca, L. Moretta, and M. C. Mingari. 2014. Development of human natural killer cells and other innate lymphoid cells. *Semin Immunol* 26: 107–13.

Moodley, Y., S. Ilancheran, C. Samuel, et al. 2010. Human amnion epithelial cell transplantation abrogates lung fibrosis and augments repair. *Am J Respir Crit Care Med* 182: 643–51.

Moodley, Y., V. Vaghjiani, J. Chan, et al. 2013. Anti-inflammatory effects of adult stem cells in sustained lung injury: A comparative study. *PLoS One* 8: e69299.

Moore, K. W., R. de Waal Malefyt, R. L. Coffman, and A. O'Garra. 2001. Interleukin-10 and the interleukin-10 receptor. *Annu Rev Immunol* 19: 683–765.

Moretta, L., E. Montaldo, P. Vacca, et al. 2014. Human natural killer cells: Origin, receptors, function, and clinical applications. *Int Arch Allergy Immunol* 164: 253–64.

Munn, D. H., M. D. Sharma, B. Baban, et al. 2005. GCN2 kinase in T cells mediates proliferative arrest and anergy induction in response to indoleamine 2,3-dioxygenase. *Immunity* 22: 633–42.

Munn, D. H., M. Zhou, J. T. Attwood, et al. 1998. Prevention of allogeneic fetal rejection by tryptophan catabolism. *Science* 281: 1191–93.

Murakami, N., and L. V. Riella. 2014. Co-inhibitory pathways and their importance in immune regulation. *Transplantation* 98: 3–14.

Murphy, S., R. Lim, H. Dickinson, et al. 2011. Human amnion epithelial cells prevent bleomycin-induced lung injury and preserve lung function. *Cell Transplant* 20: 909–23.

Murphy, S., S. Rosli, R. Acharya, et al. 2010. Amnion epithelial cell isolation and characterization for clinical use. *Curr Protoc Stem Cell Biol* Chapter 1: Unit 1E 6.

Murphy, S. V., S. C. Shiyun, J. L. Tan, et al. 2012. Human amnion epithelial cells do not abrogate pulmonary fibrosis in mice with impaired macrophage function. *Cell Transplant* 21: 1477–92.

Murray, P. J., and T. A. Wynn. 2011. Protective and pathogenic functions of macrophage subsets. *Nat Rev Immunol* 11: 723–37.

Najar, M., G. Raicevic, H. I. Boufker, et al. 2010a. Adipose-tissue-derived and Wharton's jelly-derived mesenchymal stromal cells suppress lymphocyte responses by secreting leukemia inhibitory factor. *Tissue Eng Part A* 16: 3537–46.

Najar, M., G. Raicevic, H. I. Boufker, et al. 2010b. Mesenchymal stromal cells use PGE2 to modulate activation and proliferation of lymphocyte subsets: Combined comparison of adipose tissue, Wharton's Jelly and bone marrow sources. *Cell Immunol* 264: 171–79.

Najar, M., G. Raicevic, F. Jebbawi, et al. 2012. Characterization and functionality of the CD200-CD200R system during mesenchymal stromal cell interactions with T-lymphocytes. *Immunol Lett* 146: 50–56.

Nocentini, G., and C. Riccardi. 2005. GITR: A multifaceted regulator of immunity belonging to the tumor necrosis factor receptor superfamily. *Eur J Immunol* 35: 1016–22.

Nold, M. F., C. A. Nold-Petry, G. B. Pott, et al. 2008. Endogenous IL-32 controls cytokine and HIV-1 production. *J Immunol* 181: 557–65.

Noone, C., A. Kihm, K. English, S. O'Dea, and B. P. Mahon. 2013. IFN-gamma stimulated human umbilical-tissue-derived cells potently suppress NK activation and resist NK-mediated cytotoxicity in vitro. *Stem Cells Dev* 22: 3003–14.

Onishi, R., S. Ohnishi, R. Higashi, et al. 2015. Human amnion-derived mesenchymal stem cell transplantation ameliorates dextran sulfate sodium-induced severe colitis in rats. *Cell Transplant* Mar 25 [Epub ahead of print].

Ou, D., X. Wang, D. L. Metzger, et al. 2006. Suppression of human T-cell responses to beta-cells by activation of B7-H4 pathway. *Cell Transplant* 15: 399–410.

Palucka, K. A., N. Taquet, F. Sanchez-Chapuis, and J. C. Gluckman. 1998. Dendritic cells as the terminal stage of monocyte differentiation. *J Immunol* 160: 4587–95.

Parolini, O., F. Alviano, G. P. Bagnara, et al. 2008. Concise review: Isolation and characterization of cells from human term placenta: Outcome of the First International Workshop on Placenta Derived Stem Cells. *Stem Cells* 26: 300–11.

Parolini, O., L. Souza-Moreira, F. O'Valle, et al. 2014. Therapeutic effect of human amniotic membrane-derived cells on experimental arthritis and other inflammatory disorders. *Arthritis Rheumatol* 66: 327–39.

Perrigoue, J. G., C. Zaph, K. Guild, Y. Du, and D. Artis. 2009. IL-31-IL-31R interactions limit the magnitude of Th2 cytokine-dependent immunity and inflammation following intestinal helminth infection. *J Immunol* 182: 6088–94.

The Immunomodulatory Features of Mesenchymal Stromal Cells 125

Petroff, M. G. 2005. Immune interactions at the maternal-fetal interface. *J Reprod Immunol* 68: 1–13.

Petroff, M. G., and A. Perchellet. 2010. B7 family molecules as regulators of the maternal immune system in pregnancy. *Am J Reprod Immunol* 63: 506–19.

Pianta, S., P. Bonassi Signoroni, I. Muradore, et al. 2015. Amniotic membrane mesenchymal cells-derived factors skew T cell polarization toward Treg and downregulate Th1 and Th17 cells subsets. *Stem Cell Rev* 11 (3): 394–407.

Pillay, J., V. M. Kamp, E. van Hoffen, et al. 2012. A subset of neutrophils in human systemic inflammation inhibits T cell responses through Mac-1. *J Clin Invest* 122: 327–36.

Pillay, J., T. Tak, V. M. Kamp, and L. Koenderman. 2013. Immune suppression by neutrophils and granulocytic myeloid-derived suppressor cells: Similarities and differences. *Cell Mol Life Sci* 70: 3813–27.

Poloni, A., V. Rosini, E. Mondini, et al. 2008. Characterization and expansion of mesenchymal progenitor cells from first-trimester chorionic villi of human placenta. *Cytotherapy* 10: 690–97.

Pomini, F., A. Caruso, and J. R. Challis. 1999. Interleukin-10 modifies the effects of interleukin-1 beta and tumor necrosis factor-alpha on the activity and expression of prostaglandin H synthase-2 and the NAD+-dependent 15-hydroxyprostaglandin dehydrogenase in cultured term human villous trophoblast and chorion trophoblast cells. *J Clin Endocrinol Metab* 84: 4645–51.

Portmann-Lanz, C. B., A. Schoeberlein, A. Huber, et al. 2006. Placental mesenchymal stem cells as potential autologous graft for pre- and perinatal neuroregeneration. *Am J Obstet Gynecol* 194: 664–73.

Prasanna, S. J., D. Gopalakrishnan, S. R. Shankar, and A. B. Vasandan. 2010. Pro-inflammatory cytokines, IFNgamma and TNFalpha, influence immune properties of human bone marrow and Wharton jelly mesenchymal stem cells differentially. *PLoS One* 5: e9016.

Pratama, G., V. Vaghjiani, J. Y. Tee, et al. 2011. Changes in culture expanded human amniotic epithelial cells: Implications for potential therapeutic applications. *PLoS One* 6: e26136.

Raicevic, G., M. Najar, B. Stamatopoulos, et al. 2011. The source of human mesenchymal stromal cells influences their TLR profile as well as their functional properties. *Cell Immunol* 270: 207–16.

Ribeiro, A., P. Laranjeira, S. Mendes, et al. 2013. Mesenchymal stem cells from umbilical cord matrix, adipose tissue and bone marrow exhibit different capability to suppress peripheral blood B, natural killer and T cells. *Stem Cell Res Ther* 4: 125.

Rickard, A. J., and M. J. Young. 2009. Corticosteroid receptors, macrophages and cardiovascular disease. *J Mol Endocrinol* 42: 449–59.

Roda, B., P. Reschiglian, A. Zattoni, et al. 2009. A tag-less method of sorting stem cells from clinical specimens and separating mesenchymal from epithelial progenitor cells. *Cytometry B Clin Cytom* 76: 285–90.

Roelen, D. L., B. J. van der Mast, P. S. In 't Anker, et al. 2009. Differential immunomodulatory effects of fetal versus maternal multipotent stromal cells. *Hum Immunol* 70: 16–23.

Rogers, M. L., A. Beare, H. Zola, and R. A. Rush. 2008. CD 271 (P75 neurotrophin receptor). *J Biol Regul Homeost Agents* 22: 1–6.

Romani, N., S. Gruner, D. Brang, et al. 1994. Proliferating dendritic cell progenitors in human blood. *J Exp Med* 180: 83–93.

Rosser, E. C., and C. Mauri. 2015. Regulatory B cells: Origin, phenotype, and function. *Immunity* 42: 607–12.

Rossi, D., S. Pianta, M. Magatti, P. Sedlmayr, and O. Parolini. 2012. Characterization of the conditioned medium from amniotic membrane cells: Prostaglandins as key effectors of its immunomodulatory activity. *PLoS One* 7: e46956.

Saeidi, M., A. Masoud, Y. Shakiba, et al. 2013. Immunomodulatory effects of human umbilical cord Wharton's jelly-derived mesenchymal stem cells on differentiation, maturation and endocytosis of monocyte-derived dendritic cells. *Iran J Allergy Asthma Immunol* 12: 37–49.

Sakaguchi, S., M. Miyara, C. M. Costantino, and D. A. Hafler. 2010. FOXP3+ regulatory T cells in the human immune system. *Nat Rev Immunol* 10: 490–500.

Sakuragawa, N., H. Yoshikawa, and M. Sasaki. 1992. Amniotic tissue transplantation: Clinical and biochemical evaluations for some lysosomal storage diseases. *Brain Dev* 14: 7–11.

Sallusto, F., and A. Lanzavecchia. 1994. Efficient presentation of soluble antigen by cultured human dendritic cells is maintained by granulocyte/macrophage colony-stimulating factor plus interleukin 4 and downregulated by tumor necrosis factor alpha. *J Exp Med* 179: 1109–18.

Sallusto, F., and A. Lanzavecchia. 1999. Mobilizing dendritic cells for tolerance, priming, and chronic inflammation. *J Exp Med* 189: 611–14.

Sankar, V., and R. Muthusamy. 2003. Role of human amniotic epithelial cell transplantation in spinal cord injury repair research. *Neuroscience* 118: 11–17.

Sant'Anna, L. B., A. Cargnoni, L. Ressel, G. Vanosi, and O. Parolini. 2011. Amniotic membrane application reduces liver fibrosis in a bile duct ligation rat model. *Cell Transplant* 20: 441–53.

Scaggiante, B., A. Pineschi, M. Sustersich, et al. 1987. Successful therapy of Niemann-Pick disease by implantation of human amniotic membrane. *Transplantation* 44: 59–61.

Sharpe, A. H., and G. J. Freeman. 2002. The B7-CD28 superfamily. *Nat Rev Immunol* 2: 116–26.

Shimizu, J., S. Yamazaki, T. Takahashi, Y. Ishida, and S. Sakaguchi. 2002. Stimulation of CD25(+)CD4(+) regulatory T cells through GITR breaks immunological self-tolerance. *Nat Immunol* 3: 135–42.

Shiroishi, M., K. Tsumoto, K. Amano, et al. 2003. Human inhibitory receptors Ig-like transcript 2 (ILT2) and ILT4 compete with CD8 for MHC class I binding and bind preferentially to HLA-G. *Proc Natl Acad Sci U S A* 100: 8856–61.

Shlomchik, M. J., and F. Weisel. 2012. Germinal center selection and the development of memory B and plasma cells. *Immunol Rev* 247: 52–63.

Siegmund, K., B. Ruckert, N. Ouaked, et al. 2009. Unique phenotype of human tonsillar and in vitro-induced FOXP3+CD8+ T cells. *J Immunol* 182: 2124–30.

Sojka, D. K., Z. Tian, and W. M. Yokoyama. 2014. Tissue-resident natural killer cells and their potential diversity. *Semin Immunol* 26: 127–31.

Soncini, M., E. Vertua, L. Gibelli, et al. 2007. Isolation and characterization of mesenchymal cells from human fetal membranes. *J Tissue Eng Regen Med* 1: 296–305.

Stadler, G., S. Hennerbichler, A. Lindenmair, et al. 2008. Phenotypic shift of human amniotic epithelial cells in culture is associated with reduced osteogenic differentiation in vitro. *Cytotherapy* 10: 743–52.

Strasser, A., P. J. Jost, and S. Nagata. 2009. The many roles of FAS receptor signaling in the immune system. *Immunity* 30: 180–92.

Su, Z., L. Emdad, M. Sauane, et al. 2005. Unique aspects of mda-7/IL-24 antitumor bystander activity: Establishing a role for secretion of MDA-7/IL-24 protein by normal cells. *Oncogene* 24: 7552–66.

Subrahmanyam, M. 1995. Amniotic membrane as a cover for microskin grafts. *Br J Plast Surg* 48: 477–78.

Svensson, J., M. C. Jenmalm, A. Matussek, et al. 2011. Macrophages at the fetal-maternal interface express markers of alternative activation and are induced by M-CSF and IL-10. *J Immunol* 7: 3671–82.

The Immunomodulatory Features of Mesenchymal Stromal Cells   **127**

Tan, J. L., S. T. Chan, E. M. Wallace, and R. Lim. 2014. Human amnion epithelial cells mediate lung repair by directly modulating macrophage recruitment and polarization. *Cell Transplant* 23: 319–28.

Tandon, R., N. Gupta, M. Kalaivani, et al. 2011. Amniotic membrane transplantation as an adjunct to medical therapy in acute ocular burns. *Br J Ophthalmol* 95: 199–204.

Tecchio, C., and M. A. Cassatella. 2014. Neutrophil-derived cytokines involved in physiological and pathological angiogenesis. *Chem Immunol Allergy* 99: 123–37.

Tipnis, S., C. Viswanathan, and A. S. Majumdar. 2010. Immunosuppressive properties of human umbilical cord-derived mesenchymal stem cells: Role of B7-H1 and IDO. *Immunol Cell Biol* 88: 795–806.

Tsai, P. J., H. S. Wang, G. J. Lin, et al. 2015. Undifferentiated Wharton's Jelly mesenchymal stem cell transplantation induces insulin-producing cell differentiation and suppression of T cell-mediated autoimmunity in non-obese diabetic mice. *Cell Transplant* 24: 1555–70.

Tsuji, H., S. Miyoshi, Y. Ikegami, et al. 2010. Xenografted human amniotic membrane-derived mesenchymal stem cells are immunologically tolerated and transdifferentiated into cardiomyocytes. *Circ Res* 106: 1613–23.

Tylki-Szymanska, A., D. Maciejko, M. Kidawa, U. Jablonska-Budaj, and B. Czartoryska. 1985. Amniotic tissue transplantation as a trial of treatment in some lysosomal storage diseases. *J Inherit Metab Dis* 8: 101–4.

Ueno, T., M. Y. Yeung, M. McGrath, et al. 2012. Intact B7-H3 signaling promotes allograft prolongation through preferential suppression of Th1 effector responses. *Eur J Immunol* 42: 2343–53.

Ueta, M., M. N. Kweon, Y. Sano, et al. 2002. Immunosuppressive properties of human amniotic membrane for mixed lymphocyte reaction. *Clin Exp Immunol* 129: 464–70.

van Kooten, C., and J. Banchereau. 2000. CD40-CD40 ligand. *J Leukoc Biol* 67: 2–17.

Vawda, R., and M. G. Fehlings. 2013. Mesenchymal cells in the treatment of spinal cord injury: Current & future perspectives. *Curr Stem Cell Res Ther* 8: 25–38.

Vivier, E., D. H. Raulet, A. Moretta, et al. 2011. Innate or adaptive immunity? The example of natural killer cells. *Science* 331: 44–49.

von Boehmer, H. 1994. Positive selection of lymphocytes. *Cell* 76: 219–28.

Vosdoganes, P., R. J. Hodges, R. Lim, et al. 2011. Human amnion epithelial cells as a treatment for inflammation-induced fetal lung injury in sheep. *Am J Obstet Gynecol* 205: 156.e26–33.

Vosdoganes, P., E. M. Wallace, S. T. Chan, et al. 2013. Human amnion epithelial cells repair established lung injury. *Cell Transplant* 22: 1337–49.

Waldmann, H. 1999. Transplantation tolerance-where do we stand? *Nat Med* 5: 1245–8.

Wang, D., K. Chen, W. T. Du, et al. 2010. CD14+ monocytes promote the immunosuppressive effect of human umbilical cord matrix stem cells. *Exp Cell Res* 316: 2414–23.

Wang, D., Y. R. Ji, K. Chen, et al. 2012. IL-6 production stimulated by CD14(+) monocytes-paracrined IL-1beta does not contribute to the immunosuppressive activity of human umbilical cord mesenchymal stem cells. *Cell Physiol Biochem* 29: 551–60.

Wang, H., X. Qiu, P. Ni, et al. 2014. Immunological characteristics of human umbilical cord mesenchymal stem cells and the therapeutic effects of their transplantation on hyperglycemia in diabetic rats. *Int J Mol Med* 33: 263–70.

Ward, D. J., and J. P. Bennett. 1984. The long-term results of the use of human amnion in the treatment of leg ulcers. *Br J Plast Surg* 37: 191–93.

Wegmeyer, H., A. M. Broske, M. Leddin, et al. 2013. Mesenchymal stromal cell characteristics vary depending on their origin. *Stem Cells Dev* 22: 2606–18.

Wei, J. P., M. Nawata, S. Wakitani, et al. 2009. Human amniotic mesenchymal cells differentiate into chondrocytes. *Cloning Stem Cells* 11: 19–26.

Weiss, M. L., C. Anderson, S. Medicetty, et al. 2008. Immune properties of human umbilical cord Wharton's jelly-derived cells. *Stem Cells* 26: 2865–74.

Weiss, M. L., S. Medicetty, A. R. Bledsoe, et al. 2006. Human umbilical cord matrix stem cells: Preliminary characterization and effect of transplantation in a rodent model of Parkinson's disease. *Stem Cells* 24: 781–92.

Whittle, W. L., W. Gibb, and J. R. Challis. 2000. The characterization of human amnion epithelial and mesenchymal cells: The cellular expression, activity and glucocorticoid regulation of prostaglandin output. *Placenta* 21: 394–401.

Wolbank, S., A. Peterbauer, M. Fahrner, et al. 2007. Dose-dependent immunomodulatory effect of human stem cells from amniotic membrane: A comparison with human mesenchymal stem cells from adipose tissue. *Tissue Eng* 13: 1173–83.

Wolbank, S., F. Hildner, H. Redl. et al. 2009a. Impact of human amniotic membrane preparation on release of angiogenic factors. *J Tissue Eng Regen Med* 3: 651–54.

Wolbank, S., G. Stadler, A. Peterbauer, et al. 2009b. Telomerase immortalized human amnion- and adipose-derived mesenchymal stem cells: Maintenance of differentiation and immunomodulatory characteristics. *Tissue Eng Part A* 15: 1843–54.

Wu, W., Q. Lan, H. Lu, et al. 2014. Human amnion mesenchymal cells negative co-stimulatory molecules PD-L1 expression and its capacity of modulating microglial activation of CNS. *Cell Biochem Biophys* 69: 35–45.

Yamahara, K., K. Harada, M. Ohshima, et al. 2014. Comparison of angiogenic, cytoprotective, and immunosuppressive properties of human amnion- and chorion-derived mesenchymal stem cells. *PLoS One* 9: e88319.

Yang, C. C., Y. H. Shih, M. H. Ko, et al. 2008. Transplantation of human umbilical mesenchymal stem cells from Wharton's jelly after complete transection of the rat spinal cord. *PLoS One* 3: e3336.

Yawno, T., J. Schuilwerve, T. J. Moss, et al. 2013. Human amnion epithelial cells reduce fetal brain injury in response to intrauterine inflammation. *Dev Neurosci* 35: 272–82.

Yeager, A. M., H. S. Singer, J. R. Buck, et al. 1985. A therapeutic trial of amniotic epithelial cell implantation in patients with lysosomal storage diseases. *Am J Med Genet* 22: 347–55.

Yoshinaga, S. K., J. S. Whoriskey, S. D. Khare, et al. 1999. T-cell co-stimulation through B7RP-1 and ICOS. *Nature* 402: 827–32.

Yuge, I., Y. Takumi, K. Koyabu, et al. 2004. Transplanted human amniotic epithelial cells express connexin 26 and Na-K-adenosine triphosphatase in the inner ear. *Transplantation* 77: 1452–54.

Zhang, B., T. M. Adesanya, L. Zhang, et al. 2014. Delivery of placenta-derived mesenchymal stem cells ameliorates ischemia induced limb injury by immunomodulation. *Cell Physiol Biochem* 34: 1998–2006.

Zhang, X., A. Mitsuru, K. Igura, et al. 2006. Mesenchymal progenitor cells derived from chorionic villi of human placenta for cartilage tissue engineering. *Biochem Biophys Res Commun* 340: 944–52.

Zhao, P., H. Ise, M. Hongo, et al. 2005. Human amniotic mesenchymal cells have some characteristics of cardiomyocytes. *Transplantation* 79: 528–35.

Zhou, C., B. Yang, Y. Tian, et al. 2011. Immunomodulatory effect of human umbilical cord Wharton's jelly-derived mesenchymal stem cells on lymphocytes. *Cell Immunol* 272: 33–38.

Zhou, S., J. Chen, and J. Feng. 2003. The effects of amniotic membrane on polymorphonuclear cells. *Chin Med J (Engl)* 116: 788–90.

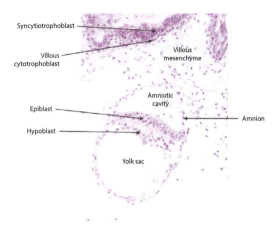

FIGURE 1.1 Hematoxylin and eosin–stained section through an embryo (Carnegie Collection #7801) presumed to be 13.5 days (Boyd and Hamilton 1970), showing the epiblast and hypoblast (which together comprise the bilaminar embryonic disc) relative to the amnion, the developing placenta (syncytiotrophoblast/cytotrophoblast), and the yolk sac. (Used with the permission of Virtual Human Embryo project [http://virtualhumanembryo.lsuhsc.edu].)

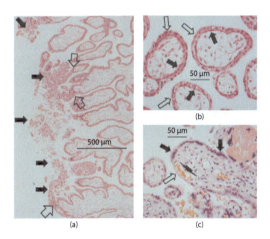

FIGURE 1.2 (a) Hematoxylin and eosin (H&E)–stained section from a placenta of approximately 5 weeks of gestation showing a region of extravillous trophoblasts from the cytotrophoblast shell (dark arrows). These extravillous trophoblasts emanate from the tips of anchoring villi (hollow arrows). (b) H&E-stained section from a placenta of approximately 6 weeks of gestation. The syncytiotrophoblast (hollow arrows) is relatively homogenous in terms of thickness and distribution of nuclei. On the fetal aspect of the syncytiotrophoblast is a more-or-less continuous layer of villous cytotrophoblasts (solid arrows) beneath which is the sparsely cellular mesenchymal core. (c) H&E-stained section from a placenta at term showing syncytiotrophoblast of varying thickness with clusters of nuclei (syncytial knots, solid arrows) and an adjacent vasculosyncytial membrane lacking nuclei (hollow arrow), allowing a fetal vessel (wide arrow) to come into proximity to the maternal blood in the intervillous space. The mesenchymal core of the villi is more densely cellular, and fetal vessels are more prominent than in early gestation villi (contrast with b).

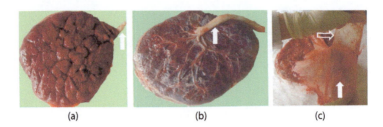

(a) (b) (c)

FIGURE 1.3 A normal term placenta viewed from (a) the maternal aspect, showing the villous surface (the umbilical cord, behind, is indicated by the arrow), and (b) the fetal aspect, with the membranes inverted to expose the pale-colored amnion that covers the entire fetal aspect of the placenta and the umbilical cord (arrow). (c) Although they fuse during gestation, even at term, the amnion (solid arrow) and smooth chorion with adherent decidua (hollow arrow) membranes can be readily separated.

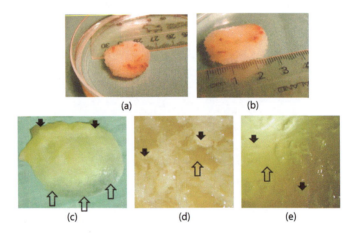

(a) (b)

(c) (d) (e)

FIGURE 1.4 The villous placenta initially forms as a sphere encompassing the embryo, but by 6–7 weeks of gestation the villi to the sides and rear of the implantation site begin regressing to form the smooth chorion. (a) A placenta of approximately 6 weeks of gestation, showing villi surrounding the embryo and (b) the same placenta as in (a) showing the surface that is obscured in (a). (c) A placenta of approximately 8 weeks of gestation showing a region where the villi are obviously regressing to form the smooth chorion at the rear and sides of the implantation site (hollow arrows), whereas the disc-like structure that will remain as the definitive placenta (chorion frondosum) can be seen at the top of the image, indicated by the solid arrows. (d and e) Images from a placenta of approximately 7 weeks of gestation showing the regressing villi (solid arrows) and the smooth chorion (hollow arrows).

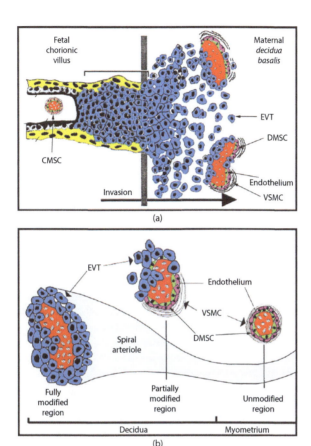

FIGURE 2.1 Uterine spiral arteriole modifications during trophoblast invasion. (a) Diagram of a uterine spiral arteriole during trophoblast invasion. The vascular chorionic mesenchymal stem cell (CMSC) niche in the chorionic villus is shown. Extravillous trophoblasts (EVTs) are shown migrating and invading into the maternal *decidua basalis*. Two partially modified spiral arterioles are shown. The decidual mesenchymal stem cells (DMSCs) in the vascular niche (shown in purple) as well as endothelial cells (green) and vascular smooth muscle cells (VSMCs) are replaced by EVTs. Invading and migrating EVTs are present in the lumen and occupy the vessel wall. (b) Spiral arteriole in various phases of modification or transformation by EVTs. In the unmodified spiral arteriole, DMSCs are present in the vascular niche, and endothelium and VSMCs are present. These cell types are replaced by invading and migrating EVTs in the fully modified arteriole. (Image [a] modified from Lindheimer, M.D. et al., 2009, *Chesley's Hypertensive Disorders in Pregnancy*, 4th ed, Academic Press, Amsterdam. Image [b] modified from Janatpour, M.J. et al., *Dev Genet* 25 (2), 146–57, 2009.)

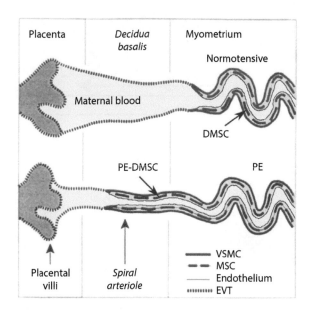

FIGURE 2.2 Illustration of the proposed increased presence of abnormal decidual mesenchymal stem cells in preeclampsia (PE) as a consequence of shallow spiral arteriole modification. In a normotensive pregnancy, modification or transformation of the spiral arterioles is deep and extends into the inner third of the myometrium. Decidual mesenchymal stem cells (DMSCs) in the vascular niche are normal. Shallow invasion of extravillous trophoblasts (EVTs) in PE results in more unremodeled spiral arteriole vessel walls present in the *decidua basalis* compared with normotensive pregnancy, and these vessels contain abnormal PE-DMSCs. (Image modified from VanWijk, M.J. et al., *Cardiovasc Res* 47 (1), 38–48, 2000.)

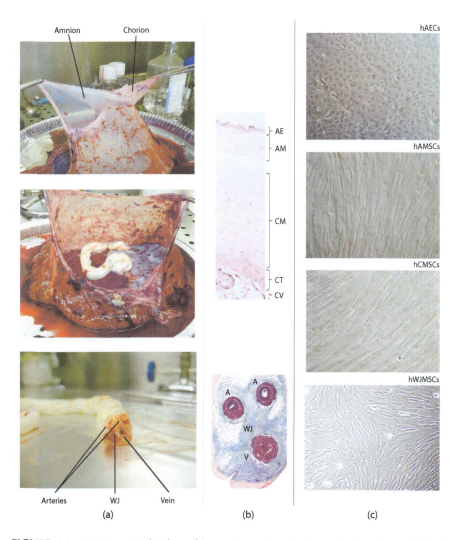

FIGURE 5.1 (a) Macroscopic view of human term placenta demonstrating the mechanical separation of amniotic membrane from chorion, and a cord section showing the umbilical vessels surrounded by Wharton's jelly. (b) Histology of fetal membranes and umbilical blood. A section of human fetal membranes (hematoxylin and eosin stain). (c) *In vitro* morphology of cells isolated from amniotic epithelium, amniotic mesoderm, chorionic mesoderm, and Wharton's jelly. AE, amniotic epithelium; AM, amniotic mesoderm; CM, chorionic mesoderm; CT, chorionic trophoblast; CV, chorionic villi. A section of human umbilical cord (Mallory stain). A, arteries; V, vein; WJ, Wharton's jelly; hAECs, human amniotic epithelial cells; hAMSCs, human amniotic membrane mesenchymal stromal cells; hCMSCs, human chorionic membrane mesenchymal stromal cells; hWJMSCs, human Wharton's jelly mesenchymal stromal cells.

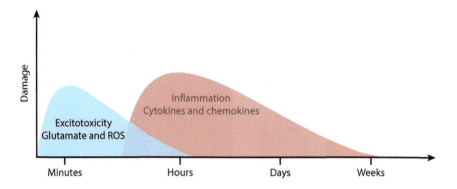

FIGURE 7.1 Time line of events after stroke. Glutamate excitotoxicity occurs within minutes after onset of stroke or traumatic brain injury, but wanes over hours after insult. Inflammation starts thereafter and can last from days to weeks, and it may even persist for months and years. ROS, reactive oxygen species.

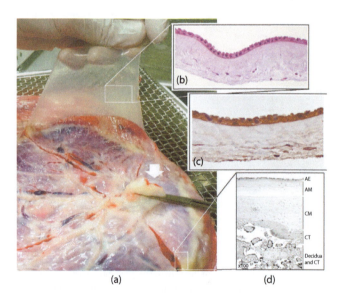

FIGURE 8.3 Human placenta and amnion tissue sections. (a) Picture of a portion of a full-term human placenta where cord tissue has been clumped cut off (arrow). The amnion tissue is manually stripped off from the fetal side. (b) Hematoxylin and eosin–stained of a section of amnion tissue. The amniotic layer is composed of a cuboidal epithelial layer and a deeper mesodermal layer where a few stromal cells are visible. (c) Staining of the amnion tissue with human leukocyte antigen-G antibody, showing positive expression in all the epithelial cells. (d) Cross-sectional representation of the human placenta. The amnion and chorion are fetal-derived membranes, whereas the decidual membrane is maternally derived. The amnion section is composed by amniotic epithelium (AE) and amniotic mesoderm (AM); the chorionic layer is comprised of a mesodermal layer (CM) and a trophoblast layer (CT).

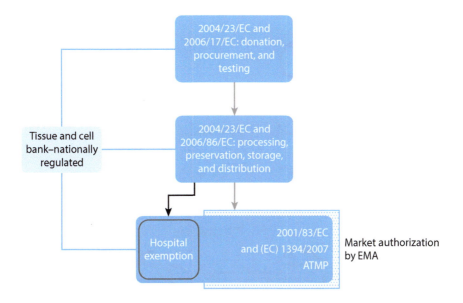

FIGURE 11.1 European Commission (EC) directives and its relation to tissues and advanced therapeutical medicinal products (ATMPs). Tissues are regulated by the EC and derivative national laws; ATMPs are regulated nationally by hospital exemption or transnationally by the European Medicines Agency.

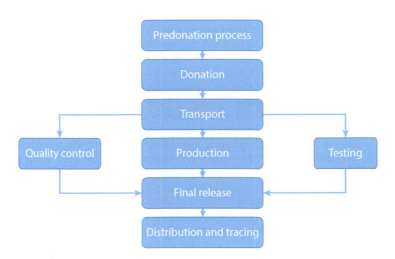

FIGURE 11.2 Design process for a tissue bank. Placental tissues and cells always require informed consent and start with the predonation steps. The general process is split into testing, production, and quality control. All information merges at the release point.

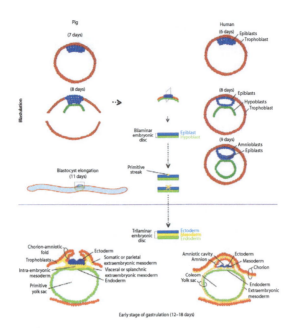

FIGURE 12.2 Comparative embryological development of the amnion in a human and an example of its development in a domestic animal. The images summarize the different kinetics and modalities of amnion formation: schematically in primates (including humans) amniotic cavity differentiates at the end of blastulation (bilaminar embryonic disc) through a process of epiblast cavitation, whereas in domestic animals it occurs later, during the early stages of gastrulation (trilaminar embryonic disc) with a folding modality. **Blastulation**. (Top) Images show the major events that occur in pig (left) and human (right) embryos during the blastulation phase when the embryonic disc is organized into a bilaminar structure (middle). **Pig embryo**. Day 7 after fertilization, embryo is a hatched blastocyst. Day 8, embryo becomes bilaminar by differentiating a flattened cell population facing the blastocyst cavity, referred to as the hypoblast (green cells), and the external multilayered group of cells, named to epiblast (blue cells). In addition, two other important events involve the trophectoderm: the trophoblast cells (red cells) covering the epiblast disappear, while the cells located at the edge of the embryonic disc start to proliferate, thus rapidly elongating the blastocyst. Day 11, the embryo appears as an elongated blastocyst ~4 cm in length with a new internal cavity delimitated by the hypoblast: the primitive yolk sac. **Human embryo**. Day 6, hatched blastocyst. Day 8, bilaminar embryonic disc covered by a persistent trophectoderm. Day 9, embryo presents two adnexa: the primitive yolk sac delimitated by the hypoblast and the amnion that differentiate through the cavitation of the multilayered epiblast. **Early stages of gastrulation**. (Bottom images) The embryonic disc becomes trilaminar through the differentiation of the mesoderm that is composed of cells that proliferate into the primitive streak and migrate (yellow cells) between the hypoblast (endoderm: green) and the epiblast (ectoderm: blue). **Pig embryo**. Days 12–18, cells of the mesoderm located at the edge of the embryonic disc, covered by the trophectoderm, form two lateral chorion-amniotic folds (top image). The folds gradually expand upward to fuse above the embryonic disc, thus delimitating the newly formed amniotic cavity (bottom image). Based on amnion formation in domestic animals, the inner epithelium of the amniotic cavity originates from the trophectoderm and from the ectoderm of the embryonic disc in the upper portion and in the floor.

7 Use of Placenta-Derived Cells in Neurological Disorders

Christopher Lawton*, Maya Elias*, Diego Lozano*, Hung Nguyen*, Stephanny Reyes*, Jaclyn Hoover*, and Cesario V. Borlongan

CONTENTS

Preface 129
7.1 Overview of Amnion Cells 130
7.2 Use of Amnion Cells for Human Disorders 130
    7.2.1 Neurological Disorders 131
    7.2.2 Cardiovascular Disease 131
    7.2.3 Pulmonary Disease 132
    7.2.4 Hepatic Disease 132
    7.2.5 Pancreatic Disease 132
    7.2.6 Muscular Disorders 133
    7.2.7 Cartilage Disorders 133
7.3 Specific Use of Amnion Cells for Targeting of Inflammation 133
7.4 Stroke As an Inflammation-Associated Disease and a Target Candidate
    for Amnion Cell Therapy 134
7.5 Concluding Remarks 136
    7.5.1 Future Directions: Stroke and Other CNS Disorders Similarly
        Characterized by Inflammation As Viable Disease Targets for
        Amnion Cell Therapy 136
    7.5.2 General Gating Items to Translate Cell Therapy from Lab to Clinic 137
References 138

PREFACE

This chapter describes the clinical potential of placenta-derived cells in regenerative medicine, specifically in neurological disorders. Human amniotic epithelial cells and human amniotic mesenchymal stromal cells express multipotency, low immunogenicity, and inflammation suppression capabilities. Transplantation of these amnion-derived cells in neurological disorders may promote re-epithelialization,

---

* These authors have contributed equally to this work.

130 Placenta: The Tree of Life

modulate differentiation and angiogenesis, and decrease inflammation. Preclinical studies have explored therapeutic interventions in neurological disorders, particularly in stroke and similar central nervous system (CNS) disorders characterized by inflammation. This chapter is primarily focused on use of human amniotic epithelial cells and human amniotic mesenchymal stromal cells in neurological disorders. Research is indicating that the inflammation-mediated mechanism of amnion cell therapy could be a promising clinical treatment for stroke. Amnion cell transplantation nonetheless warrants continued preclinical laboratory research, to promote safety and efficacy of treatment, for the eventual translation to clinical practice.

7.1 OVERVIEW OF AMNION CELLS

Cells from placental origins have shown exceptional promise for use in cell therapy for the treatment of many diseases. The amnion, the innermost fluid-filled sac surrounding the fetus during development, has increasingly become of interest in regenerative medicine. It is a transparently thin membrane consisting of two layers, an epithelial monolayer and a stromal layer (Yu et al. 2009). Because the amniotic sac is expelled with the placenta after parturition, there are no ethical grievances associated with its use, especially when considering stem cell therapy. There are two cell types that have been isolated from the amniotic membrane, namely, human amniotic epithelial cells (hAECs) and human amniotic mesenchymal stromal cells (hAMSCs) (Parolini et al. 2008). The hAECs are derived from the embryonic ectoderm and have been shown to express embryonic and pluripotent stem cell markers (Parolini et al. 2008). The hAMSCs are derived from the extraembryonic mesoderm and are phenotypically similar to mesenchymal cells from bone marrow (Kobayashi et al. 2008; Parolini et al. 2008; Portmann-Lanz et al. 2006; Soncini et al. 2007). Both hAECs and hAMSCs express glial and neuronal progenitor cell markers; *in vitro* differentiation to neuroglia phenotypes has been observed from hAMSCs (Sakuragawa et al. 1996, 2004). The low immunogenicity of amniotic cells makes them especially promising for clinical application. An immune response was not induced by allogeneic or xenogeneic transplantation of cells from the amniotic membrane (Bailo et al. 2004; Wolbank et al. 2007). Furthermore, hAECs and hAMSCs have been shown to express anti-inflammatory proteins (interleukin [IL]-1 receptor antagonist, IL-10, and tissue inhibitors of metalloproteinase 1–4), increasing their potential for use in therapy for autoimmune and degenerative disorders (Hao et al. 2000). In addition to being immunologically tolerated, there has been no tumorigenicity evident in human amnion cells transplanted into animal models (Yu et al. 2009). The absence of ethical barriers, lack of activation of an immune response, presence of inflammation suppression characteristics, and multipotency of amnion cells make them an ideal source for regenerative therapy.

7.2 USE OF AMNION CELLS FOR HUMAN DISORDERS

In recent years, cells from human amniotic membranes and amniotic fluid have demonstrated tremendous potential for clinical application in regenerative medicine. Experimental research studies revealed that amnion cell transplantation

Use of Placenta-Derived Cells in Neurological Disorders 131

promotes re-epithelialization, modulates differentiation and angiogenesis, and decreases inflammation, apoptosis, and fibrosis (Liu et al. 2008; Parolini et al. 2009, 2011; Uchida et al. 2000).

7.2.1 NEUROLOGICAL DISORDERS

Current applications of amnion cells in preclinical studies have explored therapeutic options in stroke, Parkinson's disease, and spinal cord injury (Yu et al. 2009). Sakuragawa et al. (1996) first suggested that hAECs may feature multipotentiality of neurons, astrocytes, and oligodendrocytes. In addition, in a stroke model of middle cerebral artery occlusion in rats, injection of hAECs into the dorsolateral striatum at day 1 after stroke demonstrated that hAECs may differentiate into astrocyte- and neuron-like cells (Liu et al. 2008). Moreover, hAECs derived from the human placenta can synthesize and release neurotransmitters and neurotrophic factors (Liu et al. 2008; Uchida et al. 2000). Neuroprotective and neuroregenerative properties of hAECs were further supported by preclinical studies in animal models; these models displayed the utility of these cells in CNS regeneration during acute phases of injury (Liu et al. 2008; Uchida et al. 2000). After transplantation of these cells into lesion areas of a contusion model of spinal cord injury in monkeys, robust regeneration of host axons and prevention of axotomized spinal cord neuron degeneration were observed (Sankar and Muthusamy 2003). Transplantation of hAECs into a rat model of Parkinson's disease with 6-hydroxydopamine lesions protected and enhanced the survival of dopaminergic neurons (Kakishita et al. 2003). The investigators observed an increase in nigral dopaminergic cell numbers, and the transplanted cells did not show any evidence of overgrowth after grafting. Thus, hAEC transplantation may counteract the loss of dopaminergic neurons in Parkinson's disease (Kakishita et al. 2003).

Although this chapter is primarily focused on use of hAECs/hAMSCs in neurological disorders, many disorders of non–CNS origin [e.g., cardiovascular (Moran 2014), pulmonary (Hopkins et al. 2006), hepatic (Ryan and Shawcross 2011), pancreatic (Pandey et al. 2013), muscular (Blake and Kröger 2000), and cartilage (Morawski et al. 2012)] may have subsequent neurological sequelae. Clinical applications of amnion cells in these disorders are described in detail below.

7.2.2 CARDIOVASCULAR DISEASE

Cells from the amniotic membrane may supply cardioprotective soluble factors. For example, one study investigated whether patching of human amniotic membrane cells could limit postischemic myocardial injury and cardiac dysfunction, with successful results of improved cardiac contractile function and ejection fraction in ischemic rat hearts (Cargnoni et al. 2009a). Reduction of myocardial scarring and prevention of myocardial thinning were also observed in transplantation of rat amniotic membrane–derived cells into a rat model of myocardial infarction (Fujimoto et al. 2009). However, the ability of placenta-derived cells to differentiate toward cardiomyocytes is still debated, due to unsuccessful findings in other studies (Parolini and Caruso 2011).

7.2.3 PULMONARY DISEASE

Previous efforts have demonstrated the utility of hAECs in displaying characteristics of type II pneumocytes. Moodley et al. (2010) tested the administration of hAECs as cellular therapy in a mouse model of bleomycin-induced lung inflammation and fibrosis. Primary hAECs were cultured in small airway growth medium, and these cells developed an alveolar epithelial phenotype and produced surfactant protein (Moodley et al. 2010). Moreover, lung collagen was significantly reduced by the treatment of cells, and there was also a marked reduction of several inflammatory and fibrotic cytokines (Moodley et al. 2010). A similar study was conducted of fetal membrane–derived cells on a mouse model of bleomycin-induced lung fibrosis (Cargnoni et al. 2009b). In this study, various administration routes were used: intraperitoneal or intratracheal for both xenogeneic and allogeneic cells, and intravenous for allogeneic cells (Cargnoni et al. 2009b). The authors observed decreased neutrophil infiltration and a significant reduction in the severity of bleomycin-induced lung fibrosis in those mice treated with placenta-derived cells (Cargnoni et al. 2009b).

7.2.4 HEPATIC DISEASE

One group studied the potential of cells derived from the human amniotic membrane to differentiate into hepatocytes (Tamagawa et al. 2007). They observed glycogen storage, a pivotal role of hepatocytes, by using periodic acid-Schiff staining of amniotic membrane cells (Tamagawa et al. 2007). Another study found that when cells from the rat amniotic membrane were exposed to conditions designed to induce hepatic differentiation, the cells exhibited low-density lipoprotein uptake from culture media (Marcus et al. 2008). In addition, amnion cells have clinical potential for use in biliary fibrosis. Biliary fibrosis is a chronic liver disease characterized by destruction of the bile ducts within the liver, and the only effective treatment is transplantation of the liver. However, one study demonstrated a significant reduction in the severity of bile duct ligation–induced fibrosis in rats treated with cells derived from the human amniotic membrane (Sant'Anna et al. 2011). In this study, a fragment of the human amniotic membrane was applied as a patch onto the liver surface after bile duct ligation (Sant'Anna et al. 2011). The treatment group showed only confined fibrosis at the portal/periportal area, with no signs of cirrhosis (Sant'Anna et al. 2011). The authors suggest that application of a patch of amniotic membrane cells, directly onto the liver surface, could thus protect against hepatic damage associated with fibrotic degeneration (Sant'Anna et al. 2011).

7.2.5 PANCREATIC DISEASE

Cell-based regeneration therapy with hAECs has been investigated within animal models of insulin-dependent diabetes mellitus and hyperglycemia. After several weeks of transplantation of hAECs into streptozotocin-induced diabetic mice, one study showed that hAECs could potentially differentiate into β-cells *in vivo* (Wei et al. 2003). Similar findings were found *in vitro* in a study by Hou et al. (2008), in which transplantation of hAECs into diabetic mice had reduced hyperglycemia

Use of Placenta-Derived Cells in Neurological Disorders 133

and maintained euglycemia for 30 days. Thus, differentiation of hAECs may be a valuable source for pancreatic β-cells in the treatment of insulin-dependent diabetes mellitus.

7.2.6 Muscular Disorders

Muscular dystrophy, an X-chromosome–linked recessive genetic disorder characterized by a dystrophin deficiency, currently has no effective treatment (Parolini and Caruso 2011). Kawamichi et al. (2010) conducted an *in vivo* implantation of placenta-derived cells into dystrophic muscles in a mouse model and successfully restored sarcolemmal expression of human dystrophin and laminin (Kawamichi et al. 2010). The authors suggested that this may be secondary to myogenic differentiation of the transplanted placenta-derived cells.

7.2.7 Cartilage Disorders

In cartilage repair, hAMSCs can be implanted with collagen scaffolds. One group of investigators found that positive expression of the cartilage marker genes collagen type II and aggrecan was detected after the induction of chondrogenesis with bone morphogenetic proteins in a rat model (Wei et al. 2009). Therefore, they suggested that hAMSCs implanted into bony defects have the potential to differentiate into chondrocytes, *in vitro* and *in vivo* (Wei et al. 2009).

7.3 SPECIFIC USE OF AMNION CELLS FOR TARGETING OF INFLAMMATION

Neurological disorders, such as stroke and traumatic brain injury (TBI), characterized by brain ischemia or injury from an impact to the head, leads to direct neural cell loss and necrotic cell death, as well as secondary cell death (Hernandez-Ontiveros et al. 2013). This primary injury is then followed by a secondary injury that consists of a cascade of cell death events that exacerbates the neurological damage (Tajiri et al. 2014). Inflammation has been recently implicated as primarily involved in this progressive brain damage and is responsible for major secondary cell death (Hernandez-Ontiveros et al. 2013).

Inflammation may be considered as a double-edged sword, conferring both protective and exacerbating effects after stroke and TBI. After the primary insult, an inflammatory response is triggered to repair the damaged cells and defend the injury site from pathogens (Schmidt et al. 2005). This inflammatory response is partly modulated by the immune system. Dead cells from the necrotic core and dying cells from the ischemic penumbra release signals that activate an immune response. These stressed cells release ATP, UTP, and high-mobility group protein B1, drawing immune cells into the brain parenchyma (Broughton et al. 2012). Up to this point, inflammation may be therapeutic, in that the inflammatory cells attempt to clear the dead cells. Neutrophils, monocytes, macrophages, T and B lymphocytes, and dendritic cells infiltrate across the blood–brain barrier and secrete pro-inflammatory molecules such as nitric oxide, reactive oxygen species, and matrix metalloproteinase, as well as proinflammatory cytokines, including interferon-γ, IL-6 and -17, and tumor necrosis factor (TNF) (Broughton et al. 2012). In turn, the expression of chemokines and cell adhesion molecules is upregulated, resulting in further mobilization of immune cells and glia into the intrathecal compartment (Lozano et al. 2015). This process leads to excessive inflammation, shifting from a protective nature into an exacerbating factor of neurodegeneration.

Amnion cells have proven to be useful for managing the inflammatory response. They have exhibited the capacity to directly suppress it by modulating the immune response (Yu et al. 2009). Although the underlying mechanisms are still not fully understood, it has been reported that amniotic cells secrete IL-10 and IL-6, both of which are potent anti-inflammatory cytokines (Manuelpillai et al. 2011). Furthermore, the delivery of amnion cells reportedly reduced the expression of inflammatory cytokines IL-1 and TNF-α (Manuelpillai et al. 2011). Studies have also shown their capacity to suppress T lymphocyte proliferation, secrete inhibitors of metalloproteinases, and reduce the infiltration of major histocompatibility complex class II antigen-presenting cells. Amnion cells are also capable of shifting the phenotype of activated microglia from proinflammatory M1 to anti-inflammatory M2 (Castillo-Melendez et al. 2013). All these events add up to an improved microenvironment and a limited inflammatory response (Broughton et al. 2012).

Considering the anti-inflammatory properties displayed by amnion cells, it is important to evaluate the outcome of variable administration timing. As previously stated, inflammation is a double-edged sword that is necessary to some extent for proper neural cell regeneration, but in excessive levels could lead to secondary damage and exacerbate the existing damage. A good example is the chemokine stromal cell–derived factor-1 that in the early stages of inflammation facilitates the migration of transplanted cells into the affected area (Broughton et al. 2012). If its effects are suppressed from the beginning, migration would be negatively affected, resulting in worse outcomes. Amnion cells should aid in the proinflammatory beneficial effects in the acute stages of brain insult, and also should facilitate the anti-inflammatory effects at the later stage of the disease.

7.4 STROKE AS AN INFLAMMATION-ASSOCIATED DISEASE AND A TARGET CANDIDATE FOR AMNION CELL THERAPY

Stroke is ranked fourth among the leading causes of death in the United States (Go et al. 2014). The most common type of stroke is ischemic stroke, approximately 87% of all strokes (Go et al. 2014). An ischemic stroke happens when there is a cerebral artery occlusion, preventing blood from reaching the brain or a portion of the brain. Currently, the U.S. Food and Drug Administration–approved treatment for ischemic stroke is limited to tissue plasminogen activator (tPA) and its derivatives. Despite the good outcomes, tPA treatment has a narrow therapeutic window and is not suitable for everyone due to potential serious hemorrhagic complications associated with the delayed tPA treatment, and various factors such as age, blood pressure, and glucose level (Demchuk and Bal 2012; Wardlaw et al. 2003). In addition to cell necrosis at the core, many studies have demonstrated that stroke also initiates a cascade of events that lead to a prolonged secondary inflammation

(Amantea et al. 2014; Konsman et al. 2007). In the ischemic penumbra, the region around the ischemic core, the level of oxygen is low but still sufficient for basic cell function. However, the low oxygen level concentration creates a stress on the cells in the penumbra. This stress causes the astrocytes and microglia to secrete proinflammation cytokines and chemokines. For example, TNF-α, IL-1β, IL-4, IL-6, and nitric oxide are molecules that promote inflammation (Konsman et al. 2007; Sairanen et al. 1997). The inflammation can last for days or weeks after the onset of stroke. Initially, this inflammation has some benefits to a certain extent; however, prolonged inflammation does more damage to the brain. Astrocytes and microglia not only secrete proinflammation cytokines and chemokines but also try to secrete anti-inflammation molecules and trophic factors to repair the damage. Some of the molecules are transforming growth factor beta, brain neurotrophic factor, erythropoietin, and vascular endothelial growth factor, to name a few. However, the self-repair mechanism is insufficient to repair the damage and recover function. Hence, intervention is needed and critical to improve the prognosis of stroke patients.

The inflammation occurs over a period of time, which is an ideal target for therapeutic intervention (Figure 7.1).

Recent studies have shown that stem cell therapy is a safe and effective treatment for stroke in the animal model. Increasing evidence shows that placenta-derived mesenchymal stem cells are a viable alternative source of mesenchymal stem cells for therapeutic treatment (Barlow et al. 2008). In addition, stem cells from human placenta have a high degree of plasticity (Ilancheran et al. 2009). One of the stem cells' modes of action is modulating the inflammation (Lei et al. 2014; van Velthoven et al. 2014; Wan et al. 2014). Placenta-derived mesenchymal stem cells have been shown to reduce the proinflammation cytokines and chemokines, increase anti-inflammation molecules, and increase neurotrophic factors. These multiple inflammation-relevant properties contribute to the reduction of deficits in stroke animal models.

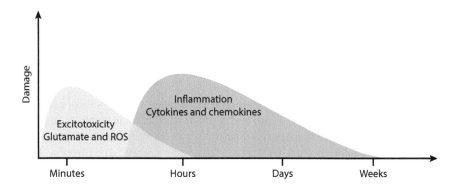

FIGURE 7.1 (**See color insert.**) Time line of events after stroke. Glutamate excitotoxicity occurs within minutes after onset of stroke or traumatic brain injury, but wanes over hours after insult. Inflammation starts thereafter and can last from days to weeks, and it may even persist for months and years. ROS, reactive oxygen species.

Although the exact therapeutic mechanism of stem cells, including those derived from the amnion, for treating stroke remains not fully understood, accumulating research evidence points to the key role of inflammation signaling pathways. Different lines of preclinical investigations suggest that amnion cell therapy is a promising method for treating stroke, likely via an inflammation-mediated mechanism.

In an *in vitro* stroke model study, Kaneko et al. (2011) demonstrated that hAECs exerted a neuroprotective effect through melatonin receptor type 1A, but not type 1B (Kaneko et al. 2011). The study also showed that the combination of hAECs and melatonin treatment increased the cells against stroke insults (Kaneko et al. 2011). In a separate study, Yu et al. (2013) showed the neuroprotection properties of dog placenta cells (DPCs) via heat shock protein 27 (Hsp27) in treating both *in vivo* and *in vitro* stroke models. The study suggested DPCs targets antioxidative stress Hsp27 through paracrine and autocrine mechanisms to afford the neuroprotective properties observed (Yu et al. 2013). The role of Hsp27 in neuroprotection has been demonstrated in many different studies (An et al. 2008; Stetler et al. 2008; Teramoto et al. 2013). In summary, both studies showed the neuroprotective effect of amnion cells in both *in vitro* and *in vivo* stroke models. The therapeutic effect is most likely due to the regulation of various inflammation control mechanisms of amnion cells.

7.5 CONCLUDING REMARKS

7.5.1 Future Directions: Stroke and Other CNS Disorders Similarly Characterized by Inflammation as Viable Disease Targets for Amnion Cell Therapy

The preceding sections provide ample scientific evidence that cells derived from human amniotic membranes and amniotic fluid demonstrate tremendous potential for clinical application in regenerative medicine. The chief areas of clinical applications being explored today include treatment for neurological disorders, and other non–CNS disorders, such as cardiovascular disease, pulmonary disease, hepatic disease, pancreatic disease, muscular disorders, and cartilage disorders that may nonetheless present with neurological complications. The specific use of amnion cells for targeting inflammation is of particular interest given the nature of stroke as an inflammation-associated disease and lack of treatment options currently available for the disease.

Although stroke is a candidate for amnion cell therapy, other CNS disorders characterized by inflammation may also be key targets for this intervention. Infectious conditions, such as encephalitis and meningitis, and progressive disorders such as multiple sclerosis, Alzheimer's disease, Parkinson's disease, amyotrophic lateral sclerosis, Creutzfeldt–Jakob disease, epilepsy, and TBI may also benefit from single or repeated treatments of amnion transplantation therapy. The scarcity of effective pharmacological treatments for stroke and other inflammatory disorders has shifted the interest of the scientific community in stem cell therapy as a potent therapeutic approach given the extensive data from animal models exemplifying promising neurological outcomes. However, as with any other cell source, injection of amnion cells still warrants rigorous and carefully designed preclinical translational research to proceed to the clinic. Critical translational research gating items include optimization of a clinically feasible routine method of delivery for stroke patients due to the fact that injections, such as intracerebral, would require routine access to suitable imaging facilities and surgical expertise (Broughton et al. 2012). Therefore, it is critical for future studies to focus on a less invasive and more clinically manageable delivery route of stem cells, such as intravenous injection (Broughton et al. 2012). In addition, safe and effective cell doses and timing of delivery should be standardized and validated in clinically relevant animal models. Equally important, as noted above, is exploiting the proinflammatory and anti-inflammatory effects of amnion cells to cater to the dynamically progressive cell death associated with stroke and relevant disorders. The endeavor of implementing large-scale stem cell therapy in the clinic remains, requiring transplant regimen optimization and investigations into mechanisms of action underlying this treatment.

7.5.2 GENERAL GATING ITEMS TO TRANSLATE CELL THERAPY FROM LAB TO CLINIC

Many clinical trials fail despite positive laboratory findings. Translating cell therapies to the clinic is no exception. To find cell therapies that are effective and safe, communication between academics, industry leaders, and government is needed. Collaboration between these groups can help researchers develop guidelines that will increase the likelihood of translating effective treatments from the bench to the clinic. For example, the advancement of stem cell therapy for stroke from the laboratory to the clinic is now guided by a set of recommendations called Stem Cell Therapeutics as an Emerging Paradigm for Stroke (STEPS) (Diamandis and Borlongan 2015). The STEPS meeting brought together leaders in stem cell research, industry, and regulatory agencies to create a common set of standards and to provide direction for future research (Wechsler et al. 2009). Although STEPS' focus was on stroke, this initiative developed guidelines that can be extended to related fields and similar initiatives are currently underway to enhance the entry of novel therapeutics in Parkinson's disease, Alzheimer's disease, epilepsy, and in several other neurodegenerative diseases (Diamandis and Borlongan 2015). The STEPS guidelines focused on the appropriate species, type of model, outcome measures and treatment protocols, imaging of cell tracking and host response, the requirements for safety indices, and the investigation of mechanisms of action. These points were determined to be crucial elements of preclinical testing that must be addressed for the effective and accurate translation of primary research into clinical practice. Addressing these key aspects of preclinical research is essential in developing placenta-derived cells as an effective and safe option for the treatment of neurological disorders. In particular, harnessing an optimal transplant regimen by targeting the inflammation signaling pathway may prove of high therapeutic value to the advancement of amnion cell therapy from the laboratory to the clinic.

REFERENCES

Amantea, D., C. Tassorelli, F. Petrelli, et al. 2014. Understanding the multifaceted role of inflammatory mediators in ischemic stroke. *Curr Med Chem* 21 (18): 2098–117.

An, J. J., Y. P. Lee, S. Y. Kim, et al. 2008. Transduced human PEP-1-heat shock protein 27 efficiently protects against brain ischemic insult. *FEBS J* 275 (6): 1296–308. doi: 10.1111/j.1742-4658.2008.06291.x.

Bailo, M., M. Soncini, E. Vertua, et al. 2004. Engraftment potential of human amnion and chorion cells derived from term placenta. *Transplantation* 78 (10): 1439–48.

Barlow, S., G. Brooke, K. Chatterjee, et al. 2008. Comparison of human placenta- and bone marrow-derived multipotent mesenchymal stem cells. *Stem Cells Dev* 17 (6): 1095–107. doi: 10.1089/scd.2007.0154.

Blake, D. J., and S. Kröger. 2000. The neurobiology of Duchenne muscular dystrophy: Learning lessons from muscle? *Trends Neurosci* 23 (3): 92–9. doi: 10.1016/S0166-2236(99)01510-6.

Broughton, B. R., R. Lim, T. V. Arumugam, et al. 2012. Post-stroke inflammation and the potential efficacy of novel stem cell therapies: Focus on amnion epithelial cells. *Front Cell Neurosci* 6: 66. doi: 10.3389/fncel.2012.00066.

Cargnoni, A., M. Di Marcello, M. Campagnol, et al. 2009a. Amniotic membrane patching promotes ischemic rat heart repair. *Cell Transplant* 18 (10): 1147–59. doi: 10.3727/096 368909x12483162196764.

Cargnoni, A., L. Gibelli, A. Tosini, et al. 2009b. Transplantation of allogeneic and xenogeneic placenta-derived cells reduces bleomycin-induced lung fibrosis. *Cell Transplant* 18 (4): 405–22. doi: 10.3727/096368909788809857.

Castillo-Melendez, M., T. Yawno, G. Jenkin, and S. L. Miller. 2013. Stem cell therapy to protect and repair the developing brain: A review of mechanisms of action of cord blood and amnion epithelial derived cells. *Front Neurosci* 7: 194. doi: 10.3389/fnins.2013.00194.

Demchuk, A. M., and S. Bal. 2012. Thrombolytic therapy for acute ischaemic stroke: What can we do to improve outcomes? *Drugs* 72 (14): 1833–45. doi: 10.2165/11635740-000000000-00000.

Diamandis, T., and C. V. Borlongan. 2015. One, two, three steps toward cell therapy for stroke. *Stroke* 46 (2): 588–91. doi: 10.1161/strokeaha.114.007105.

Fujimoto, K. L., T. Miki, L. J. Liu, et al. 2009. Naive rat amnion-derived cell transplantation improved left ventricular function and reduced myocardial scar of postinfarcted heart. *Cell Transplant* 18 (4): 477–86. doi: 10.3727/096368909788809785.

Go, A. S., D. Mozaffarian, V. L. Roger, et al. 2014. Heart disease and stroke statistics—2014 update: A report from the American Heart Association. *Circulation* 129 (3): e28–e292. doi: 10.1161/01.cir.0000441139.02102.80.

Hao, Y., D. H. Ma, D. G. Hwang, W. S. Kim, and F. Zhang. 2000. Identification of antiangiogenic and antiinflammatory proteins in human amniotic membrane. *Cornea* 19 (3): 348–52.

Hernandez-Ontiveros, D. G., N. Tajiri, S. Acosta, et al. 2013. Microglia activation as a biomarker for traumatic brain injury. *Front Neurol* 4: 30. doi: 10.3389/fneur.2013.00030.

Hopkins, R. O., S. D. Gale, and L. K. Weaver. 2006. Brain atrophy and cognitive impairment in survivors of acute respiratory distress syndrome. *Brain Injury* 20 (3): 263–71.

Hou, Y., Q. Huang, T. Liu, and L. Guo. 2008. Human amnion epithelial cells can be induced to differentiate into functional insulin-producing cells. *Acta Biochim Biophys Sin (Shanghai)* 40 (9): 830–9.

Ilancheran, S., Y. Moodley, and U. Manuelpillai. 2009. Human fetal membranes: A source of stem cells for tissue regeneration and repair? *Placenta* 30 (1): 2–10. doi: 10.1016/j.placenta.2008.09.009.

Kakishita, K., N. Nakao, N. Sakuragawa, and T. Itakura. 2003. Implantation of human amniotic epithelial cells prevents the degeneration of nigral dopamine neurons in rats with 6-hydroxydopamine lesions. *Brain Res* 980 (1): 48–56.

Kaneko, Y., T. Hayashi, S. Yu, et al. 2011. Human amniotic epithelial cells express melatonin receptor MT1, but not melatonin receptor MT2: A new perspective to neuroprotection. *J Pineal Res* 50 (3): 272–80. doi: 10.1111/j.1600-079X.2010.00837.x.

Kawamichi, Y., C. H. Cui, M. Toyoda, et al. 2010. Cells of extraembryonic mesodermal origin confer human dystrophin in the mdx model of Duchenne muscular dystrophy. *J Cell Physiol* 223 (3): 695–702. doi: 10.1002/jcp.22076.

Kobayashi, M., T. Yakuwa, K. Sasaki, et al. 2008. Multilineage potential of side population cells from human amnion mesenchymal layer. *Cell Transplant* 17 (3): 291–301.

Konsman, J. P., B. Drukarch, and A. M. Van Dam. 2007. (Peri)vascular production and action of pro-inflammatory cytokines in brain pathology. *Clin Sci (Lond)* 112 (1): 1–25. doi: 10.1042/cs20060043.

Hao L, Zou Z, Tian H, et al. 2014. Stem cell-based therapies for ischemic stroke. *BioMed Res Int* 2014: 468748. doi: 10.1155/2014/468748.

Liu, T., J. Wu, Q. Huang, et al. 2008. Human amniotic epithelial cells ameliorate behavioral dysfunction and reduce infarct size in the rat middle cerebral artery occlusion model. *Shock* 29 (5): 603–11. doi: 10.1097/SHK.0b013e318157e845.

Lozano, D., G. S. Gonzales-Portillo, S. Acosta, et al. 2015. Neuroinflammatory responses to traumatic brain injury: Etiology, clinical consequences, and therapeutic opportunities. *Neuropsychiatr Dis Treat* 11: 97–106. doi: 10.2147/ndt.s65815.

Manuelpillai, U., Y. Moodley, C. V. Borlongan, and O. Parolini. 2011. Amniotic membrane and amniotic cells: Potential therapeutic tools to combat tissue inflammation and fibrosis? *Placenta* 32 (Suppl 4): S320–5. doi: 10.1016/j.placenta.2011.04.010.

Marcus, A. J., T. M. Coyne, J. Rauch, D. Woodbury, and I. B. Black. 2008. Isolation, characterization, and differentiation of stem cells derived from the rat amniotic membrane. *Differentiation* 76 (2): 130–44. doi: 10.1111/j.1432-0436.2007.00194.x.

Moodley, Y., S. Ilancheran, C. Samuel, et al. 2010. Human amnion epithelial cell transplantation abrogates lung fibrosis and augments repair. *Am J Respir Crit Care Med* 182 (5): 643–51. doi: 10.1164/rccm.201001-0014OC.

Moran, J. F. 2014. Neurologic complications of cardiomyopathies and other myocardial disorders. In B. José and M. F. José (Eds.), *Handbook of Clinical Neurology*, pp. 111–28. Elsevier, Amsterdam, The Netherlands.

Morawski, M., G. Bruckner, T. Arendt, and R. T. Matthews. 2012. Aggrecan: Beyond cartilage and into the brain. *Int J Biochem Cell Biol* 44 (5): 690–3. doi: 10.1016/j.biocel.2012.01.010.

Pandey, M. K., P. Mittra, J. Doneria, and P. K. Maheshwari. 2013. Neurological complications in diabetic ketoacidosis—Before and after insulin therapy. *Int J Med Sci Public Health* 2 (1): 88–93.

Parolini, O., F. Alviano, G. P. Bagnara, et al. 2008. Concise review: Isolation and characterization of cells from human term placenta: Outcome of the first international Workshop on Placenta Derived Stem Cells. *Stem Cells* 26 (2): 300–11. doi: 10.1634/stemcells.2007-0594.

Parolini, O., F. Alviano, A. G. Betz, et al. 2011. Meeting report of the first conference of the International Placenta Stem Cell Society (IPLASS). *Placenta* 32 (Suppl 4): S285–90. doi: 10.1016/j.placenta.2011.04.017.

Parolini, O., and M. Caruso. 2011. Review: Preclinical studies on placenta-derived cells and amniotic membrane: An update. *Placenta* 32 (Suppl 2): S186–95. doi: 10.1016/j.placenta.2010.12.016.

Parolini, O., M. Soncini, M. Evangelista, and D. Schmidt. 2009. Amniotic membrane and amniotic fluid-derived cells: Potential tools for regenerative medicine? *Regen Med* 4 (2): 275–91. doi: 10.2217/17460751.4.2.275.

Portmann-Lanz, C. B., A. Schoeberlein, A. Huber, et al. 2006. Placental mesenchymal stem cells as potential autologous graft for pre- and perinatal neuroregeneration. *Am J Obstet Gynecol* 194 (3): 664–73. doi: 10.1016/j.ajog.2006.01.101.

Ryan, J. M., and D. L. Shawcross. 2011. Management problems in liver disease: Hepatic encephalopathy. *Medicine* 39: 617–20. doi: 10.1016/j.mpmed.2011.07.008.

Sairanen, T. R., P. J. Lindsberg, M. Brenner, and A. L. Siren. 1997. Global forebrain ischemia results in differential cellular expression of interleukin-1beta (IL-1beta) and its receptor at mRNA and protein level. *J Cereb Blood Flow Metab* 17 (10): 1107–20. doi: 10.1097/00004647-199710000-00013.

Sakuragawa, N., K. Kakinuma, A. Kikuchi, et al. 2004. Human amnion mesenchyme cells express phenotypes of neuroglial progenitor cells. *J Neurosci Res* 78 (2): 208–14. doi: 10.1002/jnr.20257.

Sakuragawa, N., R. Thangavel, M. Mizuguchi, M. Hirasawa, and I. Kamo. 1996. Expression of markers for both neuronal and glial cells in human amniotic epithelial cells. *Neurosci Lett* 209 (1): 9–12.

Sankar, V., and R. Muthusamy. 2003. Role of human amniotic epithelial cell transplantation in spinal cord injury repair research. *Neuroscience* 118 (1): 11–17.

Sant'Anna, L. B., A. Cargnoni, L. Ressel, G. Vanosi, and O. Parolini. 2011. Amniotic membrane application reduces liver fibrosis in a bile duct ligation rat model. *Cell Transplant* 20 (3): 441–53. doi: 10.3727/096368910x522252.

Schmidt, O. I., C. E. Heyde, W. Ertel, and P. F. Stahel. 2005. Closed head injury—An inflammatory disease? *Brain Res Brain Res Rev* 48 (2): 388–99. doi: 10.1016/j.brainresrev.2004.12.028.

Soncini, M., E. Vertua, L. Gibelli, et al. 2007. Isolation and characterization of mesenchymal cells from human fetal membranes. *J Tissue Eng Regen Med* 1 (4): 296–305. doi: 10.1002/term.40.

Stetler, R. A., G. Cao, Y. Gao, et al. 2008. Hsp27 protects against ischemic brain injury via attenuation of a novel stress-response cascade upstream of mitochondrial cell death signaling. *J Neurosci* 28 (49): 13038–55. doi: 10.1523/jneurosci.4407-08.2008.

Tajiri, N., S. A. Acosta, M. Shahaduzzaman, et al. 2014. Intravenous transplants of human adipose-derived stem cell protect the brain from traumatic brain injury-induced neurodegeneration and motor and cognitive impairments: Cell graft biodistribution and soluble factors in young and aged rats. *J Neurosci* 34 (1): 313–26. doi: 10.1523/jneurosci.2425-13.2014.

Tamagawa, T., S. Oi, I. Ishiwata, H. Ishikawa, and Y. Nakamura. 2007. Differentiation of mesenchymal cells derived from human amniotic membranes into hepatocyte-like cells in vitro. *Hum Cell* 20 (3): 77–84. doi: 10.1111/j.1749-0774.2007.00032.x.

Teramoto, S., H. Shimura, R. Tanaka, et al. 2013. Human-derived physiological heat shock protein 27 complex protects brain after focal cerebral ischemia in mice. *PLoS One* 8 (6): e66001. doi: 10.1371/journal.pone.0066001.

Uchida, S., Y. Inanaga, M. Kobayashi, et al. 2000. Neurotrophic function of conditioned medium from human amniotic epithelial cells. *J Neurosci Res* 62 (4): 585–90.

van Velthoven, C. T. J., F. Gonzalez, Z. S. Vexler, and D. M. Ferriero. 2014. Stem cells for neonatal stroke- the future is here. *Front Cell Neurosci* 8: 207. doi: 10.3389/fncel.2014.00207.

Wan, H., F. Li, L. Zhu, et al. 2014. Update on therapeutic mechanism for bone marrow stromal cells in ischemic stroke. *J Mol Neurosci* 52 (2): 177–85. doi: 10.1007/s12031-013-0119-0.

Wardlaw, J. M., G. Zoppo, T. Yamaguchi, and E. Berge. 2003. Thrombolysis for acute ischaemic stroke. *Cochrane Database Syst Rev* (3): CD000213. doi: 10.1002/14651858.cd000213.

Wechsler, L., D. Steindler, C. Borlongan, et al. 2009. Stem Cell Therapies as an Emerging Paradigm in Stroke (STEPS): Bridging basic and clinical science for cellular and neurogenic factor therapy in treating stroke. *Stroke* 40 (2): 510–15. doi: 10.1161/strokeaha.108.526863.

Wei, J. P., M. Nawata, S. Wakitani, et al. 2009. Human amniotic mesenchymal cells differentiate into chondrocytes. *Cloning Stem Cells* 11 (1): 19–26. doi: 10.1089/clo.2008.0027.

Wei, J. P., T. S. Zhang, S. Kawa, et al. 2003. Human amnion-isolated cells normalize blood glucose in streptozotocin-induced diabetic mice. *Cell Transplant* 12 (5): 545–52.

Wolbank, S., A. Peterbauer, M. Fahrner, et al. 2007. Dose-dependent immunomodulatory effect of human stem cells from amniotic membrane: A comparison with human mesenchymal stem cells from adipose tissue. *Tissue Eng* 13 (6): 1173–83. doi: 10.1089/ten.2006.0313.

Yu, S., N. Tajiri, N. Franzese, et al. 2013. Stem cell-like dog placenta cells afford neuroprotection against ischemic stroke model via heat shock protein upregulation. *PLoS One* 8 (9): e76329. doi: 10.1371/journal.pone.0076329.

Yu, S. J., M. Soncini, Y. Kaneko, et al. 2009. Amnion: A potent graft source for cell therapy in stroke. *Cell Transplant* 18 (2): 111–18.

8 Use of Amnion Epithelial Cells in Metabolic Liver Disorders

Roberto Gramignoli, Fabio Marongiu, and Stephen C. Strom

CONTENTS

Preface 143
8.1 Background 144
8.2 Cell-Based Therapy 145
    8.2.1 Hepatocyte Transplantation 146
    8.2.2 Stem Cell–Based Therapies 148
8.3 Amnion Epithelial Cells 148
8.4 *In Vitro* Hepatic Differentiation 151
8.5 Preclinical Models 152
8.6 Clinical Program 154
8.7 Concluding Remarks 155
References 156

PREFACE

The concept of regenerative medicine implies that the clinician works with the innate healing and regenerative process of the body. Perhaps more than with any other organ, the liver offers the greatest opportunity for regenerative medicine. Unlike other tissues, the liver has a massive capacity to regenerate after physical or chemical insult. Hepatic function is critical to survival, such that its loss often has lethal consequences.

For many years, the only option for treating chronic liver disease or metabolic defects in liver function was liver transplantation. Cellular therapy is a rapidly emerging field that holds great promise for bringing new therapeutic alternatives for life-threating and chronic diseases. More than 20 years ago, a cell-based therapeutic option for liver diseases was clinically tested: hepatocyte transplantation (HTx). Human hepatocyte infusion has been used to "bridge" or to keep the patient alive long enough to receive an organ; to regenerate liver function in acute liver failure; and primarily, to provide missing enzyme activity or function to patients with metabolic liver disease. However, a severe shortage of useful liver tissues limits a wider application of this cellular therapy, and new sources of cells are required to provide transplants to all of the patients that might benefit from this therapy.

Recent evidence reports human amnion epithelial cells (hAECs), isolated from term human placenta, have surface markers and gene expression profiles similar to those of embryonic stem cells, and more importantly, they are able to mature into functional hepatocyte and correct metabolic liver diseases.

The efficacy of hAEC transplants was assessed in two animal models of human metabolic liver diseases: an intermediate form of maple syrup urine disease (MSUD) and the phenylalanine hydroxylase deficiency model for phenylketonuria (PKU). Both conditions are 100% lethal to animals if untreated. After hAEC transplantation, survival increases to 80–100% and missing enzyme activity in the liver increases by 5–7%. Several serum and brain amino acid levels are normalized and brain neurotransmitter levels are corrected. Human DNA in the mouse liver is 5%, indicating the continued presence of human cells.

These successful results have motivated translation to cGMP isolation and banking of hAECs at the Karolinska Institutet for cellular therapies, and a phase II clinical trial is starting.

8.1 BACKGROUND

Inborn errors of metabolism usually arise from a single enzymatic defect. Abnormal storage or use of proteins, carbohydrates, lipids, or other nutrients account for a large part of the metabolic diseases. Inborn errors of metabolism can be categorized as follows:

- Disease generated by deficient hepatocytes resulting in liver-specific pathology (i.e., Crigler–Najjar syndrome)
- Disease generated by deficient hepatocytes leading to extrahepatic complications (i.e., alpha-1 antitrypsin [A1AT] deficiency)
- Disease generated extrahepatically, whose pathological manifestation results within the liver (i.e., hemochromatosis)

Medical treatment for the first two liver-related syndromes requires major invasive intervention, such as implantation of a quality graft in substitution of part of, if not the whole, native liver. During the past 50 years, the progress made in organ transplantation has revolutionized the treatment of a wide spectrum of diseases, including inborn errors of metabolism. Although liver transplantation is potentially a treatment option, patients affected by inborn errors of metabolism are commonly pediatric, and in many cases newborn babies; these babies are too small for such an invasive surgical procedure or liver graft reduction approach. In addition, the profound shortage of transplantable livers is highly affecting and limiting the whole organ replacement approach, both in pediatric and adult patients. The number of patients waiting for a liver transplant largely surpasses the supply of organs.

In inborn error deficiency, the correction can be accomplished by replacing the missing function or enzymatic activity, or part of it, rendering the patient symptom free or at least converting a life-threatening disease into a milder and more easily manageable form. Congenital liver diseases often manifest soon after birth and can range from severely debilitating to lethal, as commonly observed in patients affected by urea cycle disorders, whose ammonia-mediated neurotoxicity experienced immediately after birth can cause irreversible brain damage and lead to death. Despite routine liver transplantation and supporting medical therapies, thousands of patients currently on the waiting list (approximately 40% of listed patients each year do not receive a liver, resulting in a significant mortality; United Network for Organ Sharing source).

Liver transplantation is frequently associated with long-time recovery and postsurgical complications, such as infections, renal failure, and acute rejection, thereby significantly contributing to patient mortality and resulting in complex and expensive care. In addition, allogenic liver tissue transplantation requires life-long immunosuppression therapy. Recent estimates have calculated liver-related diseases result in costs of almost $10 billion per year in the United States alone (Everhart and Ruhl 2009).

Therefore, despite current medical and surgical therapies, there is an unmet need for more refined and widely available regenerative medicine strategies.

8.2 CELL-BASED THERAPY

A powerful shift in medical intervention may be produced by regenerative medicine. Regenerative medicine has already been shown to dramatically transform health-care, especially in hepatology, considering the liver's innate regenerative capacity. In particular, cell-based therapies are rapidly moving into clinical application and attempting to maximize the potential for repair, regeneration, or both. Cellular transplantation avoids the mortality and morbidity associated with the myriad of potential technical complications commonly described in vascularized organ transplants and in the procedure itself. Cells can be infused through minimally invasive procedures, compared to orthotopic organ transplantation. Preliminary clinical proof-of-concept for cell transplantation already exists. The routes of administration have been either local or systemic. The infusion of single-cell suspension into the circulatory system has the advantage of potentially being performed even on severely ill patients with relatively low risk. Once infused, the transplanted cells migrate and integrate into the host tissue, which in a large part of the documented cases colocalized with the liver. Nevertheless, systemic administration requires accurate evaluation to avoid entrapment of infused cells in the lung or microvasculature, resulting in embolism or thrombotic events. Cellular therapies have several additional advantages compared to the whole organ replacement: multiple infusions of cells are not only a possibility but also common practice and proposed as necessary in multiple cases; cell batches for transplant can be banked and cryopreserved for almost instant availability, avoiding concerns on procedural timing. Another significant benefit is that native organ is not removed. As proved in liver-based approaches, transplanted cells would not need to provide complete liver support, allowing temporal support and enhancement in organ regeneration in patients with fulminant liver failure. This support is of particular relevance also for patients with metabolic liver diseases, in which regeneration does not play a main role in disease correction. For example, a patient affected by OTC deficiency, a urea cycle defect, has a mutation in the enzyme ornithine transcarbamy-lase (OTC) that causes ammonia to accumulate in the blood, while the affected hepa-tocytes maintain the capacity to perform all other necessary liver functions. In the unlikely scenario of donor cellular graft failure or inefficient cell function, the native liver is retained, allowing the patient to simply return to his pretransplant condition.

8.2.1 Hepatocyte Transplantation

More than 40 years of laboratory research using preclinical models, in addition to the past 20 years of clinical procedures, have shown that diseases that are often corrected by whole liver replacement can be successfully treated by the infusion of isolated liver parenchymal cells (hepatocytes). Conceptually, liver cell transplantation (commonly referred as HTx) could provide rapid support for the failing liver by providing metabolism of toxins and secretion of proteins such as clotting factors and albumin to stabilize hemodynamic parameters. Metabolic defects in bilirubin metabolism, familial intrahepatic cholestasis, albumin secretion, copper excretion, and tyrosinemia have been corrected by HTx so far (Hansel et al. 2014b).

Since the early 1990s, when the first human hepatocyte infusions were performed (Mito et al. 1992; Strom et al. 1997b), the proof of principle for cell-based treatment as an option for patients with acquired or inherited liver diseases has been clearly established. Hepatocyte transplantation has been a useful treatment for congenital liver diseases and acute or chronic liver failure, and it still holds great promise as an alternative approach to whole organ replacement, or for bridging critical patients to organ transplantation (Allen et al. 2008; Ambrosino et al. 2005; Beck et al. 2012; Bilir et al. 2000; Darwish et al. 2004; Dhawan et al. 2004, 2005; Fisher and Strom 2006; Fox et al. 1998; Gridelli et al. 2012; Grossman et al. 1994; Habibullah et al. 1994; Lee et al. 2007; Li et al. 2012; Mito et al. 1992; Mitry et al. 2004; Muraca et al. 2002; Puppi et al. 2008; Ribes-Koninckx et al. 2012; Schneider et al. 2006; Sokal et al. 2003; Soriano et al. 2001; Stephenne et al. 2006, 2012; Strom et al. 1997a, 1997b, 1999). The reason sustaining this assumption stands on the basic fact that transplanted cells need to provide exclusively the missing function of the failing liver, while the host parenchyma is still functional in all the other hepatic functions. Therefore, identifying liver diseases that can be managed or cured by cellular transplant may help reserve livers for patients that can only be treated by whole organ transplantation, reducing the number on the waiting list and increasing the number of transplanted patients. Several of the most common liver diseases, regularly treated by liver transplantation, have also been tentatively treated by HTx (Table 8.1).

Nonetheless, despite documented improvements in the post-transplant condition of most of treated patients (Jorns et al. 2012; Puppi et al. 2012; Soltys et al. 2010), only a few centers have been able to successfully provide HTx to patients. Eighteen different groups worldwide reported performing HTx over the past 20 years (Figure 8.1). Only six groups are reported to have remained active over the past 5 years: five centers in Europe (London, UK; Brussels, Belgium; Valencia, Spain; Hannover, Germany; and Stockholm, Sweden) and one center, Children's Hospital, in the United States (Pittsburgh, PA).

The roadblocks to a widespread application of liver cell infusion can certainly be found in the current limitations to this cellular technology (Gramignoli et al. 2015). Results from human clinical trials have been inconsistent, perhaps due to the wide range in quantity and phenotype of cells transplanted (Gramignoli et al. 2014). Normally, the source of hepatocytes for clinical applications is livers unsuitable for organ transplant or livers explanted from patients with metabolic deficiencies. However, these cells are limited in number, variable in quality, and unable to proliferate *in vitro* (Bilir et al. 2000; Dhawan et al. 2010; Fox et al. 1998;

TABLE 8.1
List of Human Liver Diseases Treated by Liver Transplant and HTx, and Potentially Treatable with hAEC Transplants

Liver Disease	Liver Tx	HTx	hAEC Tx
Acute intermittent porphyria	☑	☐	☑
Alpha-1-antitrypsin	☑	☑	☑
Biliary atresia	☑	☑	☑
Complement factor deficiencies	☑	☑	☑
Crigler–Najjar syndrome	☑	☑	☑
Familial hypercholesterolemia	☑	☑	☑
Fulminant hepatic failure	☑	☑	☑
Glycogen storage diseases	☑	☑	☑
Hyperlipidemia	☑	☐	☑
Hemophilia	☑	☐	☑
Hereditary tyrosinemia type I	☑	☑	☑
Infantile Refsum's disease	☑	☑	☑
Maple syrup urine disease	☑	☐	☑
Mitochondrial defects	☑	☐	☑
Organic acidemia	☑	☐	☑
Primary oxalosis	☑	☑	☑
Phenylketonuria	☑	☑	☑
Primary sclerosing cholangitis	☑	☐	☑
Progressive familial intrahepatic cholestasis	☑	☑	☑
Urea cycle defects	☑	☑	☑

HTx, hepatocyte transplantation; hAEC; Tx, human amnion epithelial cell; transplant.

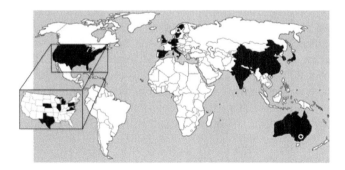

FIGURE 8.1 Worldwide distribution of centers where hepatocyte transplantation has been performed. Stars indicate the centers where clinical trials using human amnion epithelial cells (hAECs) are starting. (Courtesy of Karolinska Institute, Stockholm, Sweden; and Monash Institute of Medical Research, Melbourne, Australia.)

Gramignoli et al. 2013b; Puppi et al. 2012; Strom et al. 1997a, 1997b). Transplanted hepatocytes have resulted in partial correction of metabolic disorders, in the majority of the treated patients, failing to restore sufficient liver function to prevent eventual organ replacement (Fisher and Strom 2006; Jorns et al. 2012). Efficacy has been limited by relatively poor initial and long-term engraftment and by an inability to monitor graft function in real time, making diagnosis and treatment after rejection nearly impossible. The ability of hepatocytes to effectively repopulate a diseased liver appears to be limited to a select group of disorders in which native hepatocytes are sufficiently damaged to provide a growth advantage to the transplanted cells. Furthermore, these procedures still require immunosuppression, and immunosuppression is difficult to monitor when there is no obvious mechanism to identify a rejection episode in dispersed cell grafts.

8.2.2 Stem Cell–Based Therapies

A wider use of cell-based therapy in liver diseases will not be possible until alternative and more reliable sources of functionally proficient cells are found. Key challenges to the further development of these therapies include developing a consistent source of expandable, bankable, engraftable, and functional (liver) cells. There is a growing body of literature seeking alternative sources of abundant, high-quality hepatocytes for transplant in patients with acute liver failure or chronic liver disease and during regeneration after large hepatic resections (Duncan et al. 2009; Yu et al. 2012).

Hematopoietic stem cells, mesenchymal stromal cells, or "resident precursor cells" have also been infused in patients with different liver diseases, with inconsistent results so far. The mechanisms underlying improvement in liver function, when reported, are still unclear and largely recognized to be related to a short-term trophic effect. The most promising cell types proposed and investigated in clinical and toxicology programs are embryonic and induced pluripotent stem cells, fetal hepatoblasts, and placental cells. Several research groups have described protocols to differentiate these cells into definitive endoderm as a first step and then into cells with hepatocyte characteristics (Basma et al. 2009; Si-Tayeb et al. 2010). Although hepatocyte-like cells have been produced starting from multipotent cell sources, the achieved level of hepatic differentiation is quite primitive. In addition, the tumorigenic potential, together with some ethical concerns, has strongly limited the use of these stem cell sources. Concern over the use of pluripotent stem cells is the perceived risk of tumor formation (Amariglio et al. 2009; Andrews et al. 2005; Dlouhy et al. 2014). In the past, treatments using stem cells showed tumor cells arising from the injected graft (Amariglio et al. 2009; Dlouhy et al. 2014). A partial commitment *in vitro* before infusion has been suggested; nevertheless, the risk not all the cells will be fully differentiated remains.

8.3 AMNION EPITHELIAL CELLS

The human placenta is commonly recognized for important functions such as nutrition, respiration, and excretion, in addition to important feto-maternal tolerance. The amnion is the thin membrane portion of the placenta that surrounds the baby during development. The amniotic membrane "contains" the baby and holds the amniotic fluid; this fluid cushions the baby to protect it from mechanical injury within the womb. The placenta also forms an immunological barrier between the baby and its mother's immune system to prevent the mother from rejecting the baby as one would reject an allogeneic kidney or liver transplant without immunosuppression. The fetus is a semiallograft "tissue" (it contains genetic material from the mother as well as the father), so without immunoprotection from the maternal immune system, the baby would be easily rejected. Maintaining functional immunological tolerance of the developing embryo is crucial, and evolution evolved a system to protect the baby from immune attack to protect life.

There are a multitude of reports in the literature, some going back more than a century, describing the successful use of the amniotic membrane to treat a variety of disorders. The successful use of fragments of the placental tissue, in particular the amnion layer (Figure 8.2), and, in recent years, the importance that the different cell

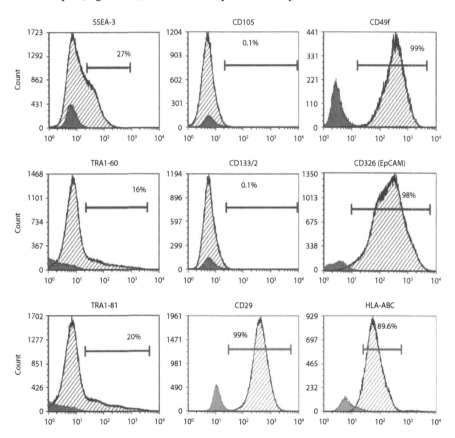

FIGURE 8.2 Flow cytometric analysis of one representative batch of human amnion epithelial cells. Comparison of several surface markers: stage-specific embryonic antigen 3 (SSEA-3), CD105 (endoglin), CD49f (α6 integrin subunit), tumor rejection antigen (TRA) 1-60; CD133/2; CD326 (EpCAM); TRA1-81; CD29 (β1 integrin subunit); and human leukocyte antigen (HLA)-ABC. The percentage of positive cells for several surface markers is relative to isotype control (gray area).

types isolated from it have gained, have encouraged clinical researchers to explore their potential effects in therapeutic applications.

With regard to their potential use in regenerative medicine, no other cell type has satisfied so many critical parameters as hAECs. Over the last decade, our group developed protocols for hAEC isolation (Miki et al. 2003, 2010), and we first reported hAECs to have stem cell characteristics (Miki et al. 2005). Surface antigens, such as stage-specific embryonic antigens (SSEAs) 3 and 4 and the tumor rejection antigens 1-60 and 1-81, commonly described on embryonic stem cells, are expressed in 10–20% of the isolated cells (Figure 8.3).

In addition to the surface markers, hAECs also express molecular markers characteristic of embryonic origin, including octamer-binding transcription factor 4 and the transcription factor Nanog (Miki et al. 2005), which is considered to be the backbone for pluripotency and self-renewal. Human amnion epithelial cells perform several important functions during development, including storage of glycogen, production and metabolism of hormones or drugs, and transport of nutrients in and out of the amniotic fluid. Probably the most interesting characteristic is their "immune-privileged" status. This feature, although not completely understood, is,

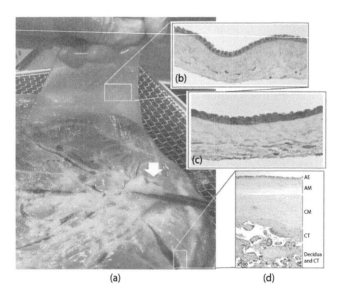

FIGURE 8.3 (See color insert.) Human placenta and amnion tissue sections. (a) Picture of a portion of a full-term human placenta where cord tissue has been clumped cut off (arrow). The amnion tissue is manually stripped off from the fetal side. (b) Hematoxylin and eosin–stained of a section of amnion tissue. The amniotic layer is composed of a cuboidal epithelial layer and a deeper mesodermal layer where a few stromal cells are visible. (c) Staining of the amnion tissue with human leukocyte antigen-G antibody, showing positive expression in all the epithelial cells. (d) Cross-sectional representation of the human placenta. The amnion and chorion are fetal-derived membranes, whereas the decidual membrane is maternally derived. The amnion section is composed by amniotic epithelium (AE) and amniotic mesoderm (AM); the chorionic layer is comprised of a mesodermal layer (CM) and a trophoblast layer (CT).

Use of Amnion Epithelial Cells in Metabolic Liver Disorders

in part, due to their unique expression of human leukocyte antigen (HLA) proteins. Rejection follows when mismatches in HLA expressions are detected by immune cells. Human amnion expresses very low levels of class I antigens; lacks expression of class II and costimulatory molecules (Banas et al. 2008; Miki et al. 2005; Parolini et al. 2008); and, interestingly, expresses high levels of HLA-G (Figure 8.2). HLA-G modulates the immune response and modulates T regulatory cell response to induce tolerance to the cell or tissue graft (Banas et al. 2008; Tee et al. 2013). Placental HLA-G expression and secretion are recognized as the prime reason that the fetus is not rejected by the mother's immune system. Human amnion epithelial cells have also been found to secrete anti-inflammatory and immunosuppressive factors that inhibited inflammation and reduced the proliferation of T and B cells *in vitro* (Li et al. 2005), in addition to macrophage inhibitory protein, Fas ligand, several interleukin molecules, and an arsenal of proregenerative cytokines (Li et al. 2005; Tee et al. 2013). Pluripotent cells such as embryonic stem cells can be expanded indefinitely *in vitro*. Amnion epithelial cells, unlike embryonic stem cells, commonly experience a drastic stop in replication after four to eight passages (Miki and Strom 2006). hAECs do not express telomerase reverse transcriptase (Miki et al. 2005). Telomerase activity is found in human embryonic stem cells and pluripotent progenitor cells, as well as in most human tumors (Chen and Chen 2011). Indeed, telomerase-expressing stem cells commonly display an unstable karyotype during proliferation and generate neoplastic grafts, often teratomas, upon injection into severe combined immunodeficiency mice (Hansel et al. 2014a). Conversely, hAECs consistently display a normal karyotype and are nontumorigenic when transplanted in animals (Marongiu et al. 2011; Miki et al. 2009). Similar characteristics, normally associated with stem cells, have also been observed in cells from other compartments of placenta (amniotic fluid and Wharton's jelly), although most of their properties suggest they are more similar to mesenchymal stromal cells than hAECs (De Coppi et al. 2007; Delo et al. 2006; In 't Anker et al. 2004; Parolini et al. 2008; Prusa et al. 2003; Tsai et al. 2004, 2006).

8.4 *IN VITRO* HEPATIC DIFFERENTIATION

Given the stem cell-like features of hAECs, we examined the ability that some or all of these cells have to produce progenitors of hepatic cells in culture. hAECs have previously shown the potential to differentiate into all three germ layers *in vitro* (Kakishita et al. 2000, 2003). We reported that under specific conditions, hAECs effectively and efficiently differentiate into cells with hepatic characteristics (Marongiu et al. 2011; Miki et al. 2009). Successful hepatic differentiation depends on both the culture substrate and the types and concentrations of growth factors added to the culture media. The ability of steroid hormone exposure to enhance hepatic differentiation has been tested, and both hydrocortisone and dexamethasone have been shown to enhance the expression of the endodermal/hepatic genes in culture (Miki et al. 2009). Markers of hepatic differentiation, such as albumin, A1AT, hepatocyte nuclear factor 4 alpha, and liver-enriched transcription factor CAAT enhancer–binding protein alpha, increased over time in culture in the presence of steroid hormones. Differentiation of hAECs to an endodermal lineage such as liver might benefit from efficient differentiation to mesendoderm first, followed by exposure to additional growth factors that would enhance endodermal formation. In several reports, activin A, a member of the transforming growth factor beta superfamily, is described as a key factor known to induce endodermal differentiation. Nevertheless, when tested on hAECs the effect of activin A was particularly inconsistent in our culture conditions, and usually resulted in a more stem cell phenotype (Marongiu et al. 2011).

One month after differentiation, structural cytoskeleton elements such as cytokeratins 8, 18, and 19, in addition several metabolic and functional genes (A1AT, albumin, several forms of cytochrome P450) as well as nuclear factors pregnane-X and constitutive-androstane receptors were detected at levels similar to those observed in human fetal livers at midgestation (18–22 weeks) (Marongiu et al. 2011). Some hepatic genes are expressed preferentially during the fetal period and decline in expression after birth. A well-known example of this type of pattern can be found in alpha-fetoprotein (AFP) and albumin. Genes in the CYP3A family show a similar pattern to AFP/albumin: fetal liver expressed predominantly CYP3A7, whereas expression of CYP3A4 occurred after birth (Noreault et al. 2005; Schuetz et al. 1994; Yang et al. 1994). Cultured hAEC-derived hepatocyte-like cells also expressed both CYP3A7 and CYP3A4 (Marongiu et al. 2011; Miki et al. 2009). However, the ratio of CYP3A4/CYP3A7 suggested that hAECs are close to fetal hepatocytes progressing toward the adult phenotype. Notably, hAEC-derived hepatocyte-like cells possessed metabolic function typical of mature hepatocyte the ability to metabolize ammonia and 17-hydroxyprogesterone caproate (normally metabolized by CYP3A7), together with inducible cytochrome P450 and phase II enzymes (Marongiu et al. 2011). The coexpression of cytokeratin 19 and AFP is reminiscent of a stage in liver development during fetal life where bipotential progenitor cells can give rise to both hepatocytes and bile duct cells, and their expression suggests that hAECs may give rise to bile ducts as well as hepatocytes. One gene has been reported to be specific for definitive endoderm, CYP7A1 (Asahina et al. 2004), an enzyme involved in the conversion of cholesterol to bile acids in hepatocytes (Ellis et al. 1998). The detection of expression of CYP7A1 in amnion-derived hepatocyte-like cells indicates that the hAECs differentiate to definitive endoderm. CYP7A1 expression is specifically downregulated by bile acid exposure along with the bile salt export protein.

8.5 PRECLINICAL MODELS

To efficiently study hAECs or any other pluripotent cells, it is important to investigate their regenerative potential *in vivo*. Undifferentiated hAECs have been described to functionally engraft into the livers of immunocompromised mouse models of liver damage, resulting in a reduction of hepatic fibrosis, inflammation, and hepatocyte apoptosis (Marongiu et al. 2011; Tee et al. 2013). When undifferentiated hAECs were transplanted into the livers of immunodeficient mice, human-specific gene expression profiles posttransplant measured levels of expression normally found in adult human liver, including nuclear hormone receptors, hepatic transport proteins, cytochrome P450 genes, plasma proteins, and metabolic enzymes causal for several

Use of Amnion Epithelial Cells in Metabolic Liver Disorders

metabolic liver diseases (Marongiu et al. 2011; Miki et al. 2009). Similar results have been described in the retrorsine model of liver repopulation where clusters of amnion epithelial cells have been shown to engraft and differentiate to mature hepatocytes in the absence of cell fusion with host-derived cells (Marongiu et al. 2011, 2015). The analysis of human-specific gene expression profiles in mouse livers posttransplant measured expression levels normally found in adult human liver for nuclear hormone receptors to hepatic transport proteins, human cytochrome P450 genes to plasma proteins, and metabolic enzymes causal for several metabolic liver diseases (Marongiu et al. 2011). Given these results, the hypothesis that hAECs, like hepatocytes, might correct metabolic liver diseases when transplanted, was tested. Two mouse models of human liver disease, MSUD and PKU, have been recipients for hAECs. In the first treated model, MSUD, a mutation inactivates the enzyme responsible for the metabolism of branched-chain amino acids in the liver (branched-chain keto-acid dehydrogenase [BCKDH] enzyme complex), resulting in high levels of brain chain amino acids in the serum and urine (Skvorak et al. 2009a). A correction of the disease has been previously possible by murine HTx (Skvorak et al. 2009b). When undifferentiated hAECs (freshly isolated or cryopreserved) rather than mouse hepatocytes were injected, similar results were reported (Skvorak et al. 2013a, 2013b). All control animals died within day 28, showing all characteristics of the untreated human disease (debilitating muscle wasting, failure in thriving, and seizures), whereas hAEC-infused mice showed characteristic improvement in survival and wild-type body weight. Treated mice displayed improved BCKDH enzymatic activity and reduced level of amino acids and metabolites, both in the brain and the blood (Skvorak et al. 2013b), in addition to complete normalization of several neurotransmitters (Skvorak et al. 2013a). Human DNA measured in the mouse liver reflected the levels of human DNA and correlated with the BCKDH level. Remarkably, the preclinical model used in this study has a normal immune system, and although immunosuppression has not been used, no evidence of rejection of hAECs was detected.

Given these results, a second human-relevant, liver-based disease was examined. PKU is a liver-based recessive genetic disease characterized by high levels of phenylalanine (Phe) in the circulation. PKU patients present a mutation in the enzyme phenylalanine hydroxylase, the enzyme responsible for Phe-to-tyrosine conversion, and this mutation results in very high levels of Phe in the blood and brain. The outcome can be severe brain dysfunction, cognitive defects, coma, and even death. PKU is not normally considered a target for liver transplants, as the risk of life-long immunosuppression must be weighed against the benefit of the organ replacement. Traditional medical therapy is palliative and includes a severely protein-restricted diet that lowers Phe levels and prevents severe mental retardation if initiated early. Nevertheless, problems with noncompliance are common, especially at adolescence and beyond, and are associated with loss of IQ points, white matter degeneration, ataxia, and seizures (Gassio et al. 2005; Greeves et al. 2000; Thompson et al. 1990). As with the MSUD mice, we transplanted freshly isolated hAECs into the liver of PKU-affected mice. We measured a complete correction of neurotransmitter amino acid (Phe, in particular) levels in the brain, and a significant correction in the serum (Strom et al. 2013). The correction of MSUD and PKU requires differentiation of hAECs to cells with a mature hepatic function.

8.6 CLINICAL PROGRAM

Given the strong preclinical data, we proposed to use hAECs for any disease considered for HTx. As reported in Table 8.1, HTx has been used for acute liver failure, fibrosis, cirrhosis, and inborn error diseases. In a clinical program, hAECs could be a useful and alternative approach for diseases that include, but will not be limited to, PKU, MSUD, urea cycle defects, Crigler–Najjar syndrome, glycogen and lysosomal storage diseases, primary hyperoxalosis, hereditary tyrosinemia type 1, porphyria, hypercholesterolemia, biliary atresia, primary sclerosing cholangitis, and other causes of cholestasis, as well as disorders affecting the metabolism of amino acid and other chemical intermediates and defects affecting the transport of molecules into or out of hepatocytes (Table 8.1). Studies in animal models of several liver-based metabolic defects have shown that disease phenotype could be changed from severe to moderate or mild by replacing about 5–10% of the missing enzyme function. Since the liver produces and secretes many proteins, such as A1AT, clotting factors, and regulators of the complement system, we propose that hAEC transplants would be useful for the treatment of hemophilia, complement factor H, and A1AT deficiencies. In addition, since hAEC transplant modulates the immune response, induces T regulatory cell response, and induces tolerance, a combined infusion of hAECs and human hepatocytes, or hAECs with whole organ transplantation could be useful to modulate the immune system and induce tolerance to the cell or organ graft in the recipient.

Although animal models of metabolic liver disease recapitulate the human processes, achieving an adequate level of engraftment is still a concern. The liver has been estimated to contain around 3–4 billion hepatocytes per kilogram of body weight. Cell dose is commonly adjusted according to the patient size, with the intent to transplant around 5% of liver mass. However 5% of liver mass cannot safely be transplanted in one single event, so the transplant is subdivide into five to eight separate infusions, over several hours or days (up to 48 h) (Gramignoli et al. 2013a; Jorns et al. 2014). The total number of infusions is based on the number of primary cells isolated from hepatic tissue and on the tolerance of the patient to infusion (avoidance of portal hypertension, portal vein thrombosis, or both). When possible, up to 2×10^8 cells/kg are commonly infused over first 24 h.

The hepatocytes are injected into the portal vein. The portal vein is accessed by umbilical vein, or surgically by a peripheral mesenteric vein. Transplantation of a larger number of cells can lead to severe portal hypertension and translocation of cells into the systemic circulation with embolization to the lungs. To avoid any risk of thrombosis, the injection rate is accurate adjusted for changes in portal pressure.

Since in all cell-based therapies it has been estimated that, once injected, up to 90% of cells are lost because of physical stress, hypoxia, anoikis, inflammation, or immunogenic rejection (Nguyen et al. 2010; Smets et al. 2002; Zvibel et al. 2002), the number of hAECs to be infused in every patient can only be estimated, at the moment. The half-life of transplanted cells is another critical factor. If the half-life is short, the therapeutic efficacy is limited to a short time. Reaching a complete, sustained correction might require repeated administrations from multiple donors.

Use of Amnion Epithelial Cells in Metabolic Liver Disorders

hAECs from more than one donor will most likely be required to provide a sufficient number of cells for transplantation to correct the disease process.

Isolated amnion-derived cells have already been successfully transplanted more than 30 years ago into normal volunteers without any signs of rejection or tumor formation (Akle et al. 1981). In addition, amnion-derived cells were injected aiming to treat Niemann–Pick and lysosomal storage patients, without immunosuppression and without any signs of rejection (Scaggiante et al. 1987; Yeager et al. 1985). Consequently, hAECs were transplanted in immunocompetent animals in preclinical models with no signs of rejection, despite the fact that immunosuppressive drugs were not provided. Based on this evidence, we have proposed to infuse hAECs into patients without providing immunosuppression.

Unlike hepatocytes, hAECs cryopreserve easily and upon thawing result in very high viability, usually in the range of 90–95%. This important feature greatly facilitates the establishment of a placental cell bank.

As a regulatory prerequisite, current good manufacturing practice (cGMP) guidelines need to be followed, as well as the use of accredited laboratory and clinical grade products and procedures. Safety for human administration requires expensive regulatory licensed and established good manufacturing-compliant facilities. Hepatocytes for use in clinical transplantation are currently isolated and prepared under cGMP guidelines in an aseptic environment and a quality control system. hAEC isolation, banking, and clinical infusion are following the path already designed for umbilical cord blood banks (Serrano-Delgado et al. 2009).

8.7 CONCLUDING REMARKS

To circumvent the number of lives lost due to waiting list times for organs and insufficient number of available livers, alternative approaches need to be developed. Cell-based therapies have shown a great deal of promise, and the progress made over the past decades via preclinical and clinical studies provides a growing rationale for their use to treat a variety of liver disorders. Cell therapy could relieve the long waiting list for an organ and allow a better use of the livers available. Research for the use of stem cells and stem-like cells in cellular therapies is mainly in preclinical and, in some cases, in early clinical stages. More than 3000 trials associated with stem cells are currently registered in the World Health Organization Clinical Trials Registry Platform (Heslop et al. 2015), and several unregulated stem cell treatments are available worldwide (Zarzeczny et al. 2014).

Gestational tissues are an abundant source of multipotent cells, with minimal ethical and legal considerations involved in their collection and use. From the perspective of donor cell availability, hAEC transplant offers several potential advantages over vascularized hepatic allografts and even hepatocyte transplantation.

Successful preclinical studies have shown that transplanted hAECs acquire sufficient hepatocyte-like functions to rescue several congenital metabolic liver diseases. Encouraged by the lack of tumorigenicity, and the well-known immunomodulatory and anti-inflammatory effects of hAECs, a clinical trial using hAECs as a substitute for human hepatocytes for metabolic disease patients is planned.

hAECs appear to be immune privileged, and they may escape detection and rejection by the recipient's immune system, even when immunosuppressive drugs are not administered. If this turns out to be true, it would be a "game changer" for cellular therapy and would reduce or even eliminate the side effects associated with immunosuppression of the patient. Current research suggests that stem cells isolated from discarded placenta may be an unlimited, noncontroversial, and safe source of cells for regenerative medicine; they also may serve as a coadjuvant therapy to organ transplant procedures, modulating the immune response of the recipient and decreasing inflammation normally associated with organ and cell transplant procedures.

REFERENCES

Akle, C. A., M. Adinolfi, K. I. Welsh, S. Leibowitz, and I. McColl. 1981. Immunogenicity of human amniotic epithelial cells after transplantation into volunteers. *Lancet* 2: 1003–5.

Allen, K. J., N. A. Mifsud, R. Williamson, P. Bertolino, and W. Hardikar. 2008. Cell-mediated rejection results in allograft loss after liver cell transplantation. *Liver Transpl* 14: 688–94.

Amariglio, N., A. Hirshberg, B. W. Scheithauer, et al. 2009. Donor-derived brain tumor following neural stem cell transplantation in an ataxia telangiectasia patient. *PLoS Med* 6: e1000029.

Ambrosino, G., S. Varotto, S. C. Strom, et al. 2005. Isolated hepatocyte transplantation for Crigler-Najjar syndrome type 1. *Cell Transplant* 14: 151–57.

Andrews, P. W., M. M. Matin, A. R. Bahrami, et al. 2005. Embryonic stem (ES) cells and embryonal carcinoma (EC) cells: Opposite sides of the same coin. *Biochem Soc Trans* 33: 1526–30.

Asahina, K., H. Fujimori, K. Shimizu-Saito, et al. 2004. Expression of the liver-specific gene Cyp7a1 reveals hepatic differentiation in embryoid bodies derived from mouse embryonic stem cells. *Genes Cells* 9: 1297–308.

Banas, R. A., C. Trumpower, C. Bentlejewski, et al. 2008. Immunogenicity and immunomodulatory effects of amnion-derived multipotent progenitor cells. *Hum Immunol* 69: 321–28.

Basma, H., A. Soto-Gutierrez, G. R. Yannam, et al. 2009. Differentiation and transplantation of human embryonic stem cell-derived hepatocytes. *Gastroenterology* 136: 990–99.

Beck, B. B., S. Habbig, K. Dittrich, et al. 2012. Liver cell transplantation in severe infantile oxalosis—A potential bridging procedure to orthotopic liver transplantation? *Nephrol Dial Transplant* 27: 2984–89.

Bilir, B. M., D. Guinette, F. Karrer, et al. 2000. Hepatocyte transplantation in acute liver failure. *Liver Transpl* 6: 32–40.

Chen, C. H., and R. J. Chen. 2011. Prevalence of telomerase activity in human cancer. *J Formos Med Assoc* 110: 275–89.

Darwish, A. A., E. Sokal, X. Stephenne, et al. 2004. Permanent access to the portal system for cellular transplantation using an implantable port device. *Liver Transpl* 10: 1213–15.

De Coppi, P., A. Callegari, A. Chiavegato, et al. 2007. Amniotic fluid and bone marrow derived mesenchymal stem cells can be converted to smooth muscle cells in the cryo-injured rat bladder and prevent compensatory hypertrophy of surviving smooth muscle cells. *J Urol* 177: 369–76.

Delo, D. M., P. De Coppi, G. Bartsch, Jr., and A. Atala. 2006. Amniotic fluid and placental stem cells. *Methods Enzymol* 419: 426–38.

Use of Amnion Epithelial Cells in Metabolic Liver Disorders 157

Dhawan, A., R. R. Mitry, and R. D. Hughes. 2005. Hepatocyte transplantation for metabolic disorders, experience at King's College hospital and review of literature. *Acta Gastroenterol Belg* 68: 457–60.

Dhawan, A., R. R. Mitry, R. D. Hughes, et al. 2004. Hepatocyte transplantation for inherited factor VII deficiency. *Transplantation* 78: 1812–14.

Dhawan, A., S. C. Strom, E. Sokal, and I. J. Fox. 2010. Human hepatocyte transplantation. *Methods Mol Biol* 640: 525–34.

Dlouhy, B. J., O. Awe, R. C. Rao, P. A. Kirby, and P. W. Hitchon. 2014. Autograft-derived spinal cord mass following olfactory mucosal cell transplantation in a spinal cord injury patient: Case report. *J Neurosurg Spine* 21: 618–22.

Duncan, A. W., C. Dorrell, and M. Grompe. 2009. Stem cells and liver regeneration. *Gastroenterology* 137: 466–81.

Ellis, E., B. Goodwin, A. Abrahamsson, et al. 1998. Bile acid synthesis in primary cultures of rat and human hepatocytes. *Hepatology* 27: 615–20.

Everhart, J. E., and C. E. Ruhl. 2009. Burden of digestive diseases in the United States part III: Liver, biliary tract, and pancreas. *Gastroenterology* 136: 1134–44.

Fisher, R. A., and S. C. Strom. 2006. Human hepatocyte transplantation: Worldwide results. *Transplantation* 82: 441–49.

Fox, I. J., J. R. Chowdhury, S. S. Kaufman, et al. 1998. Treatment of the Crigler-Najjar syndrome type I with hepatocyte transplantation. *N Engl J Med* 338: 1422–26.

Gassio, R., R. Artuch, M. A. Vilaseca, et al. Cognitive functions in classic phenylketonuria and mild hyperphenylalaninaemia: Experience in a paediatric population. *Dev Med Child Neurol* 2005; 47: 443–48.

Gramignoli, R., K. Dorko, V. Tahan, et al. 2014. Hypothermic storage of human hepatocytes for transplantation. *Cell Transplant* 23 (9): 1143–51.

Gramignoli, R., V. Tahan, K. Dorko, et al. 2013. New potential cell source for hepatocyte transplantation: Discarded livers from metabolic disease liver transplants. *Stem Cell Res* 11: 563–73.

Gramignoli, R., V. Tahan, K. Dorko, et al. 2014. Rapid-and-sensitive assessment of human hepatocyte functions. *Cell Transplant* 3 (12): 1545–56.

Gramignoli, R., M. Vosough, K. Kannisto, R. C. Srinivasan, and S. C. Strom. 2015. Clinical hepatocyte transplantation: Practical limits and possible solutions. *Eur Surg Res* 54: 162–77.

Greeves, L. G., C. C. Patterson, D. J. Carson, et al. 2000. Effect of genotype on changes in intelligence quotient after dietary relaxation in phenylketonuria and hyperphenylalaninaemia. *Arch Dis Child* 82: 216–21.

Gridelli, B., G. Vizzini, G. Pietrosi, et al. 2012. Efficient human fetal liver cell isolation protocol based on vascular perfusion for liver cell-based therapy and case report on cell transplantation. *Liver Transpl* 18: 226–37.

Grossman, M., S. E. Raper, K. Kozarsky, et al. 1994. Successful ex vivo gene therapy directed to liver in a patient with familial hypercholesterolaemia. *Nat Genet* 6: 335–41.

Habibullah, C. M., I. H. Syed, A. Qamar, Z. Taher-Uz. 1994. Human fetal hepatocyte transplantation in patients with fulminant hepatic failure. *Transplantation* 58: 951–52.

Hansel, M. C., R. Gramignoli, W. Blake, et al. 2014a. Increased reprogramming of human fetal hepatocytes compared with adult hepatocytes in feeder-free conditions. *Cell Transplant* 23 (1): 27–38.

Hansel, M. C., R. Gramignoli, K. J. Skvorak, et al. 2014b. The history and use of human hepatocytes for the treatment of liver diseases: The first 100 patients. *Curr Protoc Toxicol* 62: 14.12.11–23.

Heslop, J. A., T. G. Hammond, I. Santeramo, et al. 2015. Concise review: Workshop review: Understanding and assessing the risks of stem cell-based therapies. *Stem Cells Transl Med* 4 (4): 389–400.

In 't Anker, P. S., S. A. Scherjon, C. Kleijburg-van der Keur, et al. 2004. Isolation of mesenchymal stem cells of fetal or maternal origin from human placenta. *Stem Cells* 22: 1338–45.

Jorns, C., E. C. Ellis, G. Nowak, et al. 2012. Hepatocyte transplantation for inherited metabolic diseases of the liver. *J Intern Med* 272: 201–23.

Jorns, C., R. Gramignoli, M. Saliem, et al. 2014. Strategies for short-term storage of hepatocytes for repeated clinical infusions. *Cell Transplant* 23 (8): 1009–18.

Kakishita, K., M. A. Elwan, N. Nakao, T. Itakura, and N. Sakuragawa. 2000. Human amniotic epithelial cells produce dopamine and survive after implantation into the striatum of a rat model of Parkinson's disease: A potential source of donor for transplantation therapy. *Exp Neurol* 165: 27–34.

Kakishita, K., N. Nakao, N. Sakuragawa, and T. Itakura. 2003. Implantation of human amniotic epithelial cells prevents the degeneration of nigral dopamine neurons in rats with 6-hydroxydopamine lesions. *Brain Res* 980: 48–56.

Lee, K. W., J. H. Lee, S. W. Shin, et al. 2007. Hepatocyte transplantation for glycogen storage disease type Ib. *Cell Transplant* 16: 629–37.

Li, H., J. Y. Niederkorn, S. Neelam, et al. 2005. Immunosuppressive factors secreted by human amniotic epithelial cells. *Invest Ophthalmol Vis Sci* 46: 900–7.

Li, Z.-R., X.-H. Mao, X.-X. Hu, et al. 2012. Primary human hepatocyte transplantation in the therapy of hepatic failure: 2 cases report. *Asian Pac J Trop Med* 5 (2): 165–68.

Marongiu, F., R. Gramignoli, K. Dorko, et al. 2011. Hepatic differentiation of amniotic epithelial cells. *Hepatology* 53: 1719–29.

Marongiu, M., M. P. Serra, A. Contini, et al. 2015. Rat-derived amniotic epithelial cells differentiate into mature hepatocytes in vivo with no evidence of cell fusion. *Stem Cells Dev* 24 (12): 1429–35.

Miki, T., T. Lehmann, H. Cai, D. B. Stolz, and S. C. Strom. 2005. Stem cell characteristics of amniotic epithelial cells. *Stem Cells* 23: 1549–59.

Miki, T., T. Lehmann, H. Cai, and S. Strom. 2003. Isolation of multipotent stem cells from placenta. *Hepatology* 38: 290A.

Miki, T., F. Marongiu, K. Dorko, E. C. Ellis, and S. C. Strom. 2010. Isolation of amniotic epithelial stem cells. *Curr Protoc Stem Cell Biol* Chapter 1: Unit 1E 3.

Miki, T., F. Marongiu, E. C. Ellis, et al. 2009. Production of hepatocyte-like cells from human amnion. *Methods Mol Biol* 481: 155–68.

Miki, T., and S. C. Strom. 2006. Amnion-derived pluripotent/multipotent stem cells. *Stem Cell Rev* 2: 133–42.

Mito, M., M. Kusano, and Y. Kawaura. 1992. Hepatocyte transplantation in man. *Transplant Proc* 24: 3052–53.

Mitry, R. R., A. Dhawan, R. D. Hughes, et al. 2004. One liver, three recipients: Segment IV from split-liver procedures as a source of hepatocytes for cell transplantation. *Transplantation* 77: 1614–16.

Muraca, M., G. Gerunda, D. Neri, et al. 2002. Hepatocyte transplantation as a treatment for glycogen storage disease type 1a. *Lancet* 359: 317–18.

Nguyen, P. K., D. Nag, and J. C. Wu. 2010. Methods to assess stem cell lineage, fate and function. *Adv Drug Deliv Rev* 62: 1175–86.

Noreault, T. L., V. E. Kostrubsky, S. G. Wood, et al. 2005. Arsenite decreases CYP3A4 and RXRalpha in primary human hepatocytes. *Drug Metab Dispos* 33: 993–1003.

Parolini, O., F. Alviano, G. P. Bagnara, et al. 2008. Concise review: Isolation and characterization of cells from human term placenta: Outcome of the first international Workshop on Placenta Derived Stem Cells. *Stem Cells* 26 (2): 300–11.

Prusa, A. R., E. Marton, M. Rosner, G. Bernaschek, and M. Hengstschlager. 2003. Oct-4-expressing cells in human amniotic fluid: A new source for stem cell research? *Hum Reprod* 18: 1489–93.

Puppi, J., S. C. Strom, R. D. Hughes, et al. 2012. Improving the techniques for human hepatocyte transplantation: Report from a consensus meeting in London. *Cell Transplant* 21: 1–10.

Puppi, J., N. Tan, R. R. Mitry, et al. 2008. Hepatocyte transplantation followed by auxiliary liver transplantation—A novel treatment for ornithine transcarbamylase deficiency. *Am J Transplant* 8: 452–7.

Ribes-Koninckx, C., E. Pareja Ibars, M. A. Agrasot, et al. 2012. Clinical outcome of hepatocyte transplantation in four pediatric patients with inherited metabolic diseases. *Cell Transplant* 21 (10): 2267–82.

Scaggiante, B., A. Pineschi, M. Sustersich, et al. 1987. Successful therapy of Niemann-Pick disease by implantation of human amniotic membrane. *Transplantation* 44: 59–61.

Schneider, A., M. Attaran, P. N. Meier, et al. 2006. Hepatocyte transplantation in an acute liver failure due to mushroom poisoning. *Transplantation* 82: 1115–16.

Schuetz, J. D., D. L. Beach, and P. S. Guzelian. 1994. Selective expression of cytochrome P450 CYP3A mRNAs in embryonic and adult human liver. *Pharmacogenetics* 4: 11–20.

Serrano-Delgado, V. M., B. Novello-Garza, and E. Valdez-Martinez. 2009. Ethical issues relating the banking of umbilical cord blood in Mexico. *BMC Med Ethics* 10: 12.

Si-Tayeb, K., F. K. Noto, M. Nagaoka, et al. 2010. Highly efficient generation of human hepatocyte-like cells from induced pluripotent stem cells. *Hepatology* 51: 297–305.

Skvorak, K. J., K. Dorko, F. Marongiu, et al. 2013. Improved amino acid, bioenergetic metabolite and neurotransmitter profiles following human amnion epithelial cell transplant in intermediate maple syrup urine disease mice. *Mol Genet Metab* 109: 132–38.

Skvorak, K. J., K. Dorko, F. Marongiu, et al. 2013. Placental stem cell correction of murine intermediate maple syrup urine disease. *Hepatology* 57: 1017–23.

Skvorak, K. J., E. J. Hager, E. Arning, et al. 2009. Hepatocyte transplantation (HTx) corrects selected neurometabolic abnormalities in murine intermediate maple syrup urine disease (iMSUD). *Biochim Biophys Acta* 1792: 1004–10.

Skvorak, K. J., H. S. Paul, K. Dorko, et al. 2009. Hepatocyte transplantation improves phenotype and extends survival in a murine model of intermediate maple syrup urine disease. *Mol Ther* 17 (7): 1266–73.

Smets, F. N., Y. Chen, L. J. Wang, and H. E. Soriano. 2002. Loss of cell anchorage triggers apoptosis (anoikis) in primary mouse hepatocytes. *Mol Genet Metab* 75: 344–52.

Sokal, E. M., F. Smets, A. Bourgois, et al. 2003. Hepatocyte transplantation in a 4-year-old girl with peroxisomal biogenesis disease: Technique, safety, and metabolic follow-up. *Transplantation* 76: 735–38.

Soltys, K. A., A. Soto-Gutierrez, M. Nagaya, et al. 2010. Barriers to the successful treatment of liver disease by hepatocyte transplantation. *J Hepatol* 53: 769–74.

Soriano, H., R. Superina, G. D. Ferry, et al. 2001. Hepatocyte transplantation (HTx) in children with liver failure. *Hepatology* 34: 250A.

Stephenne, X., F. G. Debray, F. Smets, et al. 2012. Hepatocyte transplantation using the domino concept in a child with tetrabiopterin nonresponsive phenylketonuria. *Cell Transplant* 21: 2765–70.

Stephenne, X., M. Najimi, C. Sibille, et al. 2006. Sustained engraftment and tissue enzyme activity after liver cell transplantation for argininosuccinate lyase deficiency. *Gastroenterology* 130: 1317–23.

Strom, S. C., J. R. Chowdhury, and I. J. Fox. 1999. Hepatocyte transplantation for the treatment of human disease. *Semin Liver Dis* 19: 39–48.

Strom, S. C., R. A. Fisher, W. S. Rubinstein, et al. 1997. Transplantation of human hepatocytes. *Transplant Proc* 29: 2103–106.

Strom, S. C., R. A. Fisher, M. T. Thompson, et al. 1997. Hepatocyte transplantation as a bridge to orthotopic liver transplantation in terminal liver failure. *Transplantation* 63: 559–69.

Strom, S. C., K. Skvorak, R. Gramignoli, et al. 2013. Translation of amnion stem cells to the clinic. *Stem Cells Dev* 22 (Suppl 1): 96–102.

Tee, J. Y., V. Vaghjiani, Y. H. Liu, et al. 2013. Immunogenicity and immunomodulatory properties of hepatocyte-like cells derived from human amniotic epithelial cells. *Curr Stem Cell Res Ther* 8: 91–9.

Thompson, A. J., I. Smith, D. Brenton, et al. 1990. Neurological deterioration in young adults with phenylketonuria. *Lancet* 336: 602–5.

Tsai, M. S., S. M. Hwang, Y. L. Tsai, et al. 2006. Clonal amniotic fluid-derived stem cells express characteristics of both mesenchymal and neural stem cells. *Biol Reprod* 74: 545–51.

Tsai, M. S., J. L. Lee, Y. J. Chang, and S. M. Hwang. 2004. Isolation of human multipotent mesenchymal stem cells from second-trimester amniotic fluid using a novel two-stage culture protocol. *Hum Reprod* 19: 1450–56.

Yang, H. Y., Q. P. Lee, A. E. Rettie, and M. R. Juchau. 1994. Functional cytochrome P4503A isoforms in human embryonic tissues: Expression during organogenesis. *Mol Pharmacol* 46: 922–28.

Yeager, A. M., H. S. Singer, J. R. Buck, et al. 1985. A therapeutic trial of amniotic epithelial cell implantation in patients with lysosomal storage diseases. *Am J Med Genet* 22: 347–55.

Yu, Y., J. E. Fisher, J. B. Lillegard, et al. 2012. Cell therapies for liver diseases. *Liver Transpl* 18: 9–21.

Zarzeczny, A., T. Caulfield, U. Ogbogu, et al. 2014. Professional regulation: A potentially valuable tool in responding to "stem cell tourism." *Stem Cell Rep* 3: 379–84.

Zvibel, I., F. Smets, and H. Soriano. 2002. Anoikis: Roadblock to cell transplantation? *Cell Transplant* 11: 621–30.

9 The Use of Placenta-Derived Cells in Autoimmune Disorders

Antonietta R. Silini, Ornella Parolini, and Mario Delgado

CONTENTS

Preface 161
9.1 Definition and Classification of Autoimmune Diseases 162
    9.1.1 The Phenomena of Autoimmunity 162
    9.1.2 Pathogenesis of Autoimmunity 163
        9.1.2.1 Genetic Predisposition 163
        9.1.2.2 Environmental Factors 163
    9.1.3 Immune Mechanisms of Disease Induction 164
    9.1.4 Current Treatment 165
9.2 Use of Placenta-Derived Cells in Preclinical Models
of Autoimmune Disorders 166
    9.2.1 Inflammatory Bowel Disease 166
    9.2.2 Systemic Lupus Erythematosus 167
    9.2.3 Multiple Sclerosis 168
    9.2.4 Rheumatoid Arthritis 169
    9.2.5 Autoimmune Myocarditis 170
9.3 Clinical Trials 171
    9.3.1 Published Clinical Trials 172
9.4 Concluding Remarks 172
Acknowledgments 172
References 173

PREFACE

Interest in the human term placenta has increased in recent years, due mainly to its noninvasive procurement and large cell supply. In addition, cells isolated from different placental tissues share basic properties with other stem cells. Furthermore, the innate immunomodulatory and immunosuppressive functions of placental cells, and the limited expression of human leukocyte antigen (HLA) and costimulatory molecules that allow them to be used across HLA barriers, have made them unique with respect to their conventional counterparts. This chapter discusses the therapeutic potential of placental cells in autoimmune diseases. The chapter begins with

161

a general overview of the pathogenesis and the aberrant immune mechanisms in autoimmune diseases, and then discusses the relevance of placental cells in the scenario of abnormal immune processes, such as those underlying autoimmune diseases.

9.1 DEFINITION AND CLASSIFICATION OF AUTOIMMUNE DISEASES

9.1.1 THE PHENOMENA OF AUTOIMMUNITY

Every individual's immune system must effectively defend against a plethora of pathogens while simultaneously maintaining tolerance to self-antigens. The term *autoreactivity* refers to phenomena that range from a "physiological" level of self-reactivity, that is, a level fundamental in lymphocyte selection and maintenance for normal immune system homeostasis, to pathogenic autoimmunity that is associated with alterations in self-recognition control mechanisms, leading to immune-mediated dysfunction or injury (Firestein et al. 2013). After encountering an antigen, lymphocytes can either be activated, leading to an immune response, or inactivated or eliminated, leading to tolerance. Immunological tolerance is unresponsiveness to an antigen that is capable of inducing an immune response. The hallmark defect in autoimmune diseases is the breakdown of self-tolerance; this self-tolerance is mainly mediated by, but not limited to, components of the acquired immune system (B and T lymphocytes).

Currently, more than 80 diseases have a proven or strongly suspected autoimmune etiology, and all are characterized by excessive immune responses, aberrant immune responses, or both, leading to chronic inflammation, tissue destruction, and dysfunction. As a group, autoimmune diseases have been estimated to afflict 3–5% of people worldwide (Van Loveren et al. 2001).

To date, there are no consensus criteria for autoimmune diseases. A recent review that composed a comprehensive list of putative or confirmed autoimmune diseases (Hayter and Cook 2012) has estimated that there are 81 autoimmune diseases (17 putative and 67 confirmed) classified according to the Witebsky's postulates (Rose and Bona 1993). Witebsky's postulates state the following criteria: (1) direct evidence of the specific adaptive immune response directed toward the affected tissue(s) and the presence of autoantibodies or autoreactive lymphocytes that may also transfer disease to healthy individuals; (2) indirect evidence from immunization of autoantigen-inducing disease in animal models; and (3) clinical evidence demonstrating that elimination or suppression of the autoimmune response, such as through immunosuppressive interventions, prevents disease progression or improves clinical manifestation. This estimate is confirmed by the National Institutes of Health that state that there are approximately 80 autoimmune diseases based on criteria such as the breakdown of self-tolerance, presence of autoantibodies or autoreactive T cells, or infectious agents acting as triggers in disease development (Atassi and Casali 2008; Hayter and Cook 2012).

From the clinical standpoint, autoimmune diseases can be classified according to the extent of organ involvement, either organ-specific or systemic. However, a more mechanistic classification distinguishes between diseases in which there is an altered response to a particular antigen and those in which there is an alteration in the selection, regulation, or death of T or B cells (Davidson and Diamond 2001).

The Use of Placenta-Derived Cells in Autoimmune Disorders 163

Some of the most common (i.e., worldwide prevalence >25 × 10⁵) autoimmune diseases include neurological disorders such as multiple sclerosis (MS), endocrine disorders such as diabetes mellitus type 1, gastrointestinal disorders such as Crohn's disease and ulcerative colitis, cardiovascular disorders such as rheumatic fever; cutaneous disorders such as alopecia areata and vitiligo, and systemic disorders such as systemic lupus erythematosus (SLE) (Hayter and Cook 2012).

9.1.2 PATHOGENESIS OF AUTOIMMUNITY

The etiology and pathogenesis of many autoimmune diseases remain unknown. Both intrinsic (e.g., genetics, hormones, age) and environmental (e.g., infections, diet, drugs, chemicals) factors can contribute to the induction, development, and progression of autoimmune diseases (Pillai 2013).

9.1.2.1 Genetic Predisposition

Most autoimmune diseases are multigenic, whereby multiple susceptibility genes work together to produce an abnormal phenotype. In 2000, the type I diabetes locus was mapped in the major histocompatibility complex (MHC) to a 570-kilobase region (Herr et al. 2000), and other groups identified sequence variants in the NOD2 (CARD15) gene (Hugot et al. 2001) or variations in the 5q31 cytokine gene cluster (Rioux et al. 2001), which are associated with increased susceptibility to Crohn's disease. Other larger studies have identified the IDDM12/CTLA4 locus in Graves' disease (Hampe et al. 2001; Ogura et al. 2001); the PTPN22 gene in type I diabetes, rheumatoid arthritis, and SLE (Begovich et al. 2004; Kyogoku et al. 2004; Onengut-Gumuscu et al. 2004; Smyth et al. 2004); and the PDCD1 gene in SLE and rheumatoid arthritis (RA) (Lin et al. 2004; Prokunina et al. 2002). The observation that genes such as PTPN22 are common to different autoimmune diseases supports the hypothesis that some immunological pathways are common to multiple disorders, whereas other pathophysiological mechanisms are disease-specific.

9.1.2.2 Environmental Factors

Some sort of trigger, whether it be an environmental exposure or a change in the internal environment, is required for pathogenic autoreactivity, even in genetically predisposed persons. Microbial antigens can initiate autoreactivity through molecular mimicry, polyclonal activation, or the release of previously sequestered antigens (Davidson and Diamond 2001). Almost every autoimmune disease is linked to one or more infectious agents; thus, understanding of how infectious agents can contribute to autoimmunity is of considerable interest. A classic example of this relationship is the development of rheumatic fever several weeks after infection with beta hemolytic streptococcus (Vojdani 2014). Molecular resemblance between bacterial M5 protein and human alpha-myosin results in a breakdown of immunological tolerance and antibody production against the human protein in genetically susceptible individuals (Blank et al. 2002). Other infections have been associated with different autoimmune diseases, such as Epstein–Barr virus (Shimon et al. 2003), *Parvovirus* (Mori et al. 2007), and human herpes virus types 7 (Leite et al. 2010) in thyroid autoimmunity and *Porphyromonas gingivalis* (Farquharson et al. 2012),

Chlamydia (Carter et al. 2010), and *Borrelia burgdorferi* (Imai et al. 2013) in RA. Infectious agents have been shown to induce autoimmunity through molecular mimicry, epitope spreading (immune response expanded beyond original epitope recognized), standard activation, viral persistence, polyclonal activation, imbalance of immune homeostasis, and autoinflammatory activation of innate immunity (Kivity et al. 2009; Vojdani 2014). The importance of the microbiota in various autoimmune diseases is gradually becoming more evident. For example, very recently, links between alterations in microbiota and MS have been reported (Berer et al. 2011; Bhargava and Mowry 2014; Rumah et al. 2013). Studies in germ-free mice have demonstrated their resistance to developing experimental autoimmune encephalomyelitis (EAE) and reduced infiltration of T helper (Th) 1 and Th17 cells (Lee et al. 2011), which play a crucial role in EAE.

Other sorts of triggers can include hormones, such as estrogen exacerbation of SLE as demonstrated in murine models (Bynoe et al. 2000), chronic exposure to various chemicals (Bigazzi 1997), and dietary components (Pollard et al. 2010). Exposure to organic solvents as a risk factor for autoimmune diseases, such as multiple and systemic sclerosis, has also been hypothesized (Barragan-Martinez et al. 2012). Toxin- or chemical-induced autoimmunity has been attributed to mechanisms of either induced aberrant cell death, thus making hidden cellular material available to antigen-presenting cells (Germolec et al. 2012; Pollard 2012), or immune reactions to xenobiotics by covalent binding of chemicals or haptens to human proteins and formation of neoantigens (Griem et al. 1998). Another factor that can contribute to the development of autoimmunity could be structural or physical alterations in tissues, caused by injury or trauma, and ultimately inflammation that may lead to the exposure of self-antigens normally concealed from the immune system.

9.1.3 IMMUNE MECHANISMS OF DISEASE INDUCTION

A fundamental property of the "normal" immune system is self-tolerance, and a glitch in self-tolerance can result in immune reactions against self-antigens. Elucidation of the mechanisms of self-tolerance is crucial in understanding the pathogenesis of autoimmunity and the repercussions that environmental insults and infections could have at the cellular level. A comprehensive review of self-tolerance mechanisms goes beyond the scope of this chapter and only a general overview is provided.

Central tolerance is induced in self-reactive, immature lymphocytes in the thymus and bone marrow, thus acting as a safeguard to ensure that the repertoire of mature lymphocytes is inept in responding to local, self-antigens. Potent, self-antigen recognition by immature lymphocytes has several consequences. The cells may die due to a process that selects self-reactive clones (through clonal deletion or negative selection), making the cells apoptotic (Kyewski and Klein 2006; Palmer 2003; Starr et al. 2003). Or, in the case of CD4+ T cells, they can even differentiate into regulatory cells that migrate to the periphery and block immune reactions to self-antigens, through, for example, the production of inhibitory cytokines, or by direct, contact-mediated blockade of costimulatory molecules on antigen-presenting cells. In B cells, receptor editing can occur whereby the B cell receptor is rearranged so that it is no longer able to recognize self-antigen (Goodnow et al. 1990).

The Use of Placenta-Derived Cells in Autoimmune Disorders 165

Peripheral tolerance represents an additional "backup" control mechanism to prevent activation of self-reactive lymphocytes that may have escaped central tolerance (Walker and Abbas 2002). This additional control entails several different mechanisms, such as the induction of unresponsiveness (anergy), suppression (e.g., by regulatory T cells), and also cell death (Mueller 2010). A second check is also necessary to induce tolerance to "late" self-antigens that are not expressed in the thymus and bone marrow, and thus those antigens that confront the established T cell repertoire.

Abnormalities in the control mechanisms can lead to autoimmune disease. For example, defects can occur prior to lymphocyte maturation and involve negative selection or deletion, or in receptor editing for B cells (Goodnow et al. 1990). Moreover, reduced numbers of T regulatory (Treg) cells or dysfunctional Treg cells can lead to autoimmune reactions. Abnormal functioning of inhibitory receptors essential for immune homeostasis, such as CD28, CTLA-4, or PD-1, that terminate T cell responses can also favor self-reactive responses (Walker 2013). Finally, persistent activation of antigen-presenting cells, which bypass regulatory mechanisms, can also result in excessive T cell stimulation and activation.

9.1.4 Current Treatment

From a simplistic point of view, the replenishment of self-tolerance could be exploited as a therapeutic strategy for harmful immune responses, such as the self-reactive responses seen in autoimmune diseases, but also for those in allergies and transplantation rejection. However, in the past, unselective, systemic immuosuppression has been the most implemented therapeutic strategy for autoimmune diseases. Even though a vast array of immunosuppressive agents are available, they are problematic in the sense that they are often associated with significant side effects, thus rendering clinical management of the disease even more complicated.

More recently, cell-based therapeutic approaches have been tested with the aim to restore immunoregulatory networks while maintaining memory cells able to respond to pathogens. Hematopoietic stem cell transplantation is the mainstay and is commonly used to reset the deregulated immune system and create a new immunological repertoire in patients with autoimmune diseases. Another strategy has focused on restoring (rather than resetting) the immune repertoire via transfusing cells with immunomodulatory actions, such as, for example, Treg cells. Treg cells play a crucial role in the maintenance of peripheral tolerance and have been shown to modulate susceptibility to autoimmune disease (Buckner 2010).

To this end, much attention has been paid to the exploitation of the immune regulatory properties of mesenchymal stromal cells (MSCs) in treating autoimmune diseases. MSCs can be isolated from a variety of adult tissues, such adipose tissue, bone marrow, dental pulp, and peripheral blood; they also can be isolated from birth-associated tissues such as placental tissues, umbilical cord, cord blood, and amniotic fluid (reviewed in Hass et al. 2011). Indeed, MSCs have been shown to interact with a wide variety of immune cells (reviewed in Abumaree et al. 2012), and perhaps the best-characterized interactions involve their effects on T lymphocytes, including their ability to induce Treg cells (Bernardo and Fibbe 2013; Ghannam et al. 2010;

Murphy et al. 2013; Singer and Caplan 2011). In fact, the capacity of MSCs to re-educate immune cells and their relevance for autoimmune diseases has been demonstrated, as shown with MSCs from adipose tissue (Anderson et al. 2013; Gonzalez et al. 2009a, 2009b; Gonzalez-Rey et al. 2009, 2010; Ra et al. 2011) and bone marrow (Ciccocioppo et al. 2011; Duijvestein et al. 2010; Sun et al. 2009; Tyndall and Uccelli 2009; Uccelli et al. 2011).

In the last decade, MSCs from placenta have also been shown to possess interesting immunomodulatory capabilities (Abumaree et al. 2013; Caruso et al. 2012; de Girolamo et al. 2013; Evangelista et al. 2008; Manuelpillai et al. 2011; Silini et al. 2013; Soncini et al. 2007) and to induce alternatively activated regulatory phenotypes with therapeutic potential in autoimmune diseases (Parolini et al. 2014). The use of the placenta as a source of MSCs provides several advantages over "conventional" MSC sources, such as bone marrow (Murphy and Atala 2013). For example, the placenta is an ethically and legally problem-free source, attributed to its consideration as biological waste. It also represents an abundant source of MSCs that can be easily procured without invasive procedures. Placental MSCs can be easily expanded in good manufacturing practice conditions, and they can be used in the allogeneic setting, providing a ready-to-use, "off-the-shelf" therapy (Fierabracci et al. 2015). Thus, not surprisingly placental MSCs have gained considerable interest in cell therapy approaches.

The following sections provide an overview of placental MSCs in preclinical and clinical settings of autoimmune diseases. The nomenclature established by the consensus after the First International Workshop on Placenta Derived Stem Cells is adopted in this chapter (Parolini et al. 2008), and it refers to human amniotic mesenchymal stromal cells (hAMSCs) and human chorion mesenchymal stromal cells (hCMSCs). MSCs can also be isolated from human umbilical cord (UC) and are referred to as either hUC-MSCs or human Wharton's jelly (hWJ)-MSCs, and from the maternal decidua human decidual mesenchymal stromal cells (hDMSCs).

9.2 USE OF PLACENTA-DERIVED CELLS IN PRECLINICAL MODELS OF AUTOIMMUNE DISORDERS

As mentioned above, the ability of MSCs to dampen the inflammatory response makes them attractive candidates in the treatment of autoimmune diseases (Dazzi and Krampera 2011; Figueroa et al. 2012; Ma et al. 2014). Moreover, the ease in with which placenta is obtainable renders the use of MSCs from this tissue even more attractive. The rationale underlying MSC use is to provide immunosuppressive effects, and even more so, to reprogram the immune system.

9.2.1 INFLAMMATORY BOWEL DISEASE

Inflammatory bowel diseases (IBD) primarily include Crohn's disease and ulcerative colitis. The inflammation associated with Crohn's disease can be found in any part of the gastrointestinal tract and usually affects the entire bowel wall, whereas that associated with ulcerative colitis is restricted to the large intestine and is usually limited to the epithelial lining (Fakhoury et al. 2014).

The Use of Placenta-Derived Cells in Autoimmune Disorders

A recent study investigated the efficacy of hAMSCs treatment in 2,4,6-trinitrobenzene sulfonic acid (TNBS)– and dextran sodium sulfate (DSS)–induced colitis. Mice treated with TNBS or with DSS develop a severe illness characterized by bloody diarrhea, rectal prolapse, and pancolitis, accompanied by sustained weight loss resulting in high mortality. Intestinal inflammation results from impairment of the intestinal epithelial cell barrier's function, subsequent exposition of the submucosa to luminal antigens (bacteria and food), and activation of the inflammatory cells involved in innate immunity. Treatment of colitic mice with hAMSCs improved histopathological signs, reduced inflammatory markers in the colon, increased body weight recovery, and increased survival (Parolini et al. 2014). Accordingly, another study showed that MSCs from the UC were also able to improve DSS-induced colitis in mice through a decrease of colon inflammation and adjustment of Treg and Th17 cells (Li et al. 2013). An interesting study published by Gao et al. (2012) using conditioned medium (CM) prepared from WJ's-MSCs (CM-hWJ-MSCs) demonstrated that when delivered intravenously just after injury induction, CM-hWJ-MSCs were able to prevent radiation-induced intestinal injury in mouse model of intestinal damage. They showed that the therapeutic effects were mediated by the stimulation of anti-inflammatory cytokine interleukin (IL)-10, inhibition of proinflammatory cytokines, and inhibition of Th1 and Th17 responses and cytokines (Gao et al. 2012).

A different study using another birth-associated tissue, namely, amniotic fluid, demonstrated that treatment with MSCs from amniotic fluid was able to improve survival and clinical conditions of rats with established necrotizing enterocolitis (Zani et al. 2014).

9.2.2 Systemic Lupus Erythematosus

SLE is characterized by defects in apoptotic clearance, leading to the excessive deposition of apoptotic debris (Lisnevskaia et al. 2014). Antigen-presenting cells capture the nuclear particles (apoptotic debris) that macrophages fail to efficiently remove, causing the development of antinuclear antibodies through interactions with T and B cells, ultimately leading to disease (Lisnevskaia et al. 2014). A commonly used spontaneous model of lupus is the MLR/lpr mouse (Perry et al. 2011). Studies on Fas-expressing B cells and T cells from MRL/lpr mice confirmed a defect in apoptosis due to the lack of a functional Fas receptor; thus, deficiency in Fas signaling results in an SLE-like phenotype due to defective Fas-mediated apoptosis (Perry et al. 2011). MRL/lpr mice also show a remarkably high level of circulating autoantibodies. One study showed that treatment with hUC-MSCs resulted in a significant decrease in autoantibodies in MRL/lpr mice, and this treatment was also able to restore renal functions, as shown by decreased proteinuria and lower blood urea nitrogen and creatinine levels compared to control animals (Gu et al. 2010). Serum and urine monocyte chemoattractant protein (MCP)-1 (Marks et al. 2008) and high-mobility group protein B1 (Wang et al. 2001), two proinflammatory cytokines that have been suggested to play an important role in the pathogenesis of lupus in humans, were also found to decrease after treatment with hUC-MSCs. Moreover, CD4+CD25+Foxp3+ T cells were significantly increased in the spleens of MLR/lpr mice after treatment with hUC-MSCs, similar to what was observed after treatment with bone-marrow–derived (BM)-MSCs (Gu et al. 2010). Accordingly, allogeneic hUC-MSCs were shown to significantly induce CD4+CD25+ T cells from the peripheral CD4+T cells of lupus patients, while inhibiting the proliferation of CD4+ CD25–T cells (Wang et al. 2014).

9.2.3 MULTIPLE SCLEROSIS

MS is a chronic autoimmune disease of the central nervous system (CNS) that affects approximately 2.5 million people worldwide (International Multiple Sclerosis Genetics Consortium et al. 2011). The majority of patients present clinical manifestations such as the involvement of motor, sensory, and visual systems (Compston and Coles 2008). Results obtained from mouse models of MS suggest that the disease is initiated by a persistent, autoimmune attack on CNS autoantigens, such as myelin oligodendrocyte glycoprotein (MOG) (reviewed in McDonald et al. 2011).

Recently, several studies have investigated the use of the placental cells, specifically using epithelial cells isolated from the human amniotic membrane of term placenta (human amniotic epithelial cells [hAECs]) in the treatment of MS (McDonald et al. 2011). The MOG-induced murine EAE model is that which is most commonly used to study MS whereby EAE is induced by immunization with MOG peptide (33-35) in complete Freund's adjuvant with mycobacterium tuberculosis and pertussis toxin. Paralytic symptoms, such as loss of tail tone and hind legs, occur 8–14 days after immunization and mice remain chronically paralyzed (Irani 2005). One group has shown a significant reduction in severity of the disease and prolonged survival of the animals by using maternal and fetal MSCs isolated from placental tissue closest to the UC (Fisher-Shoval et al. 2012). In this study, placental MSCs were infused intravenously near the onset of EAE. Histological investigations of distribution of cells labeled with PKH26 revealed that a significant number of MSCs were present in the brain and in the host spinal cord 30 days posttransplantation, thus demonstrating their ability to home to sites of injury. The authors attributed the therapeutic effects of placental MSCs to both neuroprotection and immunomodulation in the EAE brain and spinal cord, and they speculated that the beneficial effect of the transplanted cells could be mediated by the action of the tumor necrosis factor (TNF)-α–stimulated gene/protein 6 produced at the CNS inflammatory site (Fisher-Shoval et al. 2012). This is in line with their *in vitro* findings that showed that conditioned media from lipopolysaccharide-activated astrocytes stimulated the expression of the anti-inflammatory TSG-6 in placental MSCs (Fisher-Shoval et al. 2012). In a different study that also used the MOG-induced EAE model, administration of hAECs caused a significant reduction in the infiltration of CD3+ T cells and F4/80+ monocytes/macrophages within the CNS accompanied by a reduction of myelin loss (Liu et al. 2012). The authors suggested that the observed therapeutic effects could be due to their immunosuppressive actions mediated by the production of transforming growth factor beta (TGF-β) and prostaglandin E2 observed *in vitro* (Liu et al. 2012). In a subsequent study, the same group investigated the efficacy of hAECs in established EAE after prednisolone-induced remission (Liu et al. 2014b), whereby hAECs were administered during the temporary remission phase induced by treatment with prednisolone to determine whether the mice could be maintained

The Use of Placenta-Derived Cells in Autoimmune Disorders

in remission. The authors reported that hAEC infusion was able to significantly delay relapse by 1 week and induced remission in the majority of mice that lasted over the entire study period of 5 weeks (Liu et al. 2014b). In line with reports documenting reduced numbers of peripheral Treg cells in patients with relapsing MS (Venken et al. 2008) and those showing that the transfer of CD4+CD25+ Treg cells protects mice for EAE (Kohm et al. 2002), the authors also demonstrated increased levels of CD4+CD25+, CD4+FoxP3+, and CD4+CD25+FoxP3+ Treg cells in the draining inguinal lymph nodes near the MOG injection site in hAEC-treated mice (Liu et al. 2014b). They speculated that the increased Treg cells could have a role in decreasing MOG-specific immune response. Together with observations showing lower splenocyte proliferation to the MOG peptide, the authors suggested that the therapeutic role of hAECs may lie in modulating peripheral T cell functions that could contribute the prevention of relapse in corticosteroid-remitted EAE mice. Amelioration of clinical signs, accompanied by a decrease in inflammatory cytokines in the spinal cord, and impaired peripheral MOG-specific T cell responses, were also observed by a different group that used the EAE model to investigate the efficacy of treatment with another placental cell population, namely hAMSCs (Parolini et al. 2014).

9.2.4 Rheumatoid Arthritis

RA is characterized by persistent synovial inflammation in multiple joints, more frequently involving the small joints in hands and feet (Sanmarti et al. 2013), that may ultimately lead to joint destruction and deformities. The local cell populations believed to be predominantly affected by RA are synovial and cartilage cells, and macrophage-like synovial cells are thought to be the main proinflammatory cytokine producers (Scott et al. 2010). The main autoantibodies in RA are rheumatoid factor and those against citrullinated peptides (Scott et al. 2010). The collagen-induced arthritis (CIA) mouse model is the most commonly used experimental model of RA, whereby autoimmune arthritis is induced by immunization with an emulsion of complete Freund's adjuvant and type II collagen (Brand et al. 2007). In a very recent study, the CIA model was used to investigate the effects of hAMSCs in RA (Parolini et al. 2014). Intraperitoneal injections of hAMSCs were able to attenuate established arthritis in mice, which correlated with decreased cartilage damage and bone erosion and neutrophil infiltration in the joints of treated mice. Interestingly, hAMSCs showed specific tropism for inflamed joints and lymphoid organs in arthritic mice. Studies aimed to uncover the mechanisms underlying the reduced CIA severity after hAMSC treatment showed that cells from draining lymph nodes of mice treated with hAMSCs proliferated less than those from untreated mice and that they produced lower levels of interferon (IFN)-γ–producing Th1 cells and Th17 cytokines, and high levels of IL-10 (Parolini et al. 2014). The immunosuppressive properties of hAMSCs were substantiated by *in vitro* results showing that hAMSCs were able to inhibit proliferation of peripheral blood mononuclear cells (PBMCs) isolated from patients with RA, to decrease the production of Th1 cytokines IFN-γ and TNF-α, and to induce secretion of the anti-inflammatory cytokine IL-10. Moreover, Parolini et al. (2014) observed that hAMSCs were able to induce CD4+CD25+FoxP3+ T cells in mice with CIA.

The effects of hAMSCs observed in mice with CIA and in cells from patients with RA seemed to depend primarily on the production of factors deriving from cyclooxygenase-1/2 activation and IL-10 (Parolini et al. 2014).

Another group used the same model to investigate therapeutic effects of hUC-MSCs (Liu et al. 2010). Infusion of hUC-MSCs was able to attenuate the severity of CIA and to restore normal morphology of joints with a decrease or even absence of inflammatory cell infiltration and pannus formation. This was accompanied by a decrease in circulating proinflammatory cytokines TNF-α, IL-6, and MCP-1 and an increase of anti-inflammatory cytokine IL-10. Moreover, hUC-MSCs were not detectable in the joints of mice, but they were detectable only up to 7 days postinfusion in the spleen, suggesting the paracrine action of these cells. Analysis of cytokines expressed by splenic CD4+ T cells showed that hUC-MSCs downregulated IFN-γ–producing Th1 cells and IL-17–producing Th17 cells, while upregulating IL-4–producing Th2 cells and CD4+FoxP3+ Treg cells (Liu et al. 2010). hUC-MSCs were also able to inhibit proliferation of TNF-α–stimulated fibroblast-like synoviocytes from RA patients, which was restored upon addition of IL-10, TGF-β1, and indoleamine 2,3-dioxygenase inhibitors, thus underlying a role for these factors in the immune inhibition mediated by hUC-MSCs. Other *in vitro* systems using Th17 cells in PBMCs of RA patients also showed that hUC-MSCs were able to suppress the expression of IL-17, IL-6, TNF-α, and even orphan nuclear receptor gamma, a key transcription factor for the differentiation of Th17 cells (Wang et al. 2012).

A different, interesting study showed that placental supernatants (PSs), obtained from rat placentas after removal of fetal components and external membranes and minced into small pieces and harvested for supernatant collection, were able to decrease arthritic scores in CIA rats after intraperitoneal injection. This was accompanied by a significant decrease in circulating TNF-α and IFN-γ in PS-treated rats (Cortes et al. 2008).

9.2.5 Autoimmune Myocarditis

Autoimmune responses and uncontrolled inflammation are implicated in the development of many cardiovascular diseases. Myocarditis is defined as inflammation of the myocardium with consequent myocardial injury, and in many cases its etiology is unknown. There is indirect evidence showing that several types of myocarditis are caused by autoimmunity, as seen by the association of myocarditis with other autoimmune diseases, such as lupus (Cihakova and Rose 2008). Furthermore, patients with myocarditis have been shown to produce heart-specific antibodies (Neumann et al. 1990), and immunosuppressive therapy has proven beneficial in some patients with myocarditis (Kuhl et al. 2005). Preclinical studies revealed that cardiac myosin was one of these antigens (Neumann et al. 1992) that was later found to be highly present in patients (Lauer et al. 2000).

Several mouse models that produce myocardial inflammation and elicit heart-reactive antibodies are used to study autoimmune myocarditis. The two most commonly used are disease induction by infection with Coxsackievirus B3 (Huber et al. 1998), or by immunization with cardiac myosin (Neu et al. 1987).

The Use of Placenta-Derived Cells in Autoimmune Disorders 171

Recently, significant improvements in disease were observed after allogeneic fetal membrane (FM)-derived MSCs were transplanted in a rat model of cardiac myosin–induced experimental autoimmune myocarditis (EAM) (Ishikane et al. 2010; Ohshima et al. 2012). Specifically, the authors observed significant improvements in cardiac function, as shown by improved fractional shortening, ejection fraction, and decreased wall thickness in treated animals. Furthermore, in myocardial tissues from animals treated with allogeneic FM-MSCs, there was a decreased infiltration of macrophages and IL-17–positive and IFN-γ–positive cells compared to untreated cells, and a significant decrease in peripheral Th1 and Th17 CD4+ cells (Ohshima et al. 2012). Interestingly, cell distribution studies showed that intravenously injected green fluorescent protein–tagged FM-MSCs were found for the most part in the lungs 1 day after injection, whereas cells were no longer detectable 4 weeks after injection (Ishikane et al. 2010), suggesting paracrine actions of these cells in the ability to ameliorate disease outcome.

9.3 CLINICAL TRIALS

Taking into account the beneficial effects observed in preclinical studies after transplantation of placental MSCs in animal models of autoimmune diseases, it is tempting to speculate that to some degree, these effects could be observed in humans. At the time this book was written, a search for clinical trials using "placenta-derived cells" gave rise to 25 registered clinical trials (www.clinicaltrials. gov), 8 of which were irrelevant or with unknown status. Of the 17 trials remaining, one trial was performed in patients with active RA (NCT01261403) and has been terminated, and two trials were performed in patients with Crohn's disease, the first of which is a phase I currently ongoing trial (NCT01769755), and the second was a phase II trial that has been completed (NCT01155362, no results available). All three of these trials have been or are being sponsored by Celgene Corporation. This corporation has developed and commercialized PDA001, a mesenchymal-like cell population derived from normal, full-term human placental tissue. Recently, a phase I trial was completed with the attempt to evaluate the safety of PDA001 in 12 patients with moderate-to-severe Crohn's disease refractory to oral corticosteroids and immune modulators (Mayer et al. 2013). Some of the enrolled patients continued to receive immunosuppressive therapy during the study; thus, results regarding the safety of PDA001 were difficult to interpret conclusively. PDA001 was moved to a phase II study in which the primary objective was to estimate the treatment effect of three doses of PDA001 versus placebo in patients with moderate-to-severe Crohn's disease, with the secondary objective of assessing safety and tolerability. This trial was completed in April 2014 (NCT01155362), but results have not yet been posted. Surely, lessons can be learned for the design of further clinical trials needed to obtain a clear safety profile, especially concerning the use of placental cells alone, but also in combination with standard treatments, and to provide insight into the efficacy of this treatment. Recent preclinical studies investigating the mechanisms of action of PDA001 in models of autoimmune diseases have supported previous observations that these cells are able to modulate T cell and dendritic cell differentiation (Liu et al. 2014a).

9.3.1 PUBLISHED CLINICAL TRIALS

One study analyzed treatment with hUC-MSCs in 16 patients with SLE and reported beneficial results for all patients, without adverse events (Sun et al. 2010). One million cells per kilogram were administered intravenously, with initial immunosuppression and a significant drop in clinical scores 1 month after the start of treatment was observed, and this drop persisted in 10 patients who were followed for 6 months and in the two remaining patients who completed 2-year follow-ups (Sun et al. 2010). Moreover, 3 months after treatment, 15 patients had improved renal function, as shown by a significant reduction in proteinuria and improved serum creatinine levels, and two patients remained negative for proteinuria after 2 years. Mechanistic insights showed that circulating CD4+FoxP3+ T cells increased significantly 3 months after the start of treatment in all patients. Sun et al. (2010) also noted an increase in TGF-β and IFN-γ and a decrease in IL4 in peripheral blood, and these differences were significant starting at 3 months from treatment.

Another promising recently published case study was performed in a 25-year-old male patient with MS (Hou et al. 2013). Over 4-year period, the patient received intravenous treatment, intrathecal treatment, or both of autologous BM-MSCs or allogeneic hUC-MSCs for eight treatment sessions in total. Throughout this period, the patient did not receive any other treatments. The cell therapy was well tolerated, with minor rashes and pains that were considered transient infusion-related symptoms. No new lesions were detected by magnetic resonance imaging, and many of the current lesions had resolved 7 months after the last infusion (Hou et al. 2013).

Another recent much larger study enrolled 172 patients with RA with inadequate response to traditional medication and treated them with either disease-modifying antirheumatic drugs (DMARDs) or DMARD plus hUC-MSCs (4×10^7) via intravenous infusions (Wang et al. 2013). No serious adverse effects were observed during or after infusion, and serum levels of TNF-α and IL-6 decreased significantly, accompanied by a significant increase of CD4+CD25+FoxP3+ Treg cells in peripheral blood of patients, after the first infusion. These therapeutic effects were maintained for 3–6 months without continuous infusion (Wang et al. 2013).

9.4 CONCLUDING REMARKS

Over the past decade, research on placenta as a source of stem cells has grown exponentially. The placenta harbors different tissues from which different cell types can be isolated, each with unique properties, such as the ability to curb immune responses, and perhaps other interesting properties yet to be discovered. The recent success of placental cells in clinical trials has sparked interest and questions. Surely, the immunomodulatory properties of placental cells will continue to attract attention for their potential exploitation in other diseases with underlying, dysregulated immune processes, including autoimmune diseases.

ACKNOWLEDGMENTS

We thank all the mothers who donated their baby's placenta, thus making this research possible.

REFERENCES

Abumaree, M., M. Al Jumah, R. A. Pace, and B. Kalionis. 2012. Immunosuppressive properties of mesenchymal stem cells. *Stem Cell Rev* 8 (2): 375–92. doi: 10.1007/s12015-011-9312-0.

Abumaree, M. H., M. A. Al Jumah, B. Kalionis, et al. 2013. Human placental mesenchymal stem cells (pMSCs) play a role as immune suppressive cells by shifting macrophage differentiation from inflammatory M1 to anti-inflammatory M2 macrophages. *Stem Cell Rev* 9 (5): 620–41. doi: 10.1007/s12015-013-9455-2.

Anderson, P., L. Souza-Moreira, M. Morell, et al. 2013. Adipose-derived mesenchymal stromal cells induce immunomodulatory macrophages which protect from experimental colitis and sepsis. *Gut* 62 (8): 1131–41. doi: 10.1136/gutjnl-2012-302152.

Atassi, M. Z., and P. Casali. 2008. Molecular mechanisms of autoimmunity. *Autoimmunity* 41 (2): 123–32. doi: 10.1080/08916930801929021.

Barragan-Martinez, C., C. A. Speck-Hernandez, G. Montoya-Ortiz, et al. 2012. Organic solvents as risk factor for autoimmune diseases: A systematic review and meta-analysis. *PLoS One* 7 (12): e51506. doi: 10.1371/journal.pone.0051506.

Begovich, A. B., V. E. Carlton, L. A. Honigberg, et al. 2004. A missense single-nucleotide polymorphism in a gene encoding a protein tyrosine phosphatase (PTPN22) is associated with rheumatoid arthritis. *Am J Hum Genet* 75 (2): 330–7. doi: 10.1086/422827.

Berer, K., M. Mues, M. Koutrolos, et al. 2011. Commensal microbiota and myelin autoantigen cooperate to trigger autoimmune demyelination. *Nature* 479 (7374): 538–41. doi: 10.1038/nature10554.

Bernardo, M. E., and W. E. Fibbe. 2013. Mesenchymal stromal cells: Sensors and switchers of inflammation. *Cell Stem Cell* 13 (4): 392–402. doi: 10.1016/j.stem.2013.09.006.

Bhargava, P., and E. M. Mowry. 2014. Gut microbiome and multiple sclerosis. *Curr Neurol Neurosci Rep* 14 (10): 492. doi: 10.1007/s11910-014-0492-2.

Bigazzi, P. E. 1997. Autoimmunity caused by xenobiotics. *Toxicology* 119 (1): 1–21.

Blank, M., I. Krause, M. Fridkin, et al. 2002. Bacterial induction of autoantibodies to beta2-glycoprotein-I accounts for the infectious etiology of antiphospholipid syndrome. *J Clin Invest* 109 (6): 797–804. doi: 10.1172/JCI12337.

Brand, D. D., K. A. Latham, and E. F. Rosloniec. 2007. Collagen-induced arthritis. *Nat Protoc* 2 (5): 1269–75. doi: 10.1038/nprot.2007.173.

Buckner, J. H. 2010. Mechanisms of impaired regulation by CD4(+)CD25(+)FOXP3(+) regulatory T cells in human autoimmune diseases. *Nat Rev Immunol* 10 (12): 849–59. doi: 10.1038/nri2889.

Bynoe, M. S., C. M. Grimaldi, and B. Diamond. 2000. Estrogen up-regulates Bcl-2 and blocks tolerance induction of naive B cells. *Proc Natl Acad Sci U S A* 97 (6): 2703–708. doi: 10.1073/pnas.040577497.

Carter, J. D., L. R. Espinoza, R. D. Inman, et al. 2010. Combination antibiotics as a treatment for chronic Chlamydia-induced reactive arthritis: A double-blind, placebo-controlled, prospective trial. *Arthritis Rheum* 62 (5): 1298–307. doi: 10.1002/art.27394.

Caruso, M., M. Evangelista, and O. Parolini. 2012. Human term placental cells: Phenotype, properties and new avenues in regenerative medicine. *Int J Mol Cell Med* 1 (2): 64–74.

Ciccocioppo, R., M. E. Bernardo, A. Sgarella, et al. 2011. Autologous bone marrow-derived mesenchymal stromal cells in the treatment of fistulising Crohn's disease. *Gut* 60 (6): 788–98. doi: 10.1136/gut.2010.214841.

Cihakova, D., and N. R. Rose. 2008. Pathogenesis of myocarditis and dilated cardiomyopathy. *Adv Immunol* 99: 95–114. doi: 10.1016/S0065-2776(08)00604-4.

Compston, A., and A. Coles. 2008. Multiple sclerosis. *Lancet* 372 (9648): 1502–17. doi: 10.1016/S0140-6736(08)61620-7.

Cortes, M., A. Canellada, S. Miranda, J. Dokmetjian, and T. Gentile. 2008. Placental secreted factors: Their role in the regulation of anti-CII antibodies and amelioration of collagen induced arthritis in rats. *Immunol Lett* 119 (1–2): 42–48. doi: 10.1016/j.imlet.2008.04.001.

Davidson, A., and B. Diamond. 2001. Autoimmune diseases. *N Engl J Med* 345 (5): 340–50. doi: 10.1056/NEJM200108023450506.

Dazzi, F., and M. Krampera. 2011. Mesenchymal stem cells and autoimmune diseases. *Best Pract Res Clin Haematol* 24 (1): 49–57. doi: 10.1016/j.beha.2011.01.002.

de Girolamo, L., E. Lucarelli, G. Alessandri, et al. 2013. Mesenchymal stem/stromal cells: A new "cells as drugs" paradigm. Efficacy and critical aspects in cell therapy. *Curr Pharm Des* 19 (13): 2459–73.

Duijvestein, M., A. C. Vos, H. Roelofs, et al. 2010. Autologous bone marrow-derived mesenchymal stromal cell treatment for refractory luminal Crohn's disease: Results of a phase I study. *Gut* 59 (12): 1662–69. doi: 10.1136/gut.2010.215152.

Evangelista, M., M. Soncini, and O. Parolini. 2008. Placenta-derived stem cells: New hope for cell therapy? *Cytotechnology* 58 (1): 33–42. doi: 10.1007/s10616-008-9162-z.

Fakhoury, M., R. Negrulj, A. Mooranian, and H. Al-Salami. 2014. Inflammatory bowel disease: Clinical aspects and treatments. *J Inflamm Res* 7: 113–20. doi: 10.2147/JIR. S65979.

Farquharson, D., J. P. Butcher, and S. Culshaw. 2012. Periodontitis, Porphyromonas, and the pathogenesis of rheumatoid arthritis. *Mucosal Immunol* 5 (2): 112–20. doi: 10.1038/ mi.2011.66.

Fierabracci, A., L. Lazzari, M. Muraca, and O. Parolini. 2015. How far are we from the clinical use of placental-derived mesenchymal stem cells? *Expert Opin Biol Ther* 15 (5): 613–17. doi: 10.1517/14712598.2015.1000856.

Figueroa, F. E., F. Carrion, S. Villanueva, and M. Khoury. 2012. Mesenchymal stem cell treatment for autoimmune diseases: A critical review. *Biol Res* 45 (3): 269–77. doi: 10.4067/S0716-97602012000300008.

Firestein, G. S., R. C. Budd, S. E. Gabriel, I. B. McInnes, and J. R. O'Dell. 2013. *Kelley's Textbook of Rheumatology.* 9th ed. Elsevier Saunders, PA: Philadelphia.

Fisher-Shoval, Y., Y. Barhum, O. Sadan, et al. 2012. Transplantation of placenta-derived mesenchymal stem cells in the EAE mouse model of MS. *J Mol Neurosci* 48 (1): 176–84. doi: 10.1007/s12031-012-9805-6.

Gao, Z., Q. Zhang, Y. Han, et al. 2012. Mesenchymal stromal cell-conditioned medium prevents radiation-induced small intestine injury in mice. *Cytotherapy* 14 (3): 267–73. doi: 10.3109/14653249.2011.616194.

Germolec, D., D. H. Kono, J. C. Pfau, and K. M. Pollard. 2012. Animal models used to examine the role of the environment in the development of autoimmune disease: Findings from an NIEHS Expert Panel Workshop. *J Autoimmun* 39 (4): 285–93. doi: 10.1016/ j.jaut.2012.05.020.

Ghannam, S., C. Bouffi, F. Djouad, C. Jorgensen, and D. Noel. 2010. Immunosuppression by mesenchymal stem cells: Mechanisms and clinical applications. *Stem Cell Res Ther* 1 (1): 2. doi: 10.1186/scrt2.

Gonzalez, M. A., E. Gonzalez-Rey, L. Rico, D. Buscher, and M. Delgado. 2009a. Adipose-derived mesenchymal stem cells alleviate experimental colitis by inhibiting inflammatory and autoimmune responses. *Gastroenterology* 136 (3): 978–89. doi: 10.1053/j. gastro.2008.11.041.

Gonzalez, M. A., E. Gonzalez-Rey, L. Rico, D. Buscher, and M. Delgado. 2009b. Treatment of experimental arthritis by inducing immune tolerance with human adipose-derived mesenchymal stem cells. *Arthritis Rheum* 60 (4): 1006–19. doi: 10.1002/art.24405.

Gonzalez-Rey, E., P. Anderson, M. A. Gonzalez, et al. 2009. Human adult stem cells derived from adipose tissue protect against experimental colitis and sepsis. *Gut* 58 (7): 929–39. doi: 10.1136/gut.2008.168534.

The Use of Placenta-Derived Cells in Autoimmune Disorders **175**

Gonzalez-Rey, E., M. A. Gonzalez, N. Varela, et al. 2010. Human adipose-derived mesenchymal stem cells reduce inflammatory and T cell responses and induce regulatory T cells in vitro in rheumatoid arthritis. *Ann Rheum Dis* 69 (1): 241–48. doi: 10.1136/ard.2008.101881.

Goodnow, C. C., S. Adelstein, and A. Basten. 1990. The need for central and peripheral tolerance in the B cell repertoire. *Science* 248 (4961): 1373–79.

Griem, P., M. Wulferink, B. Sachs, J. B. Gonzalez, and E. Gleichmann. 1998. Allergic and autoimmune reactions to xenobiotics: How do they arise? *Immunol Today* 19 (3): 133–41.

Gu, Z., K. Akiyama, X. Ma, et al. 2010. Transplantation of umbilical cord mesenchymal stem cells alleviates lupus nephritis in MRL/lpr mice. *Lupus* 19 (13): 1502–14. doi: 10.1177/0961203310373782.

Hampe, J., A. Cuthbert, P. J. Croucher, et al. 2001. Association between insertion mutation in NOD2 gene and Crohn's disease in German and British populations. *Lancet* 357 (9272): 1925–28. doi: 10.1016/S0140-6736(00)05063-7.

Hass, R., C. Kasper, S. Bohm, and R. Jacobs. 2011. Different populations and sources of human mesenchymal stem cells (MSC): A comparison of adult and neonatal tissue-derived MSC. *Cell Commun Signal* 9: 12. doi: 10.1186/1478-811X-9-12.

Hayter, S. M., and M. C. Cook. 2012. Updated assessment of the prevalence, spectrum and case definition of autoimmune disease. *Autoimmun Rev* 11 (10): 754–65. doi: 10.1016/j.autrev.2012.02.001.

Herr, M., F. Dudbridge, P. Zavattari, et al. 2000. Evaluation of fine mapping strategies for a multifactorial disease locus: Systematic linkage and association analysis of IDDM1 in the HLA region on chromosome 6p21. *Hum Mol Genet* 9 (9): 1291–301.

Hou, Z. L., Y. Liu, X. H. Mao, et al. 2013. Transplantation of umbilical cord and bone marrow-derived mesenchymal stem cells in a patient with relapsing-remitting multiple sclerosis. *Cell Adh Migr* 7 (5): 404–7. doi: 10.4161/cam.26941.

Huber, S. A., C. J. Gauntt, and P. Sakkinen. 1998. Enteroviruses and myocarditis: Viral pathogenesis through replication, cytokine induction, and immunopathogenicity. *Adv Virus Res* 51: 35–80.

Hugot, J. P., M. Chamaillard, H. Zouali, et al. 2001. Association of NOD2 leucine-rich repeat variants with susceptibility to Crohn's disease. *Nature* 411 (6837): 599–603. doi: 10.1038/35079107.

Imai, D., K. Holden, E. M. Velazquez, et al. 2013. Influence of arthritis-related protein (BBF01) on infectivity of Borrelia burgdorferi B31. *BMC Microbiol* 13: 100. doi: 10.1186/1471-2180-13-100.

International Multiple Sclerosis Genetics Consortium, Consortium Wellcome Trust Case Control, S. Sawcer, et al. 2011. Genetic risk and a primary role for cell-mediated immune mechanisms in multiple sclerosis. *Nature* 476 (7359): 214–19. doi: 10.1038/nature10251.

Irani, D. N. 2005. Immunological mechanisms in multiple sclerosis. *Clinical and Applied Immunology Reviews* 5 (4): 257–69.

Ishikane, S., K. Yamahara, M. Sada, et al. 2010. Allogeneic administration of fetal membrane-derived mesenchymal stem cells attenuates acute myocarditis in rats. *J Mol Cell Cardiol* 49 (5): 753–61. doi: 10.1016/j.yjmcc.2010.07.019.

Kivity, S., N. Agmon-Levin, M. Blank, and Y. Shoenfeld. 2009. Infections and autoimmunity—Friends or foes? *Trends Immunol* 30 (8): 409–14. doi: 10.1016/j.it.2009.05.005.

Kohm, A. P., P. A. Carpentier, H. A. Anger, and S. D. Miller. 2002. Cutting edge: CD4+CD25+ regulatory T cells suppress antigen-specific autoreactive immune responses and central nervous system inflammation during active experimental autoimmune encephalomyelitis. *J Immunol* 169 (9): 4712–16.

Kuhl, U., M. Pauschinger, B. Seeberg, et al. 2005. Viral persistence in the myocardium is associated with progressive cardiac dysfunction. *Circulation* 112 (13): 1965–70. doi: 10.1161/CIRCULATIONAHA.105.548156.

Kyewski, B., and L. Klein. 2006. A central role for central tolerance. *Annu Rev Immunol* 24: 571–606. doi: 10.1146/annurev.immunol.23.021704.115601.

Kyogoku, C., C. D. Langefeld, W. A. Ortmann, et al. 2004. Genetic association of the R620W polymorphism of protein tyrosine phosphatase PTPN22 with human SLE. *Am J Hum Genet* 75 (3): 504–7. doi: 10.1086/423790.

Lauer, B., M. Schannwell, U. Kuhl, B. E. Strauer, and H. P. Schultheiss. 2000. Antimyosin autoantibodies are associated with deterioration of systolic and diastolic left ventricular function in patients with chronic myocarditis. *J Am Coll Cardiol* 35 (1): 11–18.

Lee, Y. K., J. S. Menezes, Y. Umesaki, and S. K. Mazmanian. 2011. Proinflammatory T-cell responses to gut microbiota promote experimental autoimmune encephalomyelitis. *Proc Natl Acad Sci U S A* 108 (Suppl 1): 4615–22. doi: 10.1073/pnas.1000082107.

Leite, J. L., N. E. Bufalo, R. B. Santos, J. H. Romaldini, and L. S. Ward. 2010. Herpes virus type 7 infection may play an important role in individuals with a genetic profile of susceptibility to Graves' disease. *Eur J Endocrinol* 162 (2): 315–21. doi: 10.1530/EJE-09-0719.

Li, L., S. Liu, Y. Xu, et al. 2013. Human umbilical cord-derived mesenchymal stem cells downregulate inflammatory responses by shifting the Treg/Th17 profile in experimental colitis. *Pharmacology* 92 (5–6): 257–64. doi: 10.1159/000354883.

Lin, S. C., J. H. Yen, J. J. Tsai, et al. 2004. Association of a programmed death 1 gene polymorphism with the development of rheumatoid arthritis, but not systemic lupus erythematosus. *Arthritis Rheum* 50 (3): 770–75. doi: 10.1002/art.20040.

Lisnevskaia, L., G. Murphy, and D. Isenberg. 2014. Systemic lupus erythematosus. *Lancet* 384 (9957): 1878–88. doi: 10.1016/S0140-6736(14)60128-8.

Liu, W., A. Morschauser, X. Zhang, et al. 2014a. Human placenta-derived adherent cells induce tolerogenic immune responses. *Clin Trans Immunol* 3: e14. doi: 10.1038/cti.2014.5.

Liu, Y., R. Mu, S. Wang, et al. 2010. Therapeutic potential of human umbilical cord mesenchymal stem cells in the treatment of rheumatoid arthritis. *Arthritis Res Ther* 12 (6): R210. doi: 10.1186/ar3187.

Liu, Y. H., J. Chan, V. Vaghjiani, et al. 2014b. Human amniotic epithelial cells suppress relapse of corticosteroid-remitted experimental autoimmune disease. *Cytotherapy* 16 (4): 535–44. doi: 10.1016/j.jcyt.2013.10.007.

Liu, Y. H., V. Vaghjiani, J. Y. Tee, et al. 2012. Amniotic epithelial cells from the human placenta potently suppress a mouse model of multiple sclerosis. *PLoS One* 7 (4): e35758. doi: 10.1371/journal.pone.0035758.

Ma, S., N. Xie, W. Li, et al. 2014. Immunobiology of mesenchymal stem cells. *Cell Death Differ* 21 (2): 216–25. doi: 10.1038/cdd.2013.158.

Manuelpillai, U., Y. Moodley, C. V. Borlongan, and O. Parolini. 2011. Amniotic membrane and amniotic cells: Potential therapeutic tools to combat tissue inflammation and fibrosis? *Placenta* 32 (Suppl 4): S320–5. doi: 10.1016/j.placenta.2011.04.010.

Marks, S. D., S. J. Williams, K. Tullus, and N. J. Sebire. 2008. Glomerular expression of monocyte chemoattractant protein-1 is predictive of poor renal prognosis in pediatric lupus nephritis. *Nephrol Dial Transplant* 23 (11): 3521–6. doi: 10.1093/ndt/gfn270.

Mayer, L., W. M. Pandak, G. Y. Melmed, et al. 2013. Safety and tolerability of human placenta-derived cells (PDA001) in treatment-resistant Crohn's disease: A phase I study. *Inflamm Bowel Dis* 19: 754–60.

McDonald, C., C. Siatskas, and C. Bernard. 2011. The emergence of amnion epithelial stem cells for the treatment of multiple sclerosis. *Inflamm Regen* 31 (3): 256–71.

Mori, K., Y. Munakata, T. Saito, et al. 2007. Intrathyroidal persistence of human parvovirus B19 DNA in a patient with Hashimoto's thyroiditis. *J Infect* 55 (2): e29–31. doi: 10.1016/j.jinf.2007.05.173.

Mueller, D. L. 2010. Mechanisms maintaining peripheral tolerance. *Nat Immunol* 11 (1): 21–27. doi: 10.1038/ni.1817.

Murphy, M. B., K. Moncivais, and A. I. Caplan. 2013. Mesenchymal stem cells: Environmentally responsive therapeutics for regenerative medicine. *Exp Mol Med* 45: e54. doi: 10.1038/emm.2013.94.

Murphy, S. V., and A. Atala. 2013. Amniotic fluid and placental membranes: Unexpected sources of highly multipotent cells. *Semin Reprod Med* 31 (1): 62–68. doi: 10.1055/s-0032-1331799.

Neu, N., N. R. Rose, K. W. Beisel, et al. 1987. Cardiac myosin induces myocarditis in genetically predisposed mice. *J Immunol* 139 (11): 3630–36.

Neumann, D. A., C. L. Burek, K. L. Baughman, N. R. Rose, and A. Herskowitz. 1990. Circulating heart-reactive antibodies in patients with myocarditis or cardiomyopathy. *J Am Coll Cardiol* 16 (6): 839–46.

Neumann, D. A., J. R. Lane, S. M. Wulff, et al. 1992. In vivo deposition of myosin-specific autoantibodies in the hearts of mice with experimental autoimmune myocarditis. *J Immunol* 148 (12): 3806–13.

Ogura, Y., D. K. Bonen, N. Inohara, et al. 2001. A frameshift mutation in NOD2 associated with susceptibility to Crohn's disease. *Nature* 411 (6837): 603–6. doi: 10.1038/35079114.

Ohshima, M., K. Yamahara, S. Ishikane, et al. 2012. Systemic transplantation of allogenic fetal membrane-derived mesenchymal stem cells suppresses Th1 and Th17 T cell responses in experimental autoimmune myocarditis. *J Mol Cell Cardiol* 53 (3): 420–28. doi: 10.1016/j.yjmcc.2012.06.020.

Onengut-Gumuscu, S., K. G. Ewens, R. S. Spielman, and P. Concannon. 2004. A functional polymorphism (1858C/T) in the PTPN22 gene is linked and associated with type I diabetes in multiplex families. *Genes Immun* 5 (8): 678–80. doi: 10.1038/sj.gene.6364138.

Palmer, E. 2003. Negative selection—Clearing out the bad apples from the T-cell repertoire. *Nat Rev Immunol* 3 (5): 383–91. doi: 10.1038/nri1085.

Parolini, O., F. Alviano, G. P. Bagnara, et al. 2008. Concise review: Isolation and characterization of cells from human term placenta: Outcome of the first international Workshop on Placenta Derived Stem Cells. *Stem Cells* 26 (2): 300–11. doi: 10.1634/stemcells.2007-0594.

Parolini, O., L. Souza-Moreira, F. O'Valle, et al. 2014. Therapeutic effect of human amniotic membrane-derived cells on experimental arthritis and other inflammatory disorders. *Arthritis Rheumatol* 66 (2): 327–39. doi: 10.1002/art.38206.

Perry, D., A. Sang, Y. Yin, Y. Y. Zheng, and L. Morel. 2011. Murine models of systemic lupus erythematosus. *J Biomed Biotechnol* 2011: 271694. doi: 10.1155/2011/271694.

Pillai, S. 2013. Rethinking mechanisms of autoimmune pathogenesis. *J Autoimmun* 45: 97–103. doi: 10.1016/j.jaut.2013.05.003.

Pollard, K. M. 2012. Gender differences in autoimmunity associated with exposure to environmental factors. *J Autoimmun* 38 (2–3): J177–86. doi: 10.1016/j.jaut.2011.11.007.

Pollard, K. M., P. Hultman, and D. H. Kono. 2010. Toxicology of autoimmune diseases. *Chem Res Toxicol* 23 (3): 455–66. doi: 10.1021/tx9003787.

Prokunina, L., C. Castillejo-Lopez, F. Oberg, et al. 2002. A regulatory polymorphism in PDCD1 is associated with susceptibility to systemic lupus erythematosus in humans. *Nat Genet* 32 (4): 666–69. doi: 10.1038/ng1020.

Ra, J. C., I. S. Shin, S. H. Kim, et al. 2011. Safety of intravenous infusion of human adipose tissue-derived mesenchymal stem cells in animals and humans. *Stem Cells Dev* 20 (8): 1297–308. doi: 10.1089/scd.2010.0466.

Rioux, J. D., M. J. Daly, M. S. Silverberg, et al. 2001. Genetic variation in the 5q31 cytokine gene cluster confers susceptibility to Crohn disease. *Nat Genet* 29 (2): 223–28. doi: 10.1038/ng1001-223.

Rose, N. R., and C. Bona. 1993. Defining criteria for autoimmune diseases (Witebsky's postulates revisited). *Immunol Today* 14 (9): 426–30. doi: 10.1016/0167-5699(93)90244-F.

Rumah, K. R., J. Linden, V. A. Fischetti, and T. Vartanian. 2013. Isolation of *Clostridium perfringens* type B in an individual at first clinical presentation of multiple sclerosis provides clues for environmental triggers of the disease. *PLoS One* 8 (10): e76359. doi: 10.1371/journal.pone.0076359.

Sanmarti, R., V. Ruiz-Esquide, and M. V. Hernandez. 2013. Rheumatoid arthritis: A clinical overview of new diagnostic and treatment approaches. *Curr Top Med Chem* 13 (6): 698–704.

Scott, D. L., F. Wolfe, and T. W. Huizinga. 2010. Rheumatoid arthritis. *Lancet* 376 (9746): 1094–108. doi: 10.1016/S0140-6736(10)60826-4.

Shimon, I., C. Pariente, J. Shlomo-David, Z. Grossman, and J. Sack. 2003. Transient elevation of triiodothyronine caused by triiodothyronine autoantibody associated with acute Epstein-Barr-virus infection. *Thyroid* 13 (2): 211–15. doi: 10.1089/105072503321319530.

Silini, A., O. Parolini, B. Huppertz, and I. Lang. 2013. Soluble factors of amnion-derived cells in treatment of inflammatory and fibrotic pathologies. *Curr Stem Cell Res Ther* 8 (1): 6–14.

Singer, N. G., and A. I. Caplan. 2011. Mesenchymal stem cells: Mechanisms of inflammation. *Annu Rev Pathol* 6: 457–78. doi: 10.1146/annurev-pathol-011110-130230.

Smyth, D., J. D. Cooper, J. E. Collins, et al. 2004. Replication of an association between the lymphoid tyrosine phosphatase locus (LYP/PTPN22) with type 1 diabetes, and evidence for its role as a general autoimmunity locus. *Diabetes* 53 (11): 3020–23.

Soncini, M., E. Vertua, L. Gibelli, et al. 2007. Isolation and characterization of mesenchymal cells from human fetal membranes. *J Tissue Eng Regen Med* 1 (4): 296–305. doi: 10.1002/term.40.

Starr, T. K., S. C. Jameson, and K. A. Hogquist. 2003. Positive and negative selection of T cells. *Annu Rev Immunol* 21: 139–76. doi: 10.1146/annurev.immunol. 21.120601.141107.

Sun, L., K. Akiyama, H. Zhang, et al. 2009. Mesenchymal stem cell transplantation reverses multiorgan dysfunction in systemic lupus erythematosus mice and humans. *Stem Cells* 27 (6): 1421–32. doi: 10.1002/stem.68.

Sun, L., D. Wang, J. Liang, et al. 2010. Umbilical cord mesenchymal stem cell transplantation in severe and refractory systemic lupus erythematosus. *Arthritis Rheum* 62 (8): 2467–75. doi: 10.1002/art.27548.

Tyndall, A., and A. Uccelli. 2009. Multipotent mesenchymal stromal cells for autoimmune diseases: Teaching new dogs old tricks. *Bone Marrow Transplant* 43 (11): 821–28. doi: 10.1038/bmt.2009.63.

Uccelli, A., S. Morando, S. Bonanno, et al. 2011. Mesenchymal stem cells for multiple sclerosis: Does neural differentiation really matter? *Curr Stem Cell Res Ther* 6 (1): 69–72.

Van Loveren, H., J. G. Vos, D. Germolec, et al. 2001. Epidemiologic associations between occupational and environmental exposures and autoimmune disease: Report of a meeting to explore current evidence and identify research needs. *Int J Hyg Environ Health* 203 (5–6): 483–95. doi: 10.1078/1438-4639-00057.

Venken, K., N. Hellings, M. Thewissen, et al. 2008. Compromised CD4+ CD25(high) regulatory T-cell function in patients with relapsing-remitting multiple sclerosis is correlated with a reduced frequency of FOXP3-positive cells and reduced FOXP3 expression at the single-cell level. *Immunology* 123 (1): 79–89. doi: 10.1111/j.1365-2567.2007.02690.x.

Vojdani, Aristo. 2014. A potential link between environmental triggers and autoimmunity. *Autoimmune Dis* 2014: 437231. doi: 10.1155/2014/437231.

Walker, L. S. 2013. Treg and CTLA-4: Two intertwining pathways to immune tolerance. *J Autoimmun* 45: 49–57. doi: 10.1016/j.jaut.2013.06.006.

Walker, L. S., and A. K. Abbas. 2002. The enemy within: Keeping self-reactive T cells at bay in the periphery. *Nat Rev Immunol* 2 (1): 11–19. doi: 10.1038/nri701.

Wang, D., X. Feng, L. Lu, et al. 2014. A CD8 T cell/indoleamine 2,3-dioxygenase axis is required for mesenchymal stem cell suppression of human systemic lupus erythematosus. *Arthritis Rheumatol* 66 (8): 2234–45. doi: 10.1002/art.38674.

Wang, H., H. Yang, C. J. Czura, A. E. Sama, and K. J. Tracey. 2001. HMGB1 as a late mediator of lethal systemic inflammation. *Am J Respir Crit Care Med* 164 (10 Pt 1): 1768–73. doi: 10.1164/ajrccm.164.10.2106117.

Wang, L., L. Wang, X. Cong, et al. 2013. Human umbilical cord mesenchymal stem cell therapy for patients with active rheumatoid arthritis: Safety and efficacy. *Stem Cells Dev* 22 (24): 3192–202. doi: 10.1089/scd.2013.0023.

Wang, Q., X. Li, J. Luo, et al. 2012. The allogeneic umbilical cord mesenchymal stem cells regulate the function of T helper 17 cells from patients with rheumatoid arthritis in an in vitro co-culture system. *BMC Musculoskelet Disord* 13: 249. doi: 10.1186/1471-2474-13-249.

Zani, A., M. Cananzi, F. Fascetti-Leon, et al. 2014. Amniotic fluid stem cells improve survival and enhance repair of damaged intestine in necrotising enterocolitis via a COX-2 dependent mechanism. *Gut* 63 (2): 300–9. doi: 10.1136/gutjnl-2012-303735.

10 The Use of Placenta-Derived Cells in Inflammatory and Fibrotic Disorders

Euan M. Wallace, Anna Cargnoni, Rebecca Lim, Alex Hodge, and William Sievert

CONTENTS

Preface.. 181
10.1 Introduction 182
10.2 Effects of Placenta-Derived Cells and Derivatives on Lung Injury 182
    10.2.1 Human Amniotic Epithelial Cells 183
    10.2.2 Amniotic Fluid Mesenchymal Stromal Cells or Amniotic Fluid
           Stem Cells.. 185
    10.2.3 Other Cells from Human Term Placenta 186
    10.2.4 Studies Using Placental Derivatives for the Treatment
           of Lung Injury.. 187
    10.2.5 Proposed Mechanisms of Action.. 187
    10.2.6 Clinical Trials 188
10.3 Effects of Placenta-Derived Cells and Derivatives on Kidney Fibrosis....... 189
10.4 Effects of Placenta-Derived Cells and Derivatives on Liver Fibrosis 189
    10.4.1 Animal Models of Liver Inflammation and Fibrosis......................... 190
    10.4.2 Studies Using Placenta-Derived Stem Cells
           for Hepatocyte Transplantation 190
    10.4.3 Studies Using Placental Cells for the Treatment of Liver Fibrosis..... 191
10.5 Summary 194
References.. 194

PREFACE

Chronic inflammatory and fibrotic disorders are often poorly and inadequately managed by nonspecific anti-inflammatory therapies such as corticosteroids. As a result, they are never "cured," but rather only managed and, at least for conditions such as lung fibrosis and liver cirrhosis, they remain ultimately fatal. The recent development of biologics such as cytokine receptor antagonists has heralded improved therapeutic options for acute inflammatory conditions, but the lack of safe and effective

181

therapies for the chronic fibrotic disorders largely remains. This is where cell therapies and their ability to "reset" the host immune system may be particularly promising. Different cell populations from the placenta and placental membranes have been assessed in diverse experimental animals of chronic inflammatory and fibrotic diseases. In particular, amnion epithelial and mesenchymal cells appear very promising in their ability to modulate host macrophage, T cell, and fibroblast function, among other actions. In this regard, these cells that are readily and relatively cheaply derived from term placentae would be an abundant source for regenerative cell therapies for diverse and serious conditions such as idiopathic lung fibrosis and liver cirrhosis. In this chapter, we describe the various cell populations and their applications, to date, in animal models of inflammatory and fibrotic conditions. We also anticipate the next step of applying the experimental insights to clinical trials of these cells.

10.1 INTRODUCTION

Fibrosis has been described as an inappropriate and uncontrolled wound healing response. In excess, fibrosis is pathological and can critically compromise organ function. For example, fibrotic diseases such pulmonary fibrosis, liver cirrhosis, and chronic kidney disease are leading causes of morbidity and mortality worldwide. They have the potential to affect multiple tissues and organ systems. Wound healing is essentially a process where dead or damaged cells are replaced after injury. This process has two major phases that may occur in tandem: a regenerative phase where the dead or damaged cells are replaced by healthy cells of the same cell types, and a fibrotic phase where connective tissue replaces the parenchyma. These phases, although usually beneficial, can become pathogenic when either is uncontrolled. If substantial amounts of extracellular matrix (ECM) remodeling takes place, this can eventuate in irreversible scarring and potentially culminate in organ failure, whereas uncontrolled regeneration may result in neoplasia.

Pathogenic fibrosis often arises from chronic inflammation that persists for weeks, months, or even longer. In these instances, inflammation, tissue destruction, and repair processes occur simultaneously and on an ongoing basis. A common denominator across most fibrotic disorders is the persistence of a trigger, usually chronic injury, that sustains the production of growth factors, proteolytic enzymes, angiogenic factors, and fibrogenic cytokines. These come together to stimulate the excessive deposition of ECM that results in pathogenic remodeling, thus destroying tissue architecture (Tomasek et al. 2002; Wynn 2007). Although there is some recent evidence to suggest that cell therapies can mitigate or reverse fibrosis by influencing the cells directly responsible for the fibrotic response, it is generally accepted that cell therapies ameliorate fibrosis via their immunomodulatory effects. In this chapter, we discuss the application of placenta-derived stem cells in the treatment of inflammatory and fibrotic disorders across different organ systems.

10.2 EFFECTS OF PLACENTA-DERIVED CELLS AND DERIVATIVES ON LUNG INJURY

Placenta-derived cells and related derivatives, such as conditioned media generated from the *in vitro* culture of cells, have been experimentally applied to determine their effect on specific lung injuries induced in animal models. The majority of studies

The Use of Placenta-Derived Cells in Inflammatory and Fibrotic Disorders **183**

examining the effects of placenta-derived cells on lung injury and repair have been undertaken using delivery of xenogenic cells, usually human cells, in lung injury animal models. Only a handful of studies have used allogeneic (Cargnoni et al. 2014; Li et al. 2014a) or syngeneic (Garcia et al. 2013) cell transplantation. Placenta-derived cells, including cells from amniotic membrane, Wharton's jelly, amniotic fluid, and chorionic villi, as specified herein, have been administered as mono-delivery in a dose ranging from 0.1 to 4 million in mice or rat models and in a dose ranging from 120 to 180 million in sheep models (Hodges et al. 2012; Vosdoganes et al. 2011). Mostly, cells are delivered systemically (usually intravenously) (Cargnoni et al. 2009; Garcia et al. 2013; Hodges et al. 2012; Moodley et al. 2009, 2010, 2013; Vosdoganes et al. 2011; Wen et al. 2013), intraperitoneally (Cargnoni et al. 2009; Lim et al. 2013; Murphy et al. 2011a, 2012a; Tan et al. 2014, 2015; Vosdoganes et al. 2012a, 2013), or locally (intratracheal) (Cargnoni et al. 2009; Grisafi et al. 2013; Hodges et al. 2012; Li et al. 2014a; Vosdoganes et al. 2011; Zhao et al. 2014), whereas conditioned media are usually administered intrathoracically or intratracheally (Cargnoni et al. 2012, 2014; Zhao et al. 2014). In some cases, cells have been delivered more than once, or even administered for 20 and 40 days as reported by Li et al. (2014a).

10.2.1 HUMAN AMNIOTIC EPITHELIAL CELLS

Among the placenta-derived cells, human amniotic epithelial cells (hAECs) are the most studied in experimental lung injury models. Many studies have investigated the effects of timing of cell administration on the establishment and progression of lung fibrosis. For example, one study administered hAEC intravenously 24 h and 2 weeks after bleomycin instillation in severe combined immunodeficiency (SCID) mice (Moodley et al. 2010). Early administration of hAECs was able to decrease local levels of proinflammatory and profibrotic cytokines (monocyte chemoattractant protein [MCP]-1, tumor necrosis factor [TNF]-α, interleukin [IL]-1, IL-6, and transforming growth factor [TGF]-β), an observation subsequently confirmed in immune-competent mice (Murphy et al. 2011a). Both studies reported reduced lung collagen deposition after either early (Murphy et al. 2011a) or delayed (Moodley et al. 2010) hAEC administration, due to a lower number of activated myofibroblasts (Murphy et al. 2011a) or increased collagen degradation by upregulation of matrix metalloproteinase (MMP)-2 and downregulation of tissue inhibitors of metalloproteinase (TIMP)-1 and -2 (Moodley et al. 2010). Over a 4-week period, the presence of hAECs in murine lungs and the production of human surfactant proteins was observed, suggesting that hAECs could differentiate into type II pneumocytes *in vivo* (Moodley et al. 2010). Another study investigated the effects of delayed hAEC administration whereby cells were administered intraperitoneally 1 and 2 weeks postbleomycin exposure in immunecompetent mice (Vosdoganes et al. 2012a). hAEC treatment performed 2 weeks after injury was able to reduce lung tissue density, collagen content, and α-smooth muscle action (SMA) expression together with decreased lung expression of profibrotic factors such as TGF-β, platelet-derived growth factor (PDGF)-α, and PDGF-β (Vosdoganes et al. 2012a). In contrast, hAEC delivery performed after 7 days exerted no effect on lung injury. It is possible that the inflammatory milieu on day 7 was too great for the dose of hAECs to overcome or sufficiently influence. Whether a higher dose of hAECs could have been effective has not yet been reported. Interestingly, the day 14 administration of hAECs was associated with a reduction of pulmonary leukocyte infiltration, whereas the day 7 administration with an increased number of leukocytes in the lungs.

In support of these early observations, more recently there has been much evidence supporting the role of macrophages in therapeutic effects of hAECs in lung injury. Murphy and collaborators used surfactant protein C–deficient ($Sftpc^{-/-}$) mice, which are highly susceptible to pulmonary injury as a result of impairment of macrophage function, to show that the beneficial effects observed after hAEC treatment depend upon on macrophage recruitment and polarization (Murphy et al. 2012a). Specifically, hAEC treatment to bleomycin-injured $Sftpc^{-/-}$ mice did not mitigate the inflammatory and fibrotic injuries, and no preservation of lung function was observed. This was in stark contrast to hAEC-treated bleomycin-injured $Sftpc^{+/+}$ mice where injury was prevented. Furthermore, they showed a lower neutrophil infiltration, but no effect was observed on lung macrophage levels, whereas in $Sftpc^{+/+}$ mice hAEC administration was associated with increased polarization of macrophages toward alternatively activated (M2) phenotype. No hAECs were detected in the host lungs at day 7 posttransplantation (Murphy et al. 2012a).

More recently, the same group provided further evidence that interactions with macrophages are central for the reparative effects of hAECs (Tan et al. 2014). They demonstrated that intraperitoneal administration of hAECs 24 h postbleomycin reduced macrophage lung infiltration and favored polarization from classically activated macrophages (M1), predominantly present in bleomycin-untreated mice, toward the M2 phenotype, suggesting that the M1–M2 switch contributes to promote lung repair. It now appears likely that the effects of hAECs on macrophage recruitment, polarization, and function are likely to be effected both directly and via regulation of host T cells, particularly T regulatory cells (Tan et al. 2015). Alternatively, the bleomycin-induced pulmonary fibrosis model has also been used to compare the efficacy of hAECs isolated from term (37–39-gestational week) and preterm (26–34-gestational week) human placentas (Lim et al. 2013). The authors demonstrated that, in contrast with term hAECs, preterm cells were not able to improve lung fibrosis and reduce lung collagen deposition to the same extent as term hAECs, despite a reduction in activated myofibroblasts. The authors suggested that the greater reparative ability of term hAECs could be related to the higher expression of human leukocyte antigen G (HLA-G) detected in term than in preterm cells, possibly mediating immunomodulation mechanisms.

hAECs have also been used to treat models reproducing fetal or neonatal bronchopulmonary dysplasia–like injury (Vosdoganes et al. 2012b). Specifically, it has been shown that hAECs are able to prevent fetal lung injury in lipopolysaccharide (LPS)-induced chorioamnionitis (Vosdoganes et al. 2011) where hAEC given at the time of intra-amniotic LPS injection attenuated the alveolar enlargement and septa thinning. In parallel to these structural benefits, a decline in pulmonary proinflammatory cytokines (TNF-α, IL-1β, and IL-6) was observed, despite increased levels of CD45-positive leukocytes found in the lungs of hAEC-treated animals.

The Use of Placenta-Derived Cells in Inflammatory and Fibrotic Disorders **185**

In ventilation-induced fetal lung injury, hAECs were detected in low numbers in fetal lung tissues but were shown to reduce lung alterations induced by *in utero* ventilation (Hodges et al. 2012). Cell treatment reduced lung collagen and elastin content, and tissue/airspace ratio. A lower number of α-SMA–positive cells and a lower expression of proinflammatory cytokines (mainly IL-8) were observed in lungs of treated animals. Hodges et al. (2012) also demonstrated that hAECs were present in fetal lung tissues and produced surfactant proteins, and possibly adopt an alveolar lung cell phenotype. In hyperoxia-induced neonatal lung injury, intraperitoneal administration of hAECs was able to partially reduce hyperoxia-induced lung damage (Vosdoganes et al. 2013). Treated animals showed normalized body weight, reduced septa crest density, and decreased levels of proinflammatory and profibrotic cytokines (IL-1a, IL-6, TGF-β and PDGF-β), but no improvement in the hyperoxia-induced alterations of lung function and structure.

10.2.2 Amniotic Fluid Mesenchymal Stromal Cells or Amniotic Fluid Stem Cells

Cells from the amniotic fluid have also been extensively investigated in experimental lung injury models. Beneficial results have been observed by the prevention of fibrosis progression after intravenous treatment with murine amniotic fluid stem cells positive for the cell surface marker c-kit (Garcia et al. 2013). Specifically, when mice were treated with amniotic fluid stem cells (AFSCs) 2 h (early inflammatory phase) and 14 days (beginning of the fibrotic process) postbleomycin challenge, the authors observed that at both time points mice showed lower fibrosis, attenuated loss of pulmonary function, and reduced cytokine CCL2 in the bronchoalveolar lavage (BAL) and lungs. This was accompanied by a reduction in the local recruitment of immune populations expressing CCR2, such as macrophages and lymphocytes. Garcia et al. (2013) also observed a transient increase in MMP-2 expression in alveolar epithelial cells type II (AECII) isolated from animals treated with AFSCs. Considering the ability of MMP-2 to proteolytically cleave CCL2, the authors suggested that CCL2 levels were maintained low in BAL from animals treated with AFSCs through the transient action of MMP-2 on CCL2 produced during inflammatory phase (Garcia et al. 2013).

In a different study, the ability of rat amniotic fluid mesenchymal stromal cells (AF-MSCs) to alleviate emphysema-induced lung injury was investigated, based upon previous findings from the same group demonstrating the potentiality of AF-MSCs to differentiate into AECs (Li et al. 2014a). In this study, male rat AF-MSCs at passage 3 were transplanted daily for 20 or 40 days in emphysematous rats by intratracheal injection. AF-MSC–treated rats presented partially alleviated pathological characteristics of emphysema (reduced airspace and higher amount of alveoli). Furthermore, transplantation of rat AF-MSCs increased the number of cells positive for surfactant proteins A and C and reduced the count of apoptotic AECIIs.

AFSCs have also been used by others to attenuate hyperoxia-induced lung injury in neonatal rats (Grisafi et al. 2013) and in adult mice reproducing a bronchopulmonary dysplasia–like injury and an acute lung (ALI) injury (Wen et al. 2013), respectively. The intratracheal injection of AFSCs at day 7 posthyperoxia injury stimulated the neonatal recovery of alveolar growth; reduced the fibrotic areas while inducing a more homogeneous distribution of the capillary network; and decreased lung levels of IL-6, IL-1β, interferon (IFN)-γ and TGF-β (Grisafi et al. 2013). Intravenous injection of AFSCs in the ALI model was able to increase animal survival associated with a reduction of lung edema, neutrophil infiltration, apoptosis, cytokine expression (IL-1β, IL-6, and TNF-α), and early lung fibrosis detected at day 7 posthyperoxia (Wen et al. 2013). Moreover, AFSCs were present in the injured lungs 24 h postinjection and gradually decreased over time.

10.2.3 OTHER CELLS FROM HUMAN TERM PLACENTA

Besides hAECs and AF-MSC/AFSCs, other placenta-derived cells have been investigated in lung injury models, albeit less extensively. For example, one study used a mixture of human amniotic (epithelial and mesenchymal) and chorionic mesenchymal cells injected either by intratracheal or intraperitoneal route 15 min after bleomycin instillation, whereas murine cells were delivered by intrajugular injection (Cargnoni et al. 2009). Irrespective of the cell source (allogenic or xenogenic) and modes of delivery, cell treatment reduced the severity and extension of lung fibrosis and decreased neutrophil infiltration, with a low number of human donor cells present in host tissues (Cargnoni et al. 2009). Human umbilical cord mesenchymal stem cells (hUC-MSCs) derived from Wharton's jelly have also been investigated for their therapeutic potential in the bleomycin model (Moodley et al. 2009). hUC-MSCs at passages 3–6 were intravenously injected 24 h from bleomycin instillation in SCID mice. Seven, 14, and 28 days postbleomycin instillation, hUC-MSC–treated mice showed a decrease in alveolar and interstitial immune cell infiltrates, a decreased expression of proinflammatory (Macrophage inhibitory factor [MIF] and TNF-α) and profibrotic factors (TGF-β and INF-γ), and a decrease in lung fibrosis and collagen interstitial deposition. Furthermore, treated mice showed upregulation of MMP-2 and downregulation of TIMP-1, -2, -3, and -4. Interestingly, hUC-MSCs in murine lung tissues were only transiently detected 14 days after transplantation, but there was no evidence of their presence at 28 days after tail vein injection.

In a more recent study, the effects exerted by bone marrow–derived (BM) MSCs, hAMSCs, and hAECs (used at passages 0 [P0] and 5 [P5]) were compared in a double-dose bleomycin model (Moodley et al. 2013). Cells were injected intravenously 72 h after the second bleomycin dose (7 days after the first dose). The authors observed that treatment with BM-MSCs, hAMSCs, and P0-hAECs reduced lung inflammatory infiltrates, but only treatment with hAMSC reduced lung levels of IL-1, IL-6, and TNF-α, whereas only IL-6 levels were limited after treatment with BM-MSCs and P5-hAECs. Similarly, only hAMSC-treated mice showed decreased fibrosis scores, collagen content, and lung TGF-β levels and increased MMP-9 activity (Moodley et al. 2013). Furthermore, they showed increased expression of IL-1 receptor antagonist (IL-1RA) mRNA in lung tissues of mice treated with hAMSCs. The anti-inflammatory activity of IL-1RA may provide an additional explanation for the observed hAMSC effects (Moodley et al. 2013).

10.2.4 Studies Using Placental Derivatives for the Treatment of Lung Injury

Conditioned medium derived from hAMSCs (hAMSC-CM) has also been used to treat lung injury concomitant to bleomycin-induced lung fibrosis with the hypothesis that the soluble factors derived from these cells could reduce lung injury (Cargnoni et al. 2012). Indeed, reduction in fibroblast proliferation, collagen deposition, and alveolar obliteration was observed on day 14 after bleomycin instillation (Cargnoni et al. 2012). In a more recent study, the same group observed that treatment with hAMSC-CM maintained lung fibrosis at lower levels compared to controls for a duration extended to 28 days after bleomycin challenge (Cargnoni et al. 2014). No fibrosis reduction was observed when treatment was performed using CM generated from other cell types, such as human peripheral blood mononuclear cells, human T leukemia cells, and human skin fibroblasts, suggesting that hAMSCs uniquely secrete or shed active mediators of tissue repair. Furthermore, mice treated with the hAMSC-CM showed improved blood gas parameters and lower lung content of IL-6, TNF-α, macrophage inflammatory protein-1α, MCP-1, and TGF-β associated with reduced lung macrophage levels (Cargnoni et al. 2014).

CM from chorionic villi mesenchymal cells, and also the cells themselves, have been used to treat bronchiolitis obliterans in a heterotopic tracheal transplant model (Zhao et al. 2014). CM and cells were injected locally 1 and 3 days after trachea transplantation into the subcutaneous area surrounding the transplanted trachea and directly into the trachea lumen. Treatment with both CM and cells reduced luminal obliteration, improved the luminal epithelial integrity, and decreased inflammatory cell infiltration in the allografts (Zhao et al. 2014). Specifically, the authors noted decreased neutrophil and CD3+ lymphocyte infiltration, accompanied by an enhanced cellular infiltration of Foxp3+ regulatory T cells in treated animals. No transplanted cells were detected in the tracheal epithelial layer.

10.2.5 Proposed Mechanisms of Action

It is clear that placenta-derived cells, irrespective of what population of cells from the placenta from which they are derived, can both prevent lung injury and enhance lung repair, apparently irrespective of the model of lung injury. Although it may have been initially thought that the cells acted by integration into the injured respiratory epithelium and by differentiation into lung cells (Hodges et al. 2012), it is now accepted that the main mechanism of action is more likely to be the ability of the cells to modulate host immune cell responses by limiting the migration, proliferation, and activation of inflammatory cells during the early lung injury, resulting in reduced tissue fibrosis (Cargnoni et al. 2009; Murphy et al. 2012a; Tan et al. 2014, 2015). Indeed, the majority of the preclinical studies so far conducted have shown a reduction of inflammatory cell recruitment into lung parenchyma and lower levels of proinflammatory cytokines, including IL-1β, IL-6, and TNF-α, after placenta-derived cell treatment (reviewed by Manuelpillai et al. 2011).

Specifically, several studies have shown that placenta-derived cells favor the switch from macrophages with an M1 phenotype (with proinflammatory and fibrogenic activity) to an M2 phenotype (associated with fibrosis resolution) (Tan et al. 2014). The immunomodulatory ability of the placenta-derived cells was also suggested to be related to the expression or secretion of factors with immunosuppressive effects, including HLA-G (Lim et al. 2013), a nonclassical class I major histocompatibility complex mainly produced by the placenta during pregnancy and involved in maintaining fetal tolerance through the inhibition of T cell proliferation and activation and induction of regulatory T cells (Carosella 2011). Another anti-inflammatory factor whose expression is induced by the administration of placenta-derived cells, and specifically by amniotic mesenchymal cells, in a lung injury model is IL-1RA (Moodley et al. 2013). IL-1RA, an endogenous inhibitor of IL-1R1, presents a unique regulatory feature that has not been observed in other cytokines: it blocks the IL-1α and IL-1β inflammatory signaling.

Besides limiting fibrotic injury by reducing inflammatory response, placental cells have been shown to enhance lung collagen degradation, favoring fibrolysis. Indeed, administration of placental cell in animals with established fibrosis has been shown to increase lung MMP levels (Garcia et al. 2013; Moodley et al. 2009, 2010, 2013) and in parallel decrease TIMP levels (Moodley et al. 2009, 2010). Interestingly, MMPs and specifically MMP2 were suggested to even be involved in the anti-inflammatory action of placenta-derived cells (Garcia et al. 2013) by maintaining CCL2 levels low in the lungs, with consequent lower macrophage recruitment.

Although the major mechanisms of action of these cells are thought to be via paracrine actions, as evidenced by the effectiveness of conditioned media alone, that is not to say that placenta-derived cells cannot integrate into the respiratory epithelium and differentiate into lung epithelial–like cells. They can and they express lung-specific proteins, such as surfactant proteins (Carraro et al. 2008; Hodges et al. 2012; Li et al. 2014b; Moodley et al. 2010) and the cystic fibrosis transmembrane regulator (Murphy et al. 2012b). Indeed, it appears that the cells are able differentiate *in vivo* (Hodges et al. 2012) much more rapidly than they can *in vitro* (Moodley et al. 2010; Murphy et al. 2012b), suggesting opportunities for better *in vitro* cueing and for further insight into the functional regulation of these cells.

10.2.6 Clinical Trials

To date, the potential application of placenta-derived MSCs to treat patients with idiopathic pulmonary fibrosis (IPF) has been explored in only one single-center, nonrandomized phase 1b clinical trial (http://www.clinicaltrials.gov NCT01385644) performed at The Prince Charles Hospital in Brisbane, Australia (Chambers et al. 2014). Eight patients with moderately severe IPF received either 1×10^6 ($n = 4$) or 2×10^6/kg ($n = 4$) unmatched-donor, placenta-derived MSCs via a peripheral vein and were followed for 6 months. At infusion time, minimal and transient changes in peri-infusion hemodynamics and gas exchange and only minor, self-limiting adverse events were observed. No evidence of worsening fibrosis, with unchanged lung function and computed tomography scores were observed at 6 months from cell therapy, suggesting a short-term safety profile in patients with moderate-to-severe IPF (Chambers et al. 2014). This trial showed the safety of placenta-derived MSC delivery via the intravenous route, but it was not designed to assess the efficacy of

The Use of Placenta-Derived Cells in Inflammatory and Fibrotic Disorders   **189**

the treatment. These initial positive results provide an encouraging base for future trials planned to determine the true efficacy of cell therapy for fibrotic lung disease. One cautionary note regarding this trial is that the derived cells may have been of maternal, not fetal, origin.

10.3 EFFECTS OF PLACENTA-DERIVED CELLS AND DERIVATIVES ON KIDNEY FIBROSIS

Chronic kidney disease is another major public health problem, where interstitial fibrosis is one of the main pathways of disease progression, which eventually culminates in end-stage renal failure. Despite efforts in developing therapies, the numbers of patients requiring dialysis and kidney transplants continue to rise. Unsurprisingly, there has been a growing interest in applying stem cells, including placental-derived stem cells, to kidney disease. Using a mouse model of unilateral ureteral obstruction, Sun et al. (2013) showed that amniotic fluid–derived stem cells alleviated interstitial fibrosis. Treatment increased levels of vascular endothelial growth factor but lowered TGF-β1 and hypoxia-inducible factor-1α levels. These findings coincided with improved microvascular density, increased tubular epithelial cell proliferation, and reduced apoptosis. Another major cause of kidney disease worldwide is diabetic kidney disease. In a rat model of streptozotocin-induced diabetes, intravenous injection of umbilical cord blood–derived stem cells reduced urinary output, proteinuria, and renal levels of fibronectin and SMA-α (Park et al. 2012). Interestingly, exosomes released by umbilical cord MSCs have also been shown to be protective against cisplatin-induced oxidative stress and apoptosis in the kidneys of rats (Zhou et al. 2013). Indeed, it was the exosomes, rather than the stem cells themselves, that afforded renal protection against the cisplatin-related injury (Chambers et al. 2014). It is possible that the beneficial effects of cell-conditioned media in the lung were also mediated via exosomes, although this has not yet been reported.

10.4 EFFECTS OF PLACENTA-DERIVED CELLS AND DERIVATIVES ON LIVER FIBROSIS

Liver cirrhosis is characterized by extensive deposition of ECM proteins leading to distortion of the normal hepatic architecture by collagen bands coupled with the loss of substantial hepatocyte mass. Autopsy studies estimate the prevalence of cirrhosis in the general population to be 4.5–9.5% (Graudal et al. 1991). Liver cirrhosis is a significant global health burden, with more than 1 million deaths in 2010 (Mokdad et al. 2014). Mortality results from organ dysfunction and increased intrahepatic vascular resistance that culminates in liver failure, portal hypertension, and importantly, the development of hepatocellular carcinoma.

Hepatic fibrosis, which leads to cirrhosis, results from a persistent and unregulated wound healing response to many different types of injury such as excessive alcohol consumption, chronic viral hepatitis from hepatitis B or hepatitis C infection, and increasingly from steatohepatitis related to nonalcoholic fatty liver disease. Removal of the cause of injury may prevent fibrosis progression; however, this is not possible for many patients for whom the only definitive treatment remains whole organ replacement by transplantation. Unfortunately, the current demand for donor livers exceeds the supply, and the increasing number of deaths of people on the transplant waiting list represents an important unmet medical need for people with cirrhosis.

Although cirrhosis has long been considered as the end of a one-way road of progressive fibrosis, it is now clear that this process is dynamic and that fibrosis can resolve after adequate treatment or resolution of the injury stimulus, such as long-term viral suppression in hepatitis B patients (Marcellin et al. 2013). In the next sections, we examine the use of placental cells or CM from those cells in relevant cell culture and animal models of liver fibrosis.

10.4.1 ANIMAL MODELS OF LIVER INFLAMMATION AND FIBROSIS

There are several models of experimental liver inflammation and fibrosis that simulate human disease. Acute liver failure can be modeled by administration of D-galactosamine (GalN), a hepatocyte-specific transcriptional inhibitor, plus LPS that induces TNF-mediated hepatocyte necrosis that is associated with dose-dependent mortality rates. The most commonly used chronic liver disease model involves carbon tetrachloride (CCl_4) administration, usually via an intraperitoneal route. CCl_4 is metabolized by cytochrome P450 to produce a trichloromethyl radical that causes lipid peroxidation of cell membranes. Short-term CCl_4 exposure causes acute centrilobular hepatocyte necrosis, and continued dosing over 12 weeks leads to extensive parenchymal fibrosis and cirrhosis. Hepatocellular carcinoma may develop with longer exposure. A similar model involves thioacetamide administration that also causes centrilobular necrosis and hepatic fibrosis, although longer exposure is required before significant hepatic fibrosis is observed (Wallace et al. 2015). Conditions such as sclerosing cholangitis and primary biliary cirrhosis involve the biliary epithelium rather than hepatocytes. Biliary cirrhosis can be modeled by surgical ligation of the bile ducts, leading to extensive periportal inflammation and fibrosis (Tag et al. 2015). Other biliary models include genetically modified mice, such as the mdr2$^{-/-}$ mouse, that develop inflammation of bile ducts, leading to biliary cirrhosis and inflammation-induced hepatocellular carcinoma (Barashi et al. 2013).

10.4.2 STUDIES USING PLACENTA-DERIVED STEM CELLS FOR HEPATOCYTE TRANSPLANTATION

Substantial hepatocyte loss occurs both in acute liver failure and in chronic fibrotic liver disease. Evidence that hepatocyte-like cells (HLCs) can arise from placental stem cells has recently been reviewed (Vaghjiani et al. 2013). As with other lineages, it is certainly possible to induce hepatocyte differentiation from both placental MSCs and hAECs *in vitro* (Ilancheran et al. 2007; Vaghjiani et al. 2013). HLCs derived from placental stem cells exhibit characteristic hepatocyte markers such as hepatocyte nuclear factor-α and CK18, are similar in size and shape to human hepatocytes, synthesize albumin, and store glycogen. Metabolic liver diseases, due to inherited enzyme deficiency in hepatocytes, seem to be ideal models to test whether transplanted placental stem cells can differentiate into functional hepatocytes.

The Use of Placenta-Derived Cells in Inflammatory and Fibrotic Disorders   **191**

Skvorak et al. (2013) transplanted hAECs into mice with maple syrup urine disease, a condition in which deficiency of branched-chain alpha-keto-acid dehydrogenase leads to branched-chain amino acid accumulation that, without lifelong dietary branched-chain amino acids restriction, results in severe brain injury. The investigators showed that hAEC-transplanted mice had improved metabolic control and survived for 100 days compared to no survival at 30 days in untreated control animals (Skvorak et al. 2013). Despite low-level engraftment, detected only by human DNA in mouse liver, markers of hepatocyte function were identified, suggesting that hAECs could differentiate into hepatocytes. Although this finding supports the notion that hAECs can differentiate into hepatocytes in relatively uninjured liver, are such cells viable and effective in preclinical models of acute liver injury? Recent studies suggest that they are. Injection of hUC-MSCs into SCID mice with D-GalN–induced acute liver failure resulted in amelioration of liver injury and detectable hepatocyte markers up to 28 days postinfusion (Yang et al. 2015). This study suggested that the mechanism of rescue and repair is through the differentiation of injected stem cells into hepatocytes. Zhang et al. (2012a) found that hUC-MSC transplantation into SCID mice with CCl_4-induced acute liver failure indeed resulted in markers of human hepatocytes, but showed that a more important effect was downregulation of systemic inflammation and stimulation of endogenous hepatocyte regeneration. These studies provide substantial evidence that placental stem cells can differentiate into functional hepatocytes after transplantation, ameliorate acute inflammatory liver disease, and improve survival in murine models.

10.4.3 Studies Using Placental Cells for the Treatment of Liver Fibrosis

Chronic liver injury activates a wound healing process that in some patients will lead to excessive deposition of ECM proteins, primarily collagen, that disrupts the normal hepatic architecture with bands of fibrosis and regenerative nodules of hepatocytes. The net result is loss of functional hepatocyte mass and altered blood flow due to loss of normal vascular anatomy. Patients may remain asymptomatic for long durations only to develop ascites, variceal hemorrhage related to portal hypertension, hepatic encephalopathy, or hepatocellular carcinoma. The onset of ascites is the most common initial presentation with hepatic decompensation and prompts consideration of liver transplantation since the 5-year survival rate is only 50%. As mentioned, donor livers are relatively scarce and many patients will not be candidates for transplantation due to age or comorbidities. Based on animal studies, placental stem cells may be an effective cell-based therapy that will diminish inflammation and stimulate hepatic regeneration in such patients.

Currently, the majority of data on the efficacy placental stem cells in chronic fibrotic liver disease comes from animal studies (Table 10.1). Jung et al. (2009) infused hUC-MSCs intravenously into rats at 8 weeks of CCl_4-induced liver fibrosis and showed that resolution of fibrosis was greater in hUC-MSC–treated animals. Cells positive for human albumin were detected in liver specimens 4 weeks after infusion. Using the same model, Lee et al. (2010) demonstrated improved indocyanine green clearance (a marker cleared from the circulation by the liver) in addition to decreased collagen synthesis. The timing of cell transfusion is important

TABLE 10.1
Animal Models

Reference	Type of Placental Stem Cell Used/ Animal Model	Model Details	Outcome
Jung et al. 2009	Umbilical cord MSCs/Sprague–Dawley rats	$CCl_4 \times 8$ week, 1×10^6 hUC-MSCs given intravenously at 8 weeks	Improved fibrosis resolution in treated animals, decreased markers of stellate cell activation, liver cells positive for human albumin detected 4 weeks postinfusion
Lee et al. 2010	Chorionic plate–derived MSCs/Sprague–Dawley rats	$CCl_4 \times 9$ weeks, 2×10^6 hCP-MSCs given by direct hepatic injection at week 9	Improved indocyanine green clearance in treated animal, decreased α-smooth muscle actin and collagen synthesis, less bridging fibrosis in treated animals
Manuelpillai et al. 2010	hAECs/ C57BL/6 mice	$CCl_4 \times 4$ weeks, 2×10^6 hAECs by tail vein injection at 2 weeks	Reduced hepatocyte apoptosis, reduced hepatic collagen content, human albumin in mouse sera 2 weeks posttransplantation
Manuelpillai et al. 2012	hAECs/ C57BL/6 mice	$CCl_4 \times 12$ weeks, 2×10^6 hAECs by tail vein injection at week 8 or weeks 8 and 10	Reduced hepatic inflammation and fibrosis, human leukocyte antigen G+ cells detected 4 weeks posttransplantation, increased M2 macrophages
Zhang et al. 2011	Amnion MSCs/ C57BL/6J mice	$CCl_4 \times 8$ weeks, 1×10^5 hAMCs given intravenously at week 4	Decreased hepatic stellate cell activation, 35% reduction in fibrosis area, human albumin detected 4 weeks posttransplantation

hAEC, human amniotic epithelial cell; hAMC, human amnion mesenchymal cell; hCP-MSC, human chorionic plate–derived mesenchymal stromal cell; hUC-MSC, human umbilical cord mesenchymal stromal cell; MSC, mesenchymal stromal cell.

in experimental design. Rather than transfusing cells at the same time as injury commences or after cessation of the injury, administration of cells in animals with ongoing injury or established cirrhosis more closely models the clinical situation. We have injected hAECs into C57BL/6 mice at week 2 of 4-week CCl_4 administration (Manuelpillai et al. 2010) or at week 8 or at weeks 8 and 10 of 12-week CCl_4 administration (Manuelpillai et al. 2012). In the 4-week model, we detected cells positive for human HLA-G in portal triads and showed decreases in hepatocyte apoptosis, proinflammatory cytokines, and hepatic collagen content. Similar findings were identified in mice given CCl_4 for 12 weeks with the development of advanced (bridging) fibrosis. There was a significant decrease in fibrosis area, although there was no additional benefit from the double-dose strategy. Interestingly, we found increased M2 macrophages, suggesting that macrophage polarization may be a mechanism for fibrosis resolution in this model, as it is in lung injury (Murphy et al. 2012a; Tan et al. 2014). Almost identical findings in relation to reduced fibrosis area, hepatocyte apoptosis, and liver inflammation were demonstrated by Zhang et al. (2011) by using human amniotic membrane–derived MSCs. These investigators also noted decreased hepatocyte senescence and speculated that this would promote hepatocyte replication. These studies show that different types of placental stem cells infused into animal models of fibrosis and cirrhosis show similar outcomes— again, very similar to the studies of diverse cells in diverse models of lung injury. Transplanted cells can engraft and remain viable up to 4 weeks posttransplantation without signs of rejection, reduce hepatic inflammation and fibrosis, and show signs consistent with hepatocyte differentiation.

Although animal studies support the use of placental stem cells in human liver disease, relatively few studies have been done (Table 10.2). Shi et al. (2012) studied 43 patients with acute liver failure due to hepatitis B virus (HBV) infection. Twenty-four patients received hUC-MSCs and 19 received saline as controls. Remarkably, not only did the model for end-stage liver disease (MELD) score, a prognostic score for patients with end-stage liver disease used for organ allocation, improve significantly in treated patients, but overall survival at 12 weeks follow-up was significantly better in hUC-MSC–treated patients (21% died compared to 47% of untreated patients). In 30 patients with decompensated liver disease due to chronic HBV infection, Zhang et al. (2012b) found that the same regimen of hUC-MSC administration significantly improved the MELD score and resulted in significantly greater resolution of ascites compared to 15 untreated controls. Survival was not reported in this study. Although these studies are small and the patients heterogeneous, the outcomes are very encouraging and suggest that placental stem cell therapy may become part of the treatment armamentarium for patients with acute liver failure or decompensated cirrhosis. Further studies are clearly warranted.

TABLE 10.2

Human Clinical Trials

Reference	Type of Placental Stem Cell Used	No. of Patients	Outcome
Shi et al. 2012	UC-MSCs/acute liver failure from hepatitis B virus infection	0.5×10^6 UC-MSCs/kg at 0, 4, and 8 weeks, 24 treatment, 19 control	Improved liver function, lower MELD, and improved survival rates
Zhang et al. 2012b	Umbilical cord MSCs	0.5×10^6 UC-MSCs/kg at 0, 4, and 8 weeks, 30 treatment, 15 control	Improved liver function and MELD and reduced ascites

MELD, model for end-stage liver disease; UC-MSC, umbilical cord mesenchymal stromal cell.

10.5 SUMMARY

It is clear that the placenta and its associated tissues are a rich and ready source of diverse stem cells or cells with stem cell–like capabilities that have much to offer clinical regenerative medicine (Murphy et al. 2011b). Although many of the studies undertaken to date have been in experimental models only, these studies have proven efficacy, offered guidance for timing and route of administration, and offered insight into likely mechanisms of action. All of this information is critical for efficient and optimal clinical application. The clinical studies undertaken to date, albeit limited, have shown that the cells are safe and well tolerated, as was suggested nearly 35 years ago (Akle et al. 1981). Although it has certainly been a long gestation, placenta-derived cells seem to be the most promising therapy for the long list of currently incurable fibrotic diseases of the lung, liver, and kidney, among others. It is time that these cells, or their derivatives, are trialed in larger, clinical trials and made ready for clinical therapies.

REFERENCES

Akle, C. A., M. Adinolfi, K. I. Welsh, S. Leibowitz, and I. McColl. 1981. Immunogenicity of human amniotic epithelial cells after transplantation into volunteers. *Lancet* 2 (8254): 1003–5.

Barashi, N., I. Weiss, O. Wald, et al. 2013. Inflammation-induced hepatocellular carcinoma is dependent on CCR5 in mice. *Hepatology* 58: 1021–30.

Cargnoni, A., L. Gibelli, A. Tosini, et al. 2009. Transplantation of allogeneic and xenogeneic placenta-derived cells reduces bleomycin-induced lung fibrosis. *Cell Transplant* 18: 405–22.

Cargnoni, A., E. C. Piccinelli, R. Lorenzo, et al. 2013. Conditioned medium from amniotic membrane-derived cells prevents lung fibrosis and preserves blood gas exchanges in bleomycin-injured mice-specificity of the effects and insights into possible mechanisms. *Cytotherapy* 16: 17–32.

Cargnoni, A., L. Ressel, D. Rossi, et al. 2012. Conditioned medium from amniotic mesenchymal tissue cells reduces progression of bleomycin-induced lung fibrosis. *Cytotherapy* 14: 153–61.

Carosella, E. D. 2011. The tolerogenic molecule HLA-G. *Immunol Lett* 138: 22–24.

Carraro, G., L. Perin, S. Sedrakyan, et al. 2008. Human amniotic fluid stem cells can integrate and differentiate into epithelial lung lineages. *Stem Cells* 26: 2902–11.

Chambers, D. C., D. Enever, N. Ilic, et al. 2014. A phase 1b study of placenta-derived mesenchymal stromal cells in patients with idiopathic pulmonary fibrosis. *Respirology* 19: 1013–18.

Garcia, O., G. Carraro, G. Tucatel, et al. 2013. Amniotic fluid stem cells inhibit the progression of bleomycin-induced pulmonary fibrosis via CCL2 modulation in bronchoalveolar lavage. *PLoS One* 8: e71679.

Graudal, N., P. Leth, L. Mårbjerg, and A. M. Galløe. 1991. Characteristics of cirrhosis undiagnosed during life: A comparative analysis of 73 undiagnosed cases and 149 diagnosed cases of cirrhosis, detected in 4929 consecutive autopsies *J Intern Med* 230: 165–71.

Grisafi, D., M. Pozzobon, A. Dedja, et al. 2013. Human amniotic fluid stem cells protect rat lungs exposed to moderate hyperoxia. *Pediatr Pulmonol* 48: 1070–80.

Hodges, R. J., G. Jenkin, S. B. Hooper, et al. 2012. Human amnion epithelial cells reduce ventilation-induced preterm lung injury in fetal sheep. *Am J Obstet Gynecol* 206: 448. e8–448.e15.

The Use of Placenta-Derived Cells in Inflammatory and Fibrotic Disorders   **195**

Ilancheran, S., A. Michalska, G. Peh, et al. 2007. Stem cells derived from human fetal membranes display multilineage differentiation potential. *Biol Reprod* 77: 577–88.

Jung, K. H., H. Shin, S. Lee, et al. 2009. Effect of human umbilical cord blood-derived mesenchymal stem cells in a cirrhotic rat model. *Liver Int* 29: 898–909.

Lee, M., J. Jung, K. Na, et al. 2010. Anti-fibrotic effect of chorionic plate-derived mesenchymal stem cells isolated from human placenta in a rat model of ccl4-injured liver: Potential application to the treatment of hepatic diseases. *J Cell Biochem* 111: 1453–63.

Li, Y., C. Gu, J. Xu, et al. 2014a. Therapeutic effects of amniotic fluid-derived mesenchymal stromal cells on lung injury in rats with emphysema. *Respir Res* 15: 120.

Li, Y., W. Xu, J. Yan, et al. 2014b. Differentiation of human amniotic fluid-derived mesenchymal stem cells into type II alveolar epithelial cells in vitro. *Int J Mol Med* 33: 1507–13.

Lim, R., S. T. Chan, J. L. Tan, et al. 2013. Preterm human amnion epithelial cells have limited reparative potential. *Placenta* 34: 486–92.

Manuelpillai, U., D. Lourensz, V. Vaghjiani, et al. 2012. Human amniotic epithelial cell transplantation induces alternative macrophage activation and reduces established hepatic fibrosis. *PLoS One* 7: e38631.

Manuelpillai, U., Y. Moodley, C. V. Borlongan, and O. Parolini. 2011. Amniotic membrane and amniotic cells: Potential therapeutic tools to combat tissue inflammation and fibrosis? *Placenta* 32 (Suppl 4): S320–5.

Manuelpillai, U., J. Tchongue, D. Lourensz, et al. 2010. Transplantation of human amnion epithelial cells reduces hepatic fibrosis in immunocompetent CCl_4 treated mice. *Cell Transplant* 19: 1157–68.

Marcellin, P., E. Gane, M. Buti, et al. 2013. Regression of cirrhosis during tenofovir disoproxil fumarate treatment for chronic hepatitis B. *Lancet* 381: 468–75.

Mokdad, A., A. Lopez, S. Shahraz, et al. 2014. Liver cirrhosis mortality in 187 countries between 1980 and 2010: A systematic analysis. *BMC Med* 12: 145.

Moodley, Y., D. Atienza, U. Manuelpillai, et al. 2009. Human umbilical cord mesenchymal stem cells reduce fibrosis of bleomycin-induced lung injury. *Am J Pathol* 175: 303–13.

Moodley, Y., S. Ilancheran, C. Samuel, et al. 2010. Human amnion epithelial cell transplantation abrogates lung fibrosis and augments repair. *Am J Respir Crit Care Med* 182: 643–51.

Moodley, Y., V. Vaghjiani, J. Chan, et al. 2013. Anti-inflammatory effects of adult stem cells in sustained lung injury: A comparative study. *PLoS One* 8: e69299.

Murphy, S., R. Lim, H. Dickinson, et al. 2011a. Human amnion epithelial cells prevent bleomycin-induced lung injury and preserve lung function. *Cell Transplant* 20: 909–23.

Murphy, S., E. Wallace, and G. Jenkin. 2011b. Placental-derived stem cells: Potential clinical applications. In K. Appasani and R. Appasani (Eds.), *Stem Cell Biology and Regenerative Medicine*, pp. 243–63. Humana Press. NY: New York.

Murphy, S. V., R. Lim, P. Heraud, et al. 2012b. Human amnion epithelial cells induced to express functional cystic fibrosis transmembrane conductance regulator. *PLoS One* 7: e46533.

Murphy, S. V., S. C. Shiyun, J. L. Tan, et al. 2012a. Human amnion epithelial cells do not abrogate pulmonary fibrosis in mice with impaired macrophage function. *Cell Transplant* 21: 1477–92.

Park, J. H., J. Park, S. H. Hwang, et al. 2012. Delayed treatment with human umbilical cord blood-derived stem cells attenuates diabetic renal injury. *Transplant Proc* 44: 1123–26.

Shi, M., Z. Zhang, R. Xu, et al. 2012. Human mesenchymal stem cell transfusion is safe and improves liver function in acute-on-chronic liver failure patients. *Stem Cells Transl Med* 1: 725–31.

Skvorak, K., K. Dorko, F. Marongiu, et al. 2013. Placental stem cell correction of murine intermediate maple syrup urine disease. *Hepatology* 57: 1017–23.

Sun, D., L. Bu, C. Liu, et al. 2013. Therapeutic effects of human amniotic fluid-derived stem cells on renal interstitial fibrosis in a murine model of unilateral ureteral obstruction. *PLoS One* 8: e65042.

Tag, C., S. Weiskirchen, K. Hittatiya, et al. 2015. Induction of experimental obstructive cholestasis in mice. *Lab Anim* 49 (1 Suppl): 70–80.

Tan, J. L., S. T. Chan, E. M. Wallace, and R. Lim. 2014. Human amnion epithelial cells mediate lung repair by directly modulating macrophage recruitment and polarization. *Cell Transplant* 23: 319–23.

Tan, J. L., S. T. Chan, E. M. Wallace, and R. Lim. 2015. Amnion cell mediated immune modulation following bleomycin challenge: Controlling the regulatory T cell response. *Stem Cell Res Ther* 6: 8.

Tomasek, J. J., G. Gabbiani, B. Hinz, C. Chaponnier, and R. A. Brown. 2002. Myofibroblasts and mechano-regulation of connective tissue remodelling. *Nat Rev Mol Cell Biol* 3: 349–63.

Vaghjiani, V., V. Vaithilingam, B. Tuch, et al. 2013. Deriving hepatocyte-like cells from placental cells for transplantation. *Curr Stem Cell Res Ther* 8: 15–24.

Vosdoganes, P., R. J. Hodges, R. Lim, et al. 2011. Human amnion epithelial cells as a treatment for inflammation-induced fetal lung injury in sheep. *Am J Obstet Gynecol* 205: 156.e26–33.

Vosdoganes, P., R. Lim, E. Koulaeva, et al. 2013. Human amnion epithelial cells modulate hyperoxia-induced neonatal lung injury in mice. *Cytotherapy* 15: 1021–29.

Vosdoganes, P., R. Lim, T. J. M. Moss, and E. M. Wallace. 2012b. Cell therapy: A novel treatment approach for bronchopulmonary dysplasia. *Pediatrics* 130: 727–37.

Vosdoganes, P., E. M. Wallace, S. T. Chan, et al. 2012a. Human amnion epithelial cells repair established lung injury. *Cell Transplant* 22: 1337–49.

Wallace, M., K. Hamesch, M. Lunova, et al. 2015. Standard operating procedures in experimental liver research: Thioacetamide model in mice and rat. *Lab Anim* 49 (1 Suppl): 21–29.

Wen, S. T., W. Chen, H. L. Chen, et al. 2013. Amniotic fluid stem cells from EGFP transgenic mice attenuate hyperoxia-induced acute lung injury. *PLoS One* 8: e75383.

Wynn, T. A. 2007. Common and unique mechanisms regulate fibrosis in various fibroproliferative diseases. *J Clin Invest* 117: 524–29.

Yang, J., H. Cao, Q. Pa, et al. 2015. Mesenchymal stem cells from the human umbilical cord ameliorate fulminant hepatic failure and increase survival in mice. *Hepatobiliary Pancreat Dis Int* 14: 186–93.

Zhang, D., M. Jiang, and D. Miao. 2011. Transplanted human amniotic membrane-derived mesenchymal stem cells ameliorate carbon tetrachloride-induced liver cirrhosis in mouse. *PLoS One* 6: e16789.

Zhang, S., L. Chen, T. Liu, et al. 2012a. Human umbilical cord matrix stem cells efficiently rescue acute liver failure through paracrine effects rather than hepatic differentiation. *Tissue Eng A* 18: 1352–64.

Zhang, Z., H. Lin, M. Shi, et al. 2012b. Human umbilical cord mesenchymal stem cells improve liver function and ascites in decompensated liver cirrhosis patients. *J Gastroenterol Hepatol* 27 (Suppl 2): 112–20.

Zhao, Y., J. R. Gillen, D. A. Harris, et al. 2014. Treatment with placenta-derived mesenchymal stem cells mitigates development of bronchiolitis obliterans in a murine model. *J Thorac Cardiovasc Surg* 147: 1668–77.e5.

Zhou, Y., H. Xu, W. Xu, et al. 2013. Exosomes released by human umbilical cord mesenchymal stem cells protect against cisplatin-induced renal oxidative stress and apoptosis in vivo and in vitro. *Stem Cell Res Ther* 4: 34.

11 From Bench to Bedside
Strategy, Regulations, and Good Manufacturing Practice Procedures

Christian Gabriel

CONTENTS

Preface 198
11.1 Strategic Issues 198
11.2 Making a Strategic Plan 199
    11.2.1 Constraints 199
    11.2.2 Risks 199
    11.2.3 Strategic Approaches 200
    11.2.4 Financial Plan 201
    11.2.5 Staffing 202
    11.2.6 Integration 202
11.3 Execution 203
11.4 Legal Requirements 204
11.5 Predonation Phase 206
    11.5.1 Selection of Donors 207
    11.5.2 Donor Questionnaire 208
    11.5.3 Pregnancy Complications 209
    11.5.4 Informed Consent 209
11.6 Donation Procedure 210
11.7 Tissue Donation 210
11.8 Transport and Reception 211
11.9 Viral Testing 211
11.10 Production 212
11.11 Quality Control 213
11.12 Final Release 213
11.13 Summary 213
References 214

PREFACE

Transferring benchtop research to a viable product for clinical use is not an easy task. Meticulous planning to cope with strategic imperatives is one of the most important issues. Financial planning and a well-thought-out technical plan for the good manufacturing practice (GMP) facility are critical prerequisites. Furthermore, integration with stakeholders is an important aspect. The production process follows European regulatory requirements derived from GMPs, and tissue regulations have some exemptions. Careful collaboration between members of different teams, and especially those from the obstetrical department, is required. Processes follow strict rules and are divided into testing, quality control, and production; these three processes merge at the point of release, with labeling and traceability procedures.

The "bench-to-bedside" transfer of placental tissues and cells is a place where research has quite a cozy spot compared to regular production. Any researcher would be overwhelmed by the huge, confusing wave of issues relating to GMPs, sometimes in combination with economic restrictions and new management principles.

The day-to-day production process is often seen as tedious and intellectually not very challenging. Therefore, at the beginning of a promising multiyear project, a researcher might tend to lose the final route to patient-oriented support. Ignorance can be a main contributor that disrupts the dream of good research combined with jaw-dropping results leading to visions of very relevant patient outcome ameliorated with decent cash flows. It is clear that transfer of research into cell and tissue production is not as easy as it may seem, but even if this fact is understood, many contributors, even those with expertise in tissue banking and GMPs, underestimate the fact that marketing, distribution, logistics, financials, and nit-picking medical demands such as ergonomics of packaging, labeling, or user education, may be detrimental to the well-designed production process or to a product in the hospital on the market with high expectations (Sheets et al. 2015).

Why do many research projects end in the "Land of Nowhere" and turn out as unfeasible, either technically or financially, or simply by lack of capabilities? In many cases, lack of strategies and missing exploration of risks are the most likely causes of failure. High-quality research, high expectations, and even profound patient benefit are not self-selling paradigms moving a tissue or cell product sustainable to the bedside.

11.1 STRATEGIC ISSUES

There is a wide gap between research and production principles. Research may vary between experiments. In contrast, in the GMP environment, nothing can be changed without additional formal procedures (even if for a good reason), and change requires qualification and validation and, furthermore, regulatory approval. These formalities induce a quite different mindset: how process and product quality can be kept within strictly defined brackets governed by rules and specifications.

Producing cells and tissues in a tissue bank or a GMP facility requires skill and experience in steady-state procedures with a high level of robustness. Furthermore, financial strength, cost control, rigid documentation, and repeated inspections are

From Bench to Bedside 199

key characteristics of the highly regulated GMP environment. Peer review in science is not comparable with GMP inspections. An inspection is executed in a face-to-face manner by authorities who hardly know anything about the product, its features, and benefit to the patient, but they are highly experienced on standards, regulations, hygiene, design of clean rooms, and validations—factors that are not core competencies of researchers, but demonstrate the focus of determination. A GMP facility demands different requirements in regard to staff qualification, management, and skills. To bridge this gap, a strategic plan, in which all elements of infrastructure, competencies, and processes are well designed, is of paramount importance.

11.2 MAKING A STRATEGIC PLAN

A strategic plan is required to make the path from bench to bedside visible, calculated, and highly successful (Gabriel 2014). This strategic plan has to include a decision for or against building a new GMP facility. However, it may clarify whether it is better to hand off the knowledge to a GMP facility that is capable of transforming research results into a final product. This option is always one to be considered, as it has some risk of losing intellectual property, but a very low risk in the transformation process (Benninger et al. 2014).

11.2.1 CONSTRAINTS

Tissue banking or tissue engineering is a not a medical specialty nestled within a hospital facility, for example, like a blood bank. Generally, a tissue bank is located in a clinical unit that is directly using its products. For example, eye banks are commonly part of an ophthalmological department. Placental tissues and cells, however, are harvested at an obstetrical department that itself hardly has any clinical case to use these tissues. A variety of clinicians are potential users of these placental tissues and cells and need variable types of the same tissue, which in contrast to eye, heart valve, or bone banks, is derived from living donors. The location of a facility for placental cells and tissues, as well as its affiliation to a clinical unit, may vary to a wide degree and quite often is a spinoff from a university. Thus, comprehensive knowledge and standards for placental cells or tissue production are seldom available. This may pose a major constraint at the time of introduction and requires a fine-tuned adaption to clinical needs and close interaction with clinicians. Getting approval and support from clinicians is inevitable, as resources have to be directed to a facility that has no imminent benefit for patients.

11.2.2 RISKS

Large investments to build a clean room and adjacent areas have to be planned; an underestimated risk can occur even with careful planning—evidence that multiple specialists have to work together in a coordinated manner. Building such a complex facility requires highly technical, but not necessarily product-oriented, skills. A major risk is the confusion and indecisiveness that often arises in big institutions, such as public hospitals or universities, where decision making is too diffuse to manage extraordinary projects. Therefore, management of a GMP facility project has to adapt to a faster pace and a greater willingness to take risks. Assigning a centralized manager responsible to finish the project can facilitate the process.

There are no specific guidelines on how to build a tissue bank or GMP facility for placental tissues and cells. Inspection authorities may refuse approval of a facility or, even worse, set further requirements with additional, often expensive, changes in the facility. General consultants are usually not experts on clinical requirements, processes, and clinical use. The entire facility depends on experienced staff, trained in technical issues as well as in GMPs. Ultimately, selection of the wrong staff can lead to complications.

11.2.3 Strategic Approaches

Strategic planning is important in any case and should be developed in parallel with existing research as a dedicated bench-to-bedside project. It is important to develop a plan at the beginning of the research project or at least at the point when it is clear that a cell or tissue product has to be developed and produced for a clinical study. On average, the bench-to-bedside project can require up to 5 years. Close ties with clinicians and researchers should be secured from the beginning to generate an exchange of ideas and to make innovative approaches possible. This ensures support within the hospital and has persuasive effects in the hospital management. A highly professional level attracts new partners and competent staff. All stakeholders should be included in strategic planning and decisions on how to move the bench-to-bedside project forward.

However, many projects are primarily considered as experimental research projects and not directed to a continuous production for clinical studies or patient care. This may be very problematic, when a research project develops protocols with methods, reagents, and testing systems inappropriate for GMP production. Therefore, careful planning should start with the research project and a review of the process. An initial issue is directed to the envisioned product, that is, whether it is appropriate for clinical purposes, easy to produce, and poses no or low risk to patients. Selecting the best process or product requires information about clinical cases, indications, and, importantly, which alternative therapies are in use or might arise in the near future. A review of scientific evidence and alternative protocols can sometimes be more important than a review of existing protocols and their fitness for GMPs. As transfer time is quite enduring, and there is a general tendency to promote cheaper and simpler treatments, ongoing observation of comparable products, therapies, or medical procedures has to be established.

Ideally, one should start designing the GMP process of a product even when continuous financing is not secured and even if the required level of expertise may be not sufficient. It is also advisable to select procedures that can easily be performed and have a high level of reproducibility, and procedures for which the risks of adverse effects or morbidity are low. Interestingly, easily performed donation and harvesting procedures have no restriction and a high extraction probability. For example, in placental tissues, the relatively high amount of cells with high growth potential poses a good opportunity.

From Bench to Bedside

In addition, it is necessary to instruct and train employees in GMPs. Staff education is vital for the project, but it is also important to keep up with issues of tracking, a key element of GMP procedures. It is beneficial to develop a product that may be used for different therapies or that is easily transformable into different final products. Placental tissues and cells have a high potential to be used in various areas of patient care; thus, they also may be transformed into various products that must be controlled under GMP conditions.

11.2.4 Financial Plan

Understandably, one crucial issue that is often neglected is a sustainable financial plan. GMP production is costly and poses a high financial risk. One of the greatest mistakes is the underestimation of required financial support to sustain a GMP facility. Making comparisons and estimating costs from other institutions' knowledge is a feasible approach. Financial experts should cover investments, running costs of the facility, and development and validation costs (Abou-El-Enein et al. 2013). The overview should provide a guideline to estimate the final cost of the product. Furthermore, it may give an indication of how many products are required to obtain a return of investment and to maintain a constant work flow without interruptions. The highest costs in the production of placental tissues and cells derived thereof are, for example, related to product loss. These losses can be incurred either by the limitation of shelf life or very commonly by microbiological contamination of tissue sources that may easily surpass the expenses of clean rooms, and the running costs when these rooms are not in use, as well costs of maintenance and supporting staff. Usually, a GMP facility is shut down regularly once a year for maintenance. The costly shutdown results in higher staffing requirements, numerous quality checks, revalidation procedures, and exchange of devices or parts in the support systems.

To avoid these risks, cooperation with other tissue banks, core facilities, and contract manufacturers is imperative. Choosing this path reduces investments, but it may add some problems in the transfer of the project and sooner or later leads to some loss of knowledge. However, it seems that contract manufacturing with existing GMP facilities has many advantages, especially when demand is relatively low (Hourd et al. 2014b).

Final product prices may turn to skyrocketing figures, a warning signal to reconsider the project. Estimations should be seen as a framework of how to avoid risks. The framework depends on variables such as the estimations of demanded products, the number of different clinicians may order the product, the number of different products that should be produced, and especially, the costs of production. The financial plan should include the costs of often inevitable project delay and those arising from improvements or replanning and rebuilding of the facility. Technical problems with inefficient equipment (especially cooling, air conditioning) are quite common at the beginning of the production phase and can be detrimental to correct financial planning. Further problems may result from inadequate staffing or insufficient qualification of facility management. Strict regulations, manufacturers' competition, or the loss of clinical partners have to be considered in the risk management of a financial plan. In some cases, financial planning can improve and sufficient capital can be obtained if venture capital is collected to run a company. However, there are additional risks involved, such as the short-lived expectations of financial supporters and the cascade of milestones, which might distract the key persons from the general goals. Regardless, financial support is provided only if the business and income perspectives seem risk-free and palatable to entrepreneurs. If financial support is provided, transnational distribution ranges can be factors which hinder the scale-up and ultimately the probability of income. Venture capital may initially be attractive, but it is only feasible if both sides are a perfect match, and risks are extensively discussed.

11.2.5 STAFFING

An additional challenge is proper staffing in early phases (Hourd et al. 2014a). Finding trained staff, preferably with an educational track record in GMP cell and tissue production, is a nearly an out-of-reach task, especially when a research facility wants to hire a highly qualified person to set up a clinical study. Transfer of research staff into a production area is rather difficult, as the mindset required is quite different. It may be helpful to run an intermediate validation program that is very near to research tasks and allows staff to learn to work in a GMP environment. The importance of assisting personnel cannot be ignored (i.e., cleaning, hygiene, and maintenance). This group is needed not only in routine operations but also more crucially in qualification, validation, and inspection phases. The work environment should also always be taken into consideration. Repetitive work consisting of delicate procedures in a confined area without exchange of ideas and experiences with coworkers can be perceived as stressful. In addition, the production of autologous products may exert high pressure on staff, as it may be the only way to save a patient's life.

11.2.6 INTEGRATION

Tissue banks are not common to every hospital, and experience in keeping them running efficiently may be limited; therefore, integration into a network of clinicians and customers is a key issue (Hourd et al. 2014a). However, a tissue bank or GMP facility may cause changes in general management. Compared to a classical tissue bank, the donation process is totally different, as live donors are requested to donate the placenta, cord, or amniotic fluid. This implies that clinical procedures of an obstetric department are unaltered, and midwives should be included in all aspects of donation. Asking mothers for a tissue donation is a delicate situation, but midwives have the knowledge and empathy to do so. Production processes are quite different for the hospital setting not only because of different facilities but also because of their highly documented and standardized processes. Complicated logistics, new quality control procedures, and maintenance issues may give the impression to clinicians that the facility in their hospital is unproductive. Rising costs in tissue banks, either induced by regulatory authorities or by mismanagement, may be seen as a threat by employees working in hospitals. Therefore, in addition to the strategic process and its planning, integration into the existing organization is one of the key factors for success.

From Bench to Bedside 203

Placental tissues and cells encourage integration by scientific work, which should be seen as an opportunity for all persons involved. This work should be supported by management through regular communications with all stakeholders. This communication is especially needed for procedures with autologous products that may hamper or even discourage clinicians opting for this kind of treatment. Showing the direct benefit for individual patients is important and requires continuous education of clinicians about therapeutic options offered by the tissue bank.

11.3 EXECUTION

Designing a GMP facility or transferring protocols to a GMP facility requires a project team with a high level of expertise that ideally should work closely with the team validating and running the tissue or cell program. This setting fosters expertise and is ultimately advantageous to the project. GMP facilities in big organizations tend to be overrun by different opinions and failed decisions. A key person or team should be appointed with a full set of responsibilities.

Specifications of the products and the processes must be definitive before facility planning commences. Reagents, culture media, antibiotics, or tests used in research often do not fit into the GMP requirements. Meticulous reassessment of data, package inserts, and specifications may reveal flaws that might hamper the introduction of critical items used in research that are not suitable for GMP production. In this case, all items not useful for GMP procedures should be changed and revalidated.

Detailed planning starts with a set of diagrams on the product flow. These diagrams describe interactions and specifications that demonstrate where, when, and how products are manipulated, stored, discarded, or tested. Supporting this main process flow are diagrams of consumables or waste management, which should be well documented, in flow diagrams. The whole set of flow diagrams represents the master plan and a fundamental document before technical experts being to work. Replanning at this moment can avoid vast and disproportional expenses materializing later. Maintenance staff should be involved as early as possible and work with contractors and technical experts to gain important know-how on support systems, maintenance, and specifications of controlling systems. Finding limits of physical parameters such as pressure, temperature, humidity, and particle counts should also include tolerability measures and the design of maintenance schedules. Much attention should be placed on thorough and consistent documentation, especially design documents that are fundamental in qualification plans and crucial for facility acceptance by inspectors.

The first challenge arises after completion of the design phase. By showing plans and documentation in a design qualification process to national inspectors, first-hand information about the quality of the facility design can be collected. Furthermore, this information is important if systems are adequately designed and conform to national regulations and standards. The approval of a proper design qualification prevents further changes and excessive expenditures.

The last phase of execution is the construction, installation, and qualification of the GMP facility. Introducing clinicians to the facility and exchanging views and ideas are helpful. Proper sequence qualification and supporting documentation (Sensebé et al. 2013) are essential.

Installation qualification (IQ) is usually performed when equipment is installed by the vendor. Specifications set by the vendor should document correct installation and basic equipment settings, including instructions and drawings. IQ is followed by the operational qualification (OQ). OQ includes staff training, cleaning, and calibration efficacy tests, tests that determine the upper and lower limits of performance, and, if appropriately defined by risk management procedures, worst-case scenario tests in which systems are stressed to show required minimal performance. OQ is followed by performance qualification (PQ). This is the final litmus test in which equipment is tested with production material or qualified substitutes to simulate production processes. Previously established specifications of the product must be met at this point. For example, a laminar flow must be able to reduce particles in a specified time frame whenever amniotic membrane is washed or cut.

11.4 LEGAL REQUIREMENTS

Legal requirements for tissues are specific regulations that are in accordance to pharmaceutical laws (The European Commission 2013). However, tissues cannot be classified in the same way as pharmaceuticals. Issues of consistency, efficacy, and efficiency are not inherent in any tissue product as it is harvested from deceased patients or donated by individuals, healthy or not. Tissues lack many constituents of pharmaceuticals and therefore need specific adaption in which principles of GMP can be applied under certain circumstances, but meaningful GMPs may be substituted by additional safety features. Unfortunately, blood regulations appear very similar in some aspects, but European legislators missed the chance to construct an overall legal framework for all types of substances of human origin. Primarily, placental cells and tissues are legally seen as human tissues. The key directive of the European Parliament is Directive 2004/23/EC (Official Journal of the European Union 2004), and its implementation into national law in the member states of the European Union is mandatory. National lawmakers may stipulate stricter regulations, so the directive represents a minimum set of standards for tissues of human origin in Europe (Figure 11.1). National legislators are free to choose the national legal instrument, a fact that is currently causing confusion if cross-border supply is considered, because national legislation may be represented in dedicated acts for tissues or it may be included in parts of pharmaceutical acts. Specific requirements on the importance of tissues may be seen as blocking the transnational flow of tissue products. The European Union Directive includes human stem cells from peripheral blood, cord, and bone marrow; reproductive cells; fetal tissues and cells; and embryonic and adult stem cells. T is an important achievement as it enables the use of stem cells without conforming to advanced therapeutical medicinal product (ATMP) regulations and therefore opens the door for the use of mesenchymal stem cells from placental origin as tissue. Organs, blood, blood components, tissues, and cells from animal origin and those used within the same surgical procedure (one-step procedures, not applicable when using placental cells) are all excluded. Quite important is the possibility to use tissues and cells for research purposes without applying for a tissue bank license.

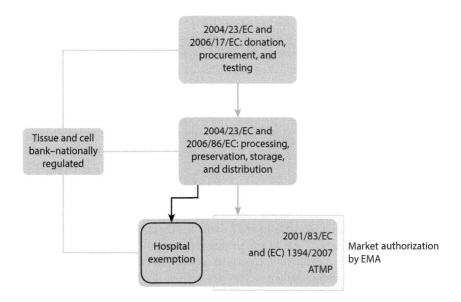

FIGURE 11.1 (See color insert.) European Commission (EC) directives and its relation to tissues and advanced therapeutical medicinal products (ATMPs). Tissues are regulated by the EC and derivative national laws; ATMPs are regulated nationally by hospital exemption or transnationally by the European Medicines Agency.

The main objective of the tissue directive is a standardized framework for quality and safety standards. All tissue banks in Europe follow the same uniform accreditation and licensing requirements, although it is well known that peculiarities of national regulators may distort the uniformity in some cases. This is in stark contrast to the approval of ATMPs on the European level. To tackle arising disparities in the accreditation process, the European Commission granted the European Union Standards and Training for the Inspection of Tissue Establishments, a project led by the Italian transplant center intended to train inspectors in Europe and find standardized definitions (The European Commission 2013). One of the major topics was the vigilance system and the notification of adverse events. Tissues, especially those that derive from cadaveric donors, pose higher risks in transmitting viral infections. Moreover, since these tissues can be split into many sections, it is important to trace all products and have a well-managed recall scheme in place. Tissues are generally harvested in a transparent system and should be accessible to all European patients in need. Therefore, nonprofit tissue donation procedures are promoted in Europe. However, this does not ban sale or production for profit. Donor identity and personal data are protected, inspections are repeated every second year, and traceability is ensured for 30 years. Unfortunately, the selection criteria for donors and their testing is not regulated in the directive, an issue that is difficult to understand in view of recent viral outbreaks and the lack of regulation to test emerging viruses such as West Nile fever virus or Chikungunya fever virus. More questionable is a recent commission directive draft for implementing West Nile fever virus testing or donor deferral (Council of the European Union 2014).

Considering placental cell products, products that are manipulated and abide by ATMP regulations, tissue regulations are fundamental because ATMPs are basically derived from tissues. Directive 2001/83/EC, amended by ATMP Regulation 1394/2007 regulates ATMPs (Official Journal of the European Union 2001, 2007). ATMP regulations cover products manipulated by gene therapy, somatic cell therapy, and tissue engineering, even when used in combination with any other medicinal product or scaffold. ATMPs require high investments, and early commercialization followed by placement on the European market (Salmikangas 2015) is a characteristic feature of these products. In fact, centralized marketing authorization guided by the European Medicines Authority (EMA) is a key element in the registration of these products. Products derived from placental tissues may scientifically be at the borderline between the definition of tissues and cells (Directive 2004/23/EC) and the ATMP regulation. The terms "substantial manipulation" and "intended use" are critical whenever a distinction is unclear. If there is no change in the biological characteristics of cells derived from placental origin, and the intended use covers the same essential function, the products may not be classified as ATMPs. In fact, this may be rare for cells derived from placental tissues. To clarify this, the EMA established the Committee for Advanced Therapies. This committee gives advice on such topics and provides information (Committee for Advanced Therapies et al. 2010; European Medicines Agency 2014).

Nonroutine products for individual patients may classify as ATMPs under the so-called hospital exemption rule, but this rule causes controversy, as the interpretation varies in member states and in professional societies (Van Wilder 2012). It is clear that ATMPs comply with the hospital exemption rule when they are prepared on a nonroutine basis according to GMPs. What is meant by a nonroutine basis is questionable, as this implies that GMPs may be missed, because a nonroutine product without validation is contradictory to GMP principles. Yet, the product can only be used in the same member state in a hospital, a fact that is circumvented by hospital providers declaring to be one hospital entity. An additional interesting fact is the requirement that the production is the full professional responsibility of a medical practitioner, who prescribes the ATMP as a custom-made product.

11.5 PREDONATION PHASE

The production of placental cells and tissues includes a predonation phase that is clearly designed for the prevention of transmissible agents, avoidance of product contamination, and proper donor selection. The process splits into three arms (Figure 11.2). One arm involves production and storage of cells and tissues in the GMP facility, the second arm involves testing, and the third arm involves quality control of the product, which may be performed either at random or on each product. All arms converge to a point where release criteria determine the further use or removal of the product. Postprocessing steps include tracking procedures, vigilance schemes, and retraction processes.

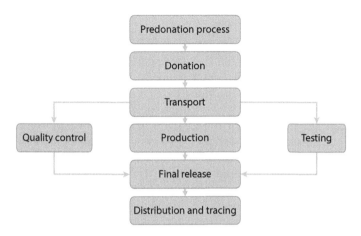

FIGURE 11.2 (See color insert.) Design process for a tissue bank. Placental tissues and cells always require informed consent and start with the predonation steps. The general process is split into testing, production, and quality control. All information merges at the release point.

11.5.1 SELECTION OF DONORS

Tissue donation generally has many constraints with regard to the safety and usability of tissues. The most frequently used tissues, such as cornea, vascular tissue, cortical bone, tendon, or cartilage tissue, are sourced from deceased donors, where it is impossible to generate safety based on thorough anamnesis or previous testing. Furthermore, infectious disease testing of cadaverous blood samples is restricted; lapse of time from death to testing, dilution of samples, and fibrinolysis are a tremendous challenge to test robustness. In comparison to tissues from cadaveric origin, sourcing placental tissues is very safe and can be performed in a timely manner. Further information is retrieved from midwives, physicians, and donors, enabling a very high level of information for donor care and product safety (Pink and Warwick 2006). It is easier to first obtain records from countries that have pregnancy and child care systems.

These systems are leading information sources when a tissue bank approaches mothers, preferably in the last trimester, and requests a tissue donation. Information generated during pregnancy (e.g., ultrasound and overall clinical chemistry; genetic screening; serologic, systemic, and infectious disease testing; and a well-documented course of the pregnancy) give enough information to include donors for further evaluation without further research. This quick look at existing data can exclude donors with a high-risk pregnancy, genetic diseases, systemic diseases, infectious complications, or viral infections. Not all genetic diseases or malformations are contraindications for a tissue donation. Autologous donation may be useful in reconstructive surgery with oronasal fistulas (Kesting et al. 2008). Some other indications may be of interest for postnatal malformations. Further questioning is required to ask specific questions and obtain informed consent. A first contact with the mother should take place well ahead of the date of birth to avoid any subjective pressure on the mother.

11.5.2 Donor Questionnaire

It is useful to design a standardized questionnaire based on regular risk assessments. A risk assessment and the regular update of donor questionnaires should be repeated regularly. A tissue bank can update its questionnaire on the basis of tissue vigilance data, usually provided by national authorities, and include new questions whenever required. One of the most important issues is the update of epidemiological data, in which persons with risk behavior may be identified, or persons coming from regions with local transmissions of emerging viruses may be excluded. A good example is the intermittent appearance of West Nile fever virus in southeastern Europe or the appearance of Chikungunya fever virus in northern Italy. National epidemiological bulletins or the European Centre for Disease Prevention and Control newsletters are valuable sources. Regular reassessment and travel anamnesis are important to exclude risks, especially for viruses not routinely included in a testing scheme. In fact, some diseases may occur in women migrating from different crisis areas of the world, where health care systems are deteriorated and chronic infections are more prevalent. In addition, the tissue bank should update a list of pharmacological agents that intentionally or unintentionally may lead to the exclusion of donors.

A face-to-face interview with the mother is required to explain the issue and enable full approval. This interview is a prerequisite by the European Directive and ideally should be performed by a knowledgeable and empathic physician who knows how to deal with additional questions from the pregnant woman. Specific care is important whenever the refusal of a tissue donation has to be explained to the mother, thereby avoiding any confusion or anxiety.

Questions include the mother's current medications, which will give an indication of any chronic disease. Chronic diseases leading to exclusion are asthma, chronic hepatitis, diabetes, inflammatory bowel disease, and any malignant or neurological disease such as multiple sclerosis. Autoimmune diseases are commonly found, and further evaluation is needed.

Of further importance is the segment in which the risk of (variant) Creutzfeld–Jacob disease (CJD) is circumscribed. The donor is asked if she has relatives with CJD, or whether she was residing for more than 6 months in Great Britain between 1980 and 1996, had treatment with growth hormones from human origin, or had a transplantation with corneal or dura mater grafts.

Infections with a tendency to chronicity and transmissibility by blood or tissues are also exclusion criteria. Tuberculosis, human T-lymphotropic virus types I and II, syphilis, Chagas disease, and malaria are usually part of a standard questionnaire. In some regions, or if a mother was traveling from a different region, it might be useful to ask about Q-fever, brucellosis, Lyme disease, babesiosis, or leishmaniosis or other tropical diseases. Cytomegalovirus (CMV) and Epstein–Barr virus infections are usually covered by prenatal care or, in some cases, by prenatal testing.

The questionnaire in regard to the evaluation of risk behavior is also important. Asking about any intravenous or intranasal drug use is as important as asking if the donor ever had sex in exchange for money or drugs. Furthermore, asking for recent

From Bench to Bedside 209

piercings or tattoos is important. However, if prenatal testing of the most relevant viruses (HIV/HBV/HCV) is done, the residual risk deems to be very low.

Additional less relevant questions might include recent symptoms of fever, bleeding, unclear lymphadenopathies, high blood pressure, diarrhea, and any sign of obvious infection.

11.5.3 PREGNANCY COMPLICATIONS

Complications during pregnancy should be considered, although they may be seen as irrelevant for the quality of placental tissue and cells. However, the well-being of the mother is of utmost importance; therefore, a review of the pregnancy must include complications. In such a case, it is advisable to forbear from donation, as there is a high probability that any additional medical intervention, as the donation procedure may be seen, leads to further distraction in a stressful situation; this is usually not accepted by midwives and physicians.

Morbus hemolyticus neonatorum may cause changes in the amniotic fluid and also some discoloration of tissues, which are refused later in the transplantation procedure. Placental insufficiency, premature placental detachment, and placenta previa lead to stressful complications coinciding with tears in amniotic membranes or disruption of the placenta. Preeclampsia and HELLP (hemolysis, elevated liver enzymes, and low platelet count) syndrome are associated with severe placental pathology. Oligohydramnios (low amniotic fluid levels) may be a sign of overdue pregnancy and is often associated with insufficiency of the placenta or with problems in the development of the kidneys or urinary tract of the fetus. Placental abnormalities may lead to intrauterine growth restriction. Cervical insufficiency is often a problem leading to preterm birth, and although treated well by cervical cerclage, it requires medication with antibiotics.

Infections during pregnancy, such as bacterial vaginosis, group B streptococcus infection, CMV infection, fifth disease, and chickenpox are a "no-go" for tissue donation.

11.5.4 INFORMED CONSENT

The optimal time to obtain informed consent for tissue donation is often during the discussion with a physician. Asking for informed consent immediately prior to or during labor is sometimes requested by inspectors, but this is unfeasible and inappropriate. First contact with the mother should be sought well ahead in the last trimester to collect enough information as possible and obtain informed consent before labor. Whenever early information and consent is given, midwives tend to accept the donation procedure.

It is advisable to provide detailed information and assess the unrestricted willingness to give away donated tissue without any reservations about its commercial exploitation, especially when donated tissue is planned for commercial use, such as stem cells for wide-scale pharmaceutical use, medicinal products extracted from placental tissue, immortalizing cells for stem cell banks, or with new stem cell therapies. Tissue donation is strictly noncommercial and any incentives for tissue donation should be avoided. However, sometimes tissues from placental origin are safe and reliable sources for commercial enterprises. A clear distinction should be drawn between the donation process and the commercial use of tissues and cells. Additional attention should be given to the circumstances when stem cells are sent to other research institutions, where there might be no regulations about how cells are handled or transferred to other entities. Special attention is also needed whenever cells are sent to different countries, where security information and genetic testing are not as restrictive as in the country where the donation takes place.

11.6 DONATION PROCEDURE

Extensive information and education of midwives and attending physicians regarding the donation procedure has to be established initially for two reasons. First, staff should get used to an additional procedure that is not in their focus, or in the interest of immediate patient care. It is recommendable to see the donation procedures through the eyes of a midwife or physician. Any distraction from the mother and the child is seen as a risk and could lead to lack of acceptance. Already established standard operating procedures (SOPs) should be amended to the local situation of a delivery room. The tissue bank should find accordance with the strict GMP setting, which is implemented in the delivery room and often felt as a culture clash. Second, in the GMP setting, no donation may start without an introduction about the documented standards and how to set up proper documentation about the product, its quality, and additional data, a requirement that is very unusual in a clinical setting. Clinical staff must get used to completing forms as required and follow the SOPs when introduced. For example, correct execution of GMPs and documented procedures is one of the most identified problems in cord blood certification systems. The tissue donation is frequently performed in conjunction with a cord blood donation (Warkentin 2003). *In utero* blood collection is especially welcomed by obstetricians, as the placenta is delivered more easily.

11.7 TISSUE DONATION

A major problem during delivery is prevention of bacterial contamination, since the delivery itself is not performed in a sterile environment. This causes some frustration, as the amount of collections required is much higher than released products. Some clinics offer the better way of tissue donation from cesarean section. Although cesarean section is often performed under circumstances in which donors may not fit well into the donor criteria, there are still many possible donations, as indications for cesarean section are lowered and many cesarean sections are planned well ahead. Tissue donation during a cesarean section is easier and the process better implemented in the operating room than in the delivery room. In addition, the rate of bacterial contamination is lower.

After delivery, the placenta should be checked for signs that would indicate that the placental tissue should be discarded, such as discolored amniotic fluid, torn amniotic membranes, or placental infarction. Sourced tissue should be placed in a sterile bag with the validated solution from the tissue bank. A sterile overwrap bag ensures that leakage is contained and the product will be still useable. A second overwrap bag is needed to transfer tissues into the clean rooms. Tissue is then placed in a precooled box that is validated for holding the temperature as long as specified—in most cases at 4°C. Optimally, a temperature logger should be packed to document any freezing of tissue, which is an overlooked hazard, if precooling or addition of cooling packs was incorrect. It may happen that the temperature of the box lowers after package and tissues are frozen slightly, thereby leading to deteriorating cell viability. If possible, cord blood samples as well as samples from the mother should be drawn, labeled correctly, and added to the tissue donation. Birth complications and details of the examinations of the placenta, amniotic membrane, and cord should be documented to avoid any confusion.

11.8 TRANSPORT AND RECEPTION

Whenever possible, the container should be transported by qualified staff to the tissue bank without delay. Depending on risk analysis, the quality of the container, precooling needs, and temperature maintenance and monitoring, a tissue bank is obliged to validate transportation under critical conditions, for example, with very high or very low environmental temperatures. These worst-case scenarios are specification brackets in which the maximum time of transport or even a maximum delay may be specified. To be safe, temperature logging is recommended. Upon reception, the tissue bank checks documents first to clarify any questionable data and, if needed, to discard the product before it is transferred to the clean room area. Samples are transferred to the testing laboratory and rechecked for correct labeling.

11.9 VIRAL TESTING

HIV antibody testing is usually performed prior to birth as a safety measure to ensure full antiviral treatment and reduce the risk of viral transmission to the child. HBV testing, usually HBsAg testing, excludes any active viral infection and its possible transmission. Newborns are generally vaccinated in countries with higher endemicity of HBV and higher mother-to-child transmission rates. However, low-level viremia may occur in anti–hepatitis B core antibody (aHBc)–positive patients and therefore testing for aHBc might give additional safety indications if HBV-nucleic acid testing (NAT) is not performed on donation samples. Hepatitis C antibody testing is not commonly performed in pregnant women, although there is a risk of transmission. This may change, because new therapies with protease inhibitors are now available and might be used in newborns in near future. Tests in the setting of a tissue and cell banks are intended to detect viruses at very low levels and are usually performed using the same validated assays used in blood banks. As cells and tissues of placental origin may be split into many parts and transplanted in many patients, it is essential to maintain the highest possible sensitivity. Therefore, single-donor nucleic acid tests of HIV, HBV, and HCV directly from cord blood are recommended, although not mandatory, in all European countries. Antibody testing of the mother's sample accomplishes the NAT results and may add safety in very rare circumstances to exclude HIV elite controllers or low-level HBV viremia.

11.10 PRODUCTION

Generally, all steps in the production process have to be validated. Preparations for the validation process can begin when documents are released, staff is trained, all equipment is qualified, tests are validated, product specifications are set, and the process is fixed. Any change after or during validation requires the activation of a complicated change management process, which is part of the GMP quality management system. A validation report is generated by a cascade of documents starting with the validation plan, the impact assessment, risk assessment, detailed testing, and production documents. Key elements are all steps reducing microbiological load, decontamination, and cleaning procedures, and all steps leading to any reduction of cell viability or loss of function of the tissues and cells. If production procedures are intended for a final devitalized product, such as placental collagen or decellularized amniotic membrane, validation should include experiments proving the purity of the product and total loss of residual cells.

After primary evaluation and reception, tissue is unpacked from the transport box in the unclassified area and then transferred in the overwrap bags to the clean room. Unpacking of the first bag begins in class C and transfer to class B areas commences. Microbiological contamination of the placenta requires frequent washing to reduce microbiological load. In addition, antibiotic decontamination is required whenever the product is used directly on patients, as for example amniotic membrane for ocular treatments (Suessner et al. 2014). In some instances, especially when using cells, antibiotics may be added to cell culture media. There is no general recommendation on which antibiotics should be used, but frequent testing of placental samples followed by the identification of the bacterial spectrum are a prerequisite to find the ideal composition of antibiotics, their optimized temperature during antibiosis, and the optimized dose. Usually, antibiosis lasts 24 h and may be set at temperatures below 37°C. Thereafter, additional washing is required to reduce residual antibiotics. Tissue samples can be cut into required sizes, and cells may be isolated by fine cutting and enzymatic digestion. This is followed by washing and purification steps. Finally, cells and ready-to-use tissue samples can be stored.

Storage temperature is highly dependent on two factors, the chosen shelf life and the turnaround of cells and tissues in the GMP facility. Determining the shelf life of a product is not easy and is commonly chosen purely by convention or convenience, as there are no regulations about the shelf life of a tissue or cell product. Validation experiments should start as soon as possible, even during the research phase. Multiple tissue samples stored at different temperatures may run through viability experiments at sequential time points to determine the loss of viable cells. Depending on the storage temperature, the addition of cryoprotectants, and the freezing conditions, the validation experiment should produce sufficient support to find the best procedure to ensure that cells and tissues meet the specifications under efficient storage conditions. A longer shelf life may be better from the economical point of view, but it may distract users if cells and tissues are not viable enough to exert their function.

From Bench to Bedside

11.11 QUALITY CONTROL

Quality control covers the whole process and should be established after thorough risk assessment. There are two basic types of quality control procedures: in-process controls and random sampling. For tissues and cells of placental origin, bacterial contamination is the major quality and safety concern. Microbiological testing is an in-process control and is complex to establish. Thorough washing of the amniotic membrane reduces the microbiological load, and antibiotics, if carefully selected and evaluated, diminish this risk. Test sensitivity and specificity have to be validated in a matrix assay with selected bacterial strains common in the vaginal microflora. Whenever antibiotics are used, the assay should follow the European Pharmacopoeia, and the validation should prove that residual antibiotics are eliminated and the microbiological assay does not significantly lose its sensitivity. In addition, cell counting and viability tests are also the most common in-process control. Testing should be performed prior to storage and after the frozen or stored product is rethawed or ready for use. The most prominent validation parameter is the loss of cells or viability.

Random quality control includes microbiological testing of clean rooms and equipment, characterization of cells, cell proliferation, and differentiation testing. Quality control in this aspect should be triggered by risk assessment of the process and validation results. Frequently, validation leads to the implementation of these quality control assays, and validation should determine the frequency of tests.

11.12 FINAL RELEASE

All processes are not synchronized, as preparation and production may lag behind testing. This lag may have the advantage to discard products with positive testing results during the production phase. All testing and quality information is used to switch a quarantined product to the released status, the most critical point in the whole process. Additional safety can be obtained through proper documentation. Each product has to be labeled and a vigilance document should be attached, when dispatched. Traceability is one of the most important issues that has to be followed by transplanting clinicians. Follow-up gives the advantage to stay in close contact and develop new products.

11.13 SUMMARY

The production process follows regulatory requirements that are generally concordant to GMPs, but tissue regulations have exemptions and very specific inclusions, such as predonation procedures. These may seem simple, but they demand careful embedding in the obstetrical department and continuous attention to emerging diseases or any changes in medical risks. Testing is performed similar to blood donation testing. The production process needs robust and carefully monitored procedures and quality control to reduce bacterial contamination by washing, antibiosis, and microbiological testing. Aside from viability testing, meticulous microbiology testing has an important function in the whole process. All processes converge to the final release point, which includes labeling and traceability procedures.

Establishing a tissue and cell bank is not an easy task, and the transition from bench to bedside requires a farsighted strategic process. The two synonymous culture changes, from research to GMP production and from the clinical setting to a production site, are accompanied by high financial risks, complexities in management, and a very long project duration. A well-designed strategic plan is fundamental to avoid early termination or long-lasting problems. Once established, the rough path from bench to bedside fulfills even the highest expectations.

REFERENCES

Abou-El-Enein, M., A. Römhild, D. Kaiser, et al. 2013. Good Manufacturing Practices (GMP) manufacturing of advanced therapy medicinal products: A novel tailored model for optimizing performance and estimating costs. *Cytotherapy* 15 (3): 362–83.

Benninger, E., P. O. Zingg, A. F. Kamath, and C. Dora. 2014. Cost analysis of fresh-frozen femoral head allografts: Is it worthwhile to run a bone bank? *Bone Joint J* 96-B (10): 1307–11.

Committee for Advanced Therapies (CAT), CAT Scientific Secretariat, C. K. Schneider, et al. 2010. Challenges with advanced therapy medicinal products and how to meet them. *Nat Rev Drug Discov* 9 (3): 195–201.

Council of the European Union. 2014. Available at http://www.parlament.gv.at/PAKT/EU/XXV/EU/03/66/EU_36647/imfname_10490619.pdf

European Medicines Agency. 2014. Reflection paper on classification of advanced therapy 5 medicinal products. Available at http://www.ema.europa.eu/docs/en_GB/document_library/Scientific_guideline/2014/06/WC500169466.pdf

Gabriel, C. 2014. How to establish and run a cell therapy unit in a blood bank. *ISBT Sci Ser* 9 (1): 155–9.

Hourd, P., A. Chandra, D. Alvey, et al. 2014a. Qualification of academic facilities for small-scale automated manufacture of autologous cell-based products. *Regen Med* 9 (6): 799–815.

Hourd, P., P. Ginty, A. Chandra, and D. J. Williams. 2014b. Manufacturing models permitting roll out/scale out of clinically led autologous cell therapies: Regulatory and scientific challenges for comparability. *Cytotherapy* 16 (8): 1033–47.

Kesting, M. R., D. J. Loeffelbein, L. Steinstraesser, et al. 2008. Cryopreserved human amniotic membrane for soft tissue repair in rats. *Ann Plast Surg* 60 (6): 684–91.

Official Journal of the European Union. 2001. Directive 2001/83/EC of zhr European Parliament and the Council. Available at http://eurlex.europa.eu/LexUriServ/LexUriServ.do?uri=OJ:L:2001:311:0067:0128:en:PDF

Official Journal of the European Union. 2004. Directive 2004/23/EC of the European Parliament and of the council of 31 March 2004. Available at http://eur-lex.europa.eu/legal-content/EN/TXT/PDF/?uri=CELEX:32004L0023&qid=1428183929851&from=EN

Official Journal of the European Union. 2007. Regulation (EC) No 1394/2007 of the European Parliament and of the council of 13 November 2007. Available at http://ec.europa.eu/health/files/eudralex/vol-1/reg_2007_1394/reg_2007_1394_en.pdf

Salmikangas, P., M. Menezes-Ferreira, I. Reischl, et al. 2015. Manufacturing, characterization and control of cell-based medicinal products: Challenging paradigms toward commercial use. *Regen Med.* 10 (1): 65–78.

Sensebé, L., M. Gadelorge, and S. Fleury-Cappellesso. 2013. Production of mesenchymal stromal/stem cells according to good manufacturing practices: A review. *Stem Cell Res Ther* 4 (3): 66.

Sheets, R. L., V. Rangavajhula, J. K. Pullen, et al. 2015. Now that you want to take your HIV/AIDS vaccine/biological product research concept into the clinic: What are the "cGMP"? *Vaccine* 33 (15): 1757–66.

Suessner, S., S. Hennerbichler, S. Schreiberhuber, D. Stuebl, and C. Gabriel. 2014. Validation of an alternative microbiological method for tissue products. *Cell Tissue Bank* 15 (2): 277–86.

The European Commission. 2013. European Union standards and training in the inspection of tissue establishments. Available at http://ec.europa.eu/chafea/documents/health/conference_27-28_06_2013/EUSTITE_-_European_Union_Standards_and_Training_in_the_Inspection_of_Tissue_Establishments.pdf

Van Wilder, P. 2012. Advanced therapy medicinal products and exemptions to the regulation 1394/2007: How confident can we be? An exploratory analysis. *Front Pharmacol* 3: 12.

Warkentin, P. 2003. Voluntary accreditation of cellular therapies: Foundation for the Accreditation of Cellular Therapy (FACT). *Cytotherapy* 5 (4): 299–305.

12 Applications of Placenta-Derived Cells in Veterinary Medicine

Barbara Barboni*, Valentina Russo*, Paolo Berardinelli, Aurelio Muttini, and Mauro Mattioli

CONTENTS

Preface	218
12.1 Cell-Based Regenerative Medicine in Veterinary Science	218
12.1.1 Therapeutic Potential of Cell-Based Approaches in Veterinary Practice	218
12.1.2 Preclinical Animal Models for Regenerative Medicine	220
12.2 Placental Tissues as Source of Stem/Progenitor Cells for Veterinary Regenerative Medicine	221
12.2.1 Comparative Placenta Adnexa Embryology	223
12.2.2 Placenta-Derived Cell Characterization in Domestic Animals	226
12.2.2.1 Amniotic-Derived Tissue and Cells	226
12.2.2.2 UC-Derived Cells	238
12.2.2.3 Umbilical Cord Blood Mesenchymal Stem Cells	238
12.2.2.4 Umbilical Cord Matrix Mesenchymal Stem Cells	246
12.3 Placenta Cell-Based Regenerative Medicine in Domestic Animals	247
12.3.1 Preclinical Studies in Domestic Animal Models	247
12.3.1.1 Spinal Cord Injury	260
12.3.1.2 Myocardial Infarction	260
12.3.1.3 Wound Healing	260
12.3.1.4 Tendon Injuries	260
12.3.1.5 Bone Defects	261
12.3.1.6 Cartilage Defects	261
12.3.1.7 Joints	261
12.3.1.8 Prenatal Diseases	262
12.3.2 Clinical Application of PCs in Veterinary Regenerative Medicine	262
12.3.2.1 Tendinopathy	262
12.3.2.2 Ophthalmology	267
12.4 Concluding Remarks and Future Perspectives	267
Acknowledgments	268
References	268

* These authors have contributed equally to this work.

PREFACE

The field of regenerative medicine is moving toward clinical practice. In this context, domestic animals cover a dual role, acting as both patients and valuable translational models.

The absence of regulatory and ethical guidance has encouraged rapid translation of unproven stem cell protocols in veterinary medicine. However, evidenced-based preclinical and clinical trials of stem cell therapies performed in domestic animal models are critical in advancing stem cell therapies. To this end, placenta-derived stem cells have been recently investigated in domestic animals, and a large amount of information has been obtained in regard to cell origin, isolation, enrichment, and processing. Aside from their high *in vitro* plasticity, four main characteristics of placenta-derived cells make them attractive candidates for therapeutic approaches: stemness features common to cells from large-sized mammals, low immunogenicity and immunomodulatory properties, multilineage regenerative capacity, and the successful engraftment and long-term survival in various host tissues after auto- or allo/xenotransplantation.

Preclinical studies performed on experimental models, as well as clinical trials designed to treat spontaneous diseases, have demonstrated beneficial regenerative effects, particularly those exerted by amniotic-derived cells. These effects arise from a mutual tissue-specific cell differentiation (tendon- and bone-derived lineage cells), and paracrine secretion of bioactive molecules in host tissues that ultimately drive crucial repairing processes (e.g., anti-inflammatory, antifibrotic, angiogenic, and neurogenic).

The knowledge acquired thus far on the mechanisms of action of placental-derived stem cells and on their effectiveness and safety in animal models highlights promising perspectives for the treatment of musculoskeletal disorders and widespread common and incurable pathologies affecting companion animals and humans.

12.1 CELL-BASED REGENERATIVE MEDICINE IN VETERINARY SCIENCE

Stem cell–based regenerative medicine progresses rapidly in veterinary practice. One reason is that domestic animals play a dual role in this dynamic field of science. As patients, they are waiting to receive efficient stem cell–based therapies for a wide range of pathological conditions that still lack effective drug or surgical solutions. For research purposes, domestic animals have a central role in advancing stem cell research since they are valuable in addressing the unresolved challenges related to the application of cell-based approaches in regenerative medicine applied to humans (Frey-Vasconcells et al. 2012; Cibelli et al. 2013).

12.1.1 THERAPEUTIC POTENTIAL OF CELL-BASED APPROACHES IN VETERINARY PRACTICE

Clinical stem cell studies are largely documented in domestic animals, and in particular they have focused on treatment of equine and canine tendon ligaments (Smith et al. 2003; Richardson et al. 2007; Nixon et al. 2008; de Mattos Carvalho

Applications of Placenta-Derived Cells in Veterinary Medicine 219

et al. 2011; Muttini et al. 2012) and cartilage/joint diseases (Brehm et al. 2006; Xiang et al. 2006; Black et al. 2007, 2008; Wilke et al. 2007; Ahern et al. 2009; Chiang and Jiang 2009; Fortier et al. 2010; Carrade et al. 2011; Guercio et al. 2012; Sandoval et al. 2013; Cuervo et al. 2014; Vilar et al. 2014). More recently, stem cell–based therapies have been proposed for a broad spectrum of other pathologies, such as neuromuscular injuries, cardiac dysfunction, and bone and wound healing defects (Kraus and Kirker-Head 2006; Ribitsch et al. 2010; Fortier and Travis 2011; Gade et al. 2012; Volk and Theoret 2013).

The cell types receiving the most attention so far are mesenchymal stem cells (MSCs) isolated from bone marrow (BM) (Martin et al. 2002; Smith et al. 2003; Volk et al. 2005, 2012; Pacini et al. 2007; Vidal et al. 2008, 2012; Frisbie et al. 2009; Crovace et al. 2010; Fortier et al. 2010; Toupadakis et al. 2010; Ahern et al. 2011; Kisiel et al. 2012; Webb et al. 2012), adipose tissue (AT) (Koerner et al. 2006; Black et al. 2007, 2008; Richardson et al. 2007; Neupane et al. 2008; Vidal et al. 2008; Toupadakis et al. 2010; Ahern et al. 2011; de Mattos Carvalho et al. 2011; Guercio et al. 2012; Kisiel et al. 2012; Webb et al. 2012; Cuervo et al. 2014; Vilar et al. 2014), and to a lesser extent from umbilical cord (UC) (Weiss and Deryl 2006; Arufe et al. 2011; Kang et al. 2013; Van Loon et al. 2014), placenta adnexa (Parolini et al. 2008, 2009; Cremonesi et al. 2011; Parolini and Caruso 2011) and peripheral blood (Koerner et al. 2006; Ahern et al. 2011).

In particular, the autologous use of AT has been common in the veterinary clinic. In contrast, the promising induced pluripotent stem cells (iPSCs) (Cebrian-Serrano et al. 2013) and the spermatogonial stem cells (Fortier and Travis 2011), although extensively investigated in animal models, have not yet moved into veterinary practice where, instead, some relevant reports involving the use of embryonic stem cells (ESCs) (Guest et al. 2010; Watts et al. 2011) and amniotic-derived cells (ACs) (Barboni et al. 2012a, 2012b, 2014; Lange-Consiglio et al. 2012; Mattioli et al. 2012; Lange-Consiglio et al. 2013a, 2013b; Muttini et al. 2013) have recently been published.

The first report of cell-based therapies aimed at joint regeneration in horses dates back to the mid-1990s when a research team from Cornell University created a bank of chondrocytes (Nixon et al. 1992) that were allogenically transplanted for resurfacing experimental articular cartilage defects (Hendrickson et al. 1994). Currently, the use of adult and fetal MSCs represents the most proposed cell source after their validation through a long series of experimental investigations in rodents, and in medium- and large-sized mammals (Ribitsch et al. 2010; Fortier and Travis 2011; Gade et al. 2012; Volk and Theoret 2013). However, in the absence of regulatory guidance, cell-based protocols in veterinary practice have exploded in an uncontrolled with a strong interest of animal owners, leading a rapid translation of unapproved protocols to clinical practice. Several private companies, spinoffs, and university departments are engaged in a widespread stem cells service aimed to provide cells isolated from patient's tissue samples with or without an amplification step, or, alternatively, to sell kits that allow for in-house cell isolation from tissues (Vet-stem; VetBiologics; UCDAVIS; Riddle and Rood; RENOVOCyte; Biologics Medivet; Celavet; Therapies, ART Advanced Regenerative; Laboratories, Fat-Stem). Such a service, originated in North America, has now extended to Asia (South Korea Histostem Co. Ltd.) and

Europe (Belgium Fat-Stem Laboratories) and supports cell-based treatments to thousands of animals (Fortier and Travis 2011; Cyranoski 2013).

However, this empirical use of cell products applied in a variety of pathological conditions mainly in horses, dogs, and cats has not really enhanced knowledge on of the properties and mechanisms of these innovative therapeutic procedures for the care of animals. The major clinical outcomes generated by this widespread practice are represented by anecdotal and case reports. Although the animal cell products have been commercially available since 2003, few studies have documented the scientific improvement promoted by the injection of autologous cells collected by AT. In particular, two double-blinded controlled and multicenter studies performed on 21 (Black et al. 2007) and 39 dogs (Cuervo et al. 2014) affected by coxofemoral osteoarthritis (OA), and two clinical trials involving 14 recruited dogs with humeroradial joint OA (Black et al. 2008) and 10 dogs with severe hip OA (Vilar et al. 2014) have been published. The guidance void, and in parallel the widespread availability of cell products available in the United States, have fueled a large debate that involves the Veterinary Scientific Association and the Food and Drug Administration. Currently, the regulatory agency that has jurisdiction on development and approval of veterinary products has neither halted nor approved any veterinary stem cell therapy. Stakeholders now expect guidance from the regulatory agencies that should streamline the approval of cell treatments but also take into account the specificity of the veterinary field and the limited economic resources related to animal health and advances in veterinary medical care.

12.1.2 PRECLINICAL ANIMAL MODELS FOR REGENERATIVE MEDICINE

Regenerative medicine is moving toward translation to clinical practice and is becoming increasingly dependent on animal models and on the information generated regarding the potential efficacy and safety of stem cell–based technologies.

Companion (dog, cat, and horse) and farm animals (sheep, goat, bovine, and pig), often improperly referred to as large animal models, have been demonstrated to have an enhanced ability over laboratory mammals to predict clinical efficacy of new medical devices, pharmacological therapies, and cell-organ–based therapies (Dehoux and Gianello 2007).

In particular, in regenerative medicine, companion animals have been instrumental in advancing hematopoietic stem cell therapy for malignancies. The huge clinical impact in human medicine of this cell-based therapy would not have been achieved without the translational preclinical studies performed in canine models that initially allowed us to develop this therapy (Deeg and Storb 1994; Thomas 1999; Thomas and Storb 1999) and then optimize the technique over four decades (Sykes 2009; Trobridge and Kiem 2010).

Although genetically altered mice maintain a central role in the study of stem cell biology, no rodent model has biomedical properties more comparable to the biology and anatomy of humans. Compared to humans, the medium-sized mammals (pig, sheep, goat, and dog) share a similar basal cell metabolism rate, a longer life span, and comparable organ size morphology and physiology (Wagner and Storb 1996; Bruns et al. 2000; Wang 2006; McCarty et al. 2009; Parker et al. 2010).

Applications of Placenta-Derived Cells in Veterinary Medicine

Unlike laboratory rodents, the health conditions of companion and farm animals are under lifestyle influences. They are outbred and thereby continually exposed environmental factors that underlie several diseases (e.g., cancer, diabetes) or traumatic defects. For example, horses and dogs are often engaged in athletic or working careers. This increases the incidence of the long-standing musculoskeletal disorders in these species that continues to be a therapeutic, diagnostic, and clinical challenge for medicine. Another relevant aspect that increases the scientific interest in nonrodent animal models is that they suffer from a variety of naturally occurring diseases able to reproduce human phenotype and etiology. In particular, some companion and farm animals are currently used to study human genetic diseases. Indeed, they may be spontaneously affected by genetic disorders induced by a single gene defect, or due to the complex interaction between gene expression and environmental conditions (Lunney 2007; Kuzmuk and Schook 2011; Switonski 2014). Alternatively, genetically induced defects are also obtained in nonrodent animals to reproduce invalidating degenerative disorders (e.g., Alzheimer's and Huntington's diseases, cystic fibrosis, muscular dystrophy) by targeting specific genomic sites (Volk and Theoret 2013). Thus, the use of companion and farm animals may have a tremendous potential for validating and advancing the crucial field of regenerative medicine.

12.2 PLACENTAL TISSUES AS SOURCE OF STEM/PROGENITOR CELLS FOR VETERINARY REGENERATIVE MEDICINE

Placenta-derived cells (PCs) can be considered a relatively new entry within the large field of cell-based regenerative medicine. Looking at the relevant literature (Figure 12.1), the biology and properties of cells isolated from placental tissues started to generate a consistent number of publications one decade ago (2004–2005), even though the amniotic membrane (AM) has a long story in human therapy (DeRoth 1940; Colocho et al. 1974; Gruss and Jirsch 1978; Trelford-Sauder et al. 1978; Trelford and Trelford-Sauder 1979; Faulk et al. 1980; Ward and Bennett 1984; Ward et al. 1989; Young et al. 1991; Subrahmanyam 1995; Azuara-Blanco et al. 1999; Ravishanker et al. 2003; Mermet et al. 2007).

Placenta-related research, however, rapidly evolved when the organ, essential to support fetal development and to predict genetic diseases of the newborn, started to be considered as a source of stem/progenitor cells. Advances in the biology and *in vitro* protocols apply mainly to human PCs and have rapidly opened a role for them as an alternative stem cell source. In particular, placenta has become a valid and alternative tissue to collect a large amount of MSCs.

Encouraging results have been obtained more recently from studying PCs in domestic animal models. However, even though considerable information on animal placenta tissues (Parolini et al. 2009; Cremonesi et al. 2011) is now available, there are still knowledge gaps and technical concerns regarding their comparative biology. Importantly, although the placentas of eutherian mammals share common physiological and functional features, there are key differences in terms of embryogenesis, phenotype, and cell function that must be taken into account before comparing results and translating them among species to humans.

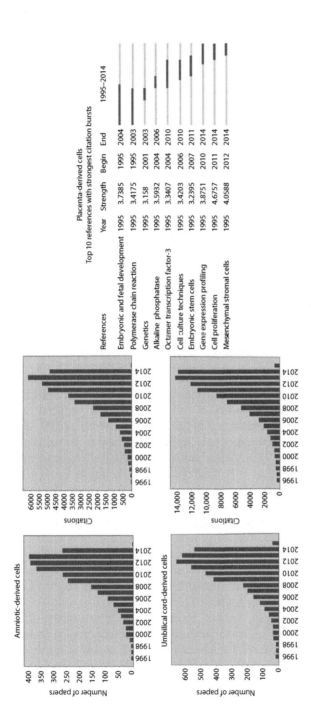

FIGURE 12.1 Histograms on the left show the number of papers published per year on amniotic-derived cells (top) and umbilical cord–derived cells (bottom), and histograms on the right illustrate the number of citations per year for the same topics. The data are elaborated from Web of Science Core Collection (http://thomsonreuters.com/web-of-science-core-collection/) by using key words "amniotic-derived cells" and "umbilical cord–derived cells." The diagram on the right shows the citation bursts (black bar) from 1995 to 2014 for "placenta-derived cells." The key words are listed in the references column and the bursting time (beginning and end) of the relative papers. During the first decade (up to 2003–2004), the most cited papers on placenta-derived cells had key words relating to the use of placental tissue/cells as diagnostic tools, whereas more recently, they refer to the use of placenta as a source of stem/progenitor cells. The data, referring to the 50 most cited papers per year, have been elaborated with CiteSpace (http://cluster.cis.drexel.edu/~cchen/citespace/).

12.2.1 Comparative Placenta Adnexa Embryology

The early stages of embryo development during which mammal fetal adnexa are differentiated take place through a complex series of processes that are frequently species specific (Barone 2003; Sinowatz 2009; Evans and de Lahunta 2013).

Postfertilization, the zygote develops into the embryo through subsequent mitosis that increases cell number without any overall volume gain for the presence of the external zona pellucida (ZP). Cleavages occur during the transport of the embryo through the female genital tract in a species-specific manner (Table 12.1).

In most of the mammals, the tubaric period ends at the morula stage. Compaction of the morula, preliminary to blastulation, occurs mainly in the uterus. The process leads to the formation of a unilaminar disc named the blastocyst that displays a central fluid-filled cavity, the blastocyst cavity, and an inner cell mass (ICM) that differentiates peripherally. The blastocyst expands under the osmotic pressure inside until the rupture of the surrounding ZP via hatching. In humans and primates, the embryo implants into the uterine mucosa, whereas in domestic animals placentation occurs at a much more advanced stage. The ICM then becomes bilaminar (bilaminar embryonic disc) around the time of blastocyst, hatching by differentiating a flattened cell population facing the blastocyst cavity, referred as hypoblast, and an external multilayered group of cells, named the epiblast (Figure 12.2). The hypoblast gradually forms a complete inner lining beneath the epiblast and the trophectoderm delimitating the primitive yolk sac. The hypoblast forms the inner epithelium of the yolk sac, and depending on the species, may become engaged in placentation. Differently from humans, the bilaminar embryos of domestic animals lose the trophectoderm that covers the epiblast, which consequently remains exposed to the uterine environment (Figure 12.2). The hatching blastocyst, which retains the spherical shape in humans, becomes gradually elongated in domestic animals (Figure 12.2). The formation of the amnion in humans occurs earlier (at the end of blastulation) than in domestic mammal embryos (at an early stage of gastrulation), and through different modalities. Indeed, the formation occurs in primates through cavitation, whereas in domestic animals it follows a more complex folding process.

Cavitation begins in the bilaminar embryonic disc where several gaps that progressively converge to form a central cavity start to appear inside the epiblast: the amniotic cavity. The epiblast-derived cells that constitute the innermost amniotic epithelium layer are named amnioblasts.

In contrast, the folding modality that characterizes the differentiation of amnion in domestic animals occurs in the early stage of gastrulation when the embryo becomes a trilaminar structure (trilaminar embryonic disc). The two layers located at the edge of the embryonic disc form the lateral chorion-amniotic folds and gradually expand upward and fuse. The union of the trophectoderm and the extraembryonic mesoderm will form the outer layers of the embryonic part of the placenta, named the chorion. The site where the chorion-amniotic folds meet and fuse is known as the mesamnion. In the horse and carnivores, the mesamnion disappears during fetal development, leaving no connection between the amnion and chorion. As a result, foals, pups, and kittens are born covered by an intact amnion that can be easily collected during delivery. In contrast, in the pig and ruminants,

TABLE 12.1

Major Developmental Events Occurring during the Early Stages of Embryogenesis

| Species | Embryo Stage at Arrival in Uterus | Time Expressed in Days after Fertilization | | | | | |
		Blastocyst	Gastrulation	Primitive Yolk Sac	Amnion/Chorion	Allantoids	Placentation
Cattle	8–16 cells	7–8	14	13	15 (development of amniotic folds)	18	19
Sheep	8–16 cells	6–7	12	10	14 (amniotic folds)	18	15
Pig	4–8 cells	5–6	12	—	17–18 (complete amnion cavity)	15	17
Horse	Morula/developing into blastocyst	6	16	10	20–21 (complete amnion cavity)	28	—
Dog	Blastocyst	8	16	—	21 (complete amnion cavity)	21	17
Cat	Blastocyst	6	13	—	21 (development of amniotic folds)	23	13
Human	Morula	6	21 (complete gastrulation)	8	8 (complete amnion cavity)	16	7–9

Note: The data are according to Barone (2003), Hyttel et al. (2010), and Evans and de Lahunta (2013).

Applications of Placenta-Derived Cells in Veterinary Medicine

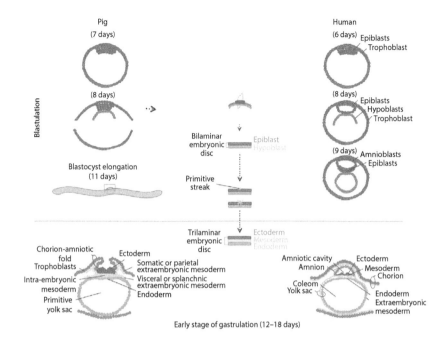

FIGURE 12.2 (See color insert.) Comparative embryological development of the amnion in a human and an example of its development in a domestic animal. The images summarize the different kinetics and modalities of amnion formation: schematically in primates (including humans) amniotic cavity differentiates at the end of blastulation (bilaminar embryonic disc) through a process of epiblast cavitation, whereas in domestic animals it occurs later, during the early stages of gastrulation (trilaminar embryonic disc) with a folding modality. **Blastulation.** (Top) Images show the major events that occur in pig (left) and human (right) embryos during the blastulation phase when the embryonic disc is organized into a bilaminar structure (middle). **Pig embryo.** Day 7 after fertilization, embryo is a hatched blastocyst. Day 8, embryo becomes bilaminar by differentiating a flattened cell population facing the blastocyst cavity, referred to as the hypoblast (green cells), and the external multilayered group of cells, named to epiblast (blue cells). In addition, two other important events involve the trophectoderm: the trophoblast cells (red cells) covering the epiblast disappear, while the cells located at the edge of the embryonic disc start to proliferate, thus rapidly elongating the blastocyst. Day 11, the embryo appears as an elongated blastocyst ~4 cm in length with a new internal cavity delimitated by the hypoblast: the primitive yolk sac. **Human embryo.** Day 6, hatched blastocyst. Day 8, bilaminar embryonic disc covered by a persistent trophectoderm. Day 9, embryo presents two adnexa: the primitive yolk sac delimitated by the hypoblast and the amnion that differentiate through the cavitation of the multilayered epiblast. **Early stages of gastrulation.** (Bottom images) The embryonic disc becomes trilaminar through the differentiation of the mesoderm that is composed of cells that proliferate into the primitive streak and migrate (yellow cells) between the hypoblast (endoderm: green) and the epiblast (ectoderm: blue). **Pig embryo.** Days 12–18, cells of the mesoderm located at the edge of the embryonic disc, covered by the trophectoderm, form two lateral chorion-amniotic folds (top image). The folds gradually expand upward to fuse above the embryonic disc, thus delimitating the newly formed amniotic cavity (bottom image). Based on amnion formation in domestic animals, the inner epithelium of the amniotic cavity originates from the trophectoderm and from the ectoderm of the embryonic disc in the upper portion and in the floor.

the mesamnion persists; as a result, the amnion is torn during parturition and offspring are generally born without covering membranes.

Instead, there are no major differences during the formation of the UC. In placental mammals, the UC develops as a stalk/conduit between the embryo/fetus and the placenta. During prenatal development, the UC is physiologically and genetically part of the fetus and normally contains two arteries and one vein. The cord is covered by a layer of ectoderm that is continuous with that of the amnion, and its various constituents are enveloped by an embryonic gelatinous tissue, named Wharton's jelly.

12.2.2 Placenta-Derived Cell Characterization in Domestic Animals

Unlike other stem cells (e.g., hematopoietic and ESCs), PCs have recently started to be tested in humans. Despite the extensive bibliography available, however, human PCs results are not easily comparable, especially considering the current lack of common protocols and the wide spectrum of cell origins and applications (Parolini et al. 2008, 2009). This is further complicated in animal-derived PCs where cell can be collected from different species and at different gestational stages. Another methodological aspect that makes domestic animal cell research challenging is the absence of commercially species-specific reagents, a source of the frequent discrepancy between results. For all these reasons, a careful update on the state of the art of domestic animal PCs is crucial to begin to focus on conserved and species-specific biological characteristics and to better understand their effects in preclinical settings.

12.2.2.1 Amniotic-Derived Tissue and Cells

12.2.2.1.1 Amniotic Membrane

Although there are relatively few reports in veterinary literature, most of them describe the successful use of AM for equine and canine corneal or conjunctival epithelium reconstruction (Barros et al. 1998, 2005; Lassaline et al. 2005; Ollivier et al. 2006; Kim et al. 2009; Plummer et al. 2009; Wichayacoop et al. 2009). Alternative clinical uses of AM have recently been proposed for endometrial cell replenishment when low proliferation is associated with pregnancy failure (Corradetti et al. 2014), as well as for the repair of full-thickness cartilage defects (Garcia et al. 2014). Even if the underlying mechanisms remain to be clarified, the anti-inflammatory genomic profile (Corradetti et al. 2014) and the immunomodulatory activity (Lange-Consiglio et al. 2013b; Barboni et al. 2014) may represent two conserved and facilitator mechanisms, respectively, that can explain the positive clinical outcomes of AM in these different diseases.

12.2.2.1.2 Amniotic Epithelial Cells

Amniotic epithelial cells (AECs) derived from different domestic animals have not been extensively studied to date except in ovine (oAEC), feline (fAEC), and equine (eAEC) models.

Despite the late embryonic origin of domestic animal amnion, AECs retain early stem cell markers except for c-Kit (Table 12.2), similar to humans (Miki et al. 2005; Parolini et al. 2008, 2009). In detail, SOX2 and NANOG are largely expressed in

TABLE 12.2

Comparison of Key Features of Domestic Animal Amniotic-Derived Cells

Cell Type	Species	Surface Markers	Expression of Pluripotent/ Embryonic Markers	Immunological Features	Differentiation Potential
AEC	Cat	[1]*CD29⁻ CD44⁺ CD73⁻ CD90⁻ CD166⁺* [1]*CD34⁻*	[1]TRA-1-60⁺ SSEA4⁺ [1]*OCT 4⁺ NANOG⁺*	[1]*MHC-I⁺ MHC-II⁻*	[1]Mesodermal differentiation: osteo, chondro, adipogenic [1]Endodermal differentiation: neurogenic
	Horse	[2]*CD29⁺ CD44⁺ CD166⁺ CD105⁺* [2]*CD34⁻*	[2]*ALP⁺ OCT4⁺*	[2]*MHC-I⁺ MHC-II⁻*	[2]Mesodermal differentiation: osteo, chondro, adipogenic [2]Endodermal differentiation: neurogenic
	Sheep	[3,4,5,6,7,8,9]CD29⁺ CD49f⁺ CD166⁺ [3,4,5,6,7,8,9]CD14⁻ CD31⁻ CD45⁻	[10]SSEA1⁺ SSEA3⁺ SSEA4⁺ TRA-1-160⁺ TRA-1-81⁺ [3,5,6,7,8]c-Kit⁻ [3,4,5,6,7,8,9]NANOG⁺ SOX2⁺ TERT⁺ [3,4,6,7,10,8]*OCT4⁺ NANOG⁺ SOX2⁺* [10]*REX1⁺* [3,4,6,7,8]*TERT⁺*	[4,5,9]MHC-I⁺ MHC-II⁻ [4]*In vitro* inhibition of PBMCs proliferation [3,7,9]*In vivo* inhibition of leukocyte infiltration [3,7,8]Long-term engraftment in tendons [6,9]Long-term engraftment in bone	Mesodermal differentiation: [3,4,5,6,9]osteo, [4,9]adipo [3,5,7,9]tenogenic [4,5,10]Endodermal differentiation: neurogenic [5]Ectodermal differentiation: hepatogenic [9] Expression of bone-related protein (OCN) in PKH26-transplanted AECs [7]Ovine COL1 mRNA recorded in repaired equine tendon after transplantation of ovine AECs [3]Secretion of COL1 in PKH26-AEC transplanted in tendon defect

(Continued)

TABLE 12.2 *(Continued)*

Comparison of Key Features of Domestic Animal Amniotic-Derived Cells

Cell Type	Species	Surface Markers	Expression of Pluripotent/ Embryonic Markers	Immunological Features	Differentiation Potential
	Buffalo		[11]*ALP+* [11]*OCT4+ NANOG+ SOX2+*		[11]Mesodermal differentiation: osteogenic
AMSC	Cat	[12]CD44+ [12]CD73− [13]CD73+/− [12,13]CD90+ [12]CD105+ [12]CD14− [12,13]CD34−CD45−[13]CD79a−			[12,13]Mesodermal differentiation: osteo, chondro, adipogenic [13]No teratoma formation
	Dog	[14]CD90+ CD105+ [15,16]*CD29+ CD44+ CD184+* [14]CD34− CD45− [15,16]*CD34− CD45−* [14]*CD3− CD11c− CD 28− CD38− CD62L−* [14]*CD41a−*	[15,16,14]*OCT4+* [15,16]*NANOG−* [14]*NANOG+* [14]*SOX2+ KLF4+*	[15,16]*DLA-DRA1− DLA-DQ1− DLA-79− DLA-88−*	Mesodermal differentiation: [4,15,14]osteo, [15,14]chondro, [15,14]adipogenic [16,14]Endodermal differentiation: neurogenic
	Horse	[17,18,19]CD 44+ CD90+ [20,17]CD105+ [2,20,17]*CD29+ CD44+ CD166+* [2,20,17,21]CD105+	[2,20,17]TRA-1-60+ SSEA4+ [2]SSEA3+ [2,20,17,19]OCT4+ [22]c-Kit+ [17]cMyc+ [20]ALP+ [21]*OCT4+ NANOG+ SOX2+*	[19]MHC-I− MHC-II− [2,20,17]*MHC-I+ MHC-II−* [22]*In vitro* inhibition of PBMC Proliferation	[2,20,18,19,21]Mesodermal differentiation: osteo, chondro, adipogenic [2,20]Endodermal differentiation: neurogenic

(Continued)

Applications of Placenta-Derived Cells in Veterinary Medicine 229

TABLE 12.2 (Continued)
Comparison of Key Features of Domestic Animal Amniotic-Derived Cells

Cell Type	Species	Surface Markers	Expression of Pluripotent/ Embryonic Markers	Immunological Features	Differentiation Potential
		[19]CD14+ [18]CD34- [18]CD31- [2,20,17,21]CD34- [18]CD19- CD20- CD28- CD38- CD62L- CD62P- CD200- [18]CD31- CD41a-			
	Pig	[23]CD29+ CD44+ [23]CD45-		[23]SLA1- SLAII-DR- SLAII-DQ-	[23]Switch-on of cardiomyocytes markers (troponins I and T) detected in green fluorescent protein–transplanted porcine AMSCs [24]Mesodermal differentiation: osteo, myogenic
	Sheep		[24]OCT4+, TERT+		
	Bovine	[25]CD29+ CD44+ CD73+ CD105- CD166+ [25]CD14- CD34+ CD45-	[25]OCT4+ Myc+	[25]MHC-I+ MHC-II-	[25]Mesodermal differentiation: osteo, adipogenic [25]Endodermal differentiation: neurogenic
	Buffalo		[26]ALP+ [26]OCT4+ SOX2+ NANOG+ SCS+		

(Continued)

TABLE 12.2 *(Continued)*

Comparison of Key Features of Domestic Animal Amniotic-Derived Cells

Cell Type	Species	Surface Markers	Expression of Pluripotent/ Embryonic Markers	Immunological Features	Differentiation Potential
AFMSC	Cat	[12]CD44+ CD73+/− CD90+ CD105+ [12]CD14− CD34+/− CD45−			[12]Mesodermal differentiation: osteo, chondro, adipogenic
	Dog	[27,28]CD29+ CD44+ CD90+ [15]*CD29− CD44+ CD184−* [27,28]CD34− [15]*CD34− CD45−*	[27]OCT4+ SOX2+ NANOG+ [27,28]*OCT4+ NANOG+* [27,28]*SOX2+* [15]*NANOG−*	[15]*DLA-DRA1+ DLA79+* [15]*DLA-DQA1− DLA88−*	Mesodermal differentiation: [15,27,29]osteo, [27,29] chondro, [15,27,29]adipogenic [15,28,29]Endodermal differentiation: neurogenic [27]Ectodermal differentiation: hepatogenic
	Horse	[30]CD29+ [31,30,32]CD44+ [30]CD73(clone5F/B9)− [30]CD73(cloneAD2)+ [31]CD73− [31,30,33,32] CD90+ [31,33,34] CD105+ [30,31]CD14− [31,30,33,32] CD34−[30,31]CD45− [30]*CD14−*[30,33]*CD34−*[30] *CD45−*	[33]OCT4+ [33] TRA-1-60+/− SSEA4+	[33]*ELA-ABC+ ELA-DR−*	[31,30,33,34,32]Mesodermal differentiation: osteo, chondro, adipo, [30]tenogenic

(Continued)

TABLE 12.2 (Continued)
Comparison of Key Features of Domestic Animal Amniotic-Derived Cells

Cell Type	Species	Surface Markers	Expression of Pluripotent/Embryonic Markers	Immunological Features	Differentiation Potential
	Pig	[32]CD19-CD20-CD28-CD38- CD62L-CD62P-CD200- [32]CD31-CD41a- [35,36]CD44+ [36]CD73+ [37,36]CD90+ [37]CD105- [36]CD105+ [35]CD166+ [35]CD90+ [36]CD14+/- [37,35,36] CD34-CD45- [35]CD54- [37]vWF+ [35]CD45-	[38,36]SSEA4+/- [36] SSEA1-Tra-1-60+ Tra-1-81+ [38,37]ckit- [38] OCT4+/- [36]OCT4+ [38,36]NANOG+ [36] TERT+ [38,37,39,41]OCT4+ THY+ [37,41]SOX2+ [36]c-Kit+	[36] HLA-ABC+ [38]Survival into ischemic myocardium after xenotransplantation in mice	Mesodermal differentiation: [37,39,40]osteo, [36,37,40]adipo, [39]chondro, [42,38,36]myogenic, [38,37,40]endothelial lineage differentiation [35,38,39,40]Endodermal differentiation: neurogenic
	Sheep	[38]CD31+ [42,43,44,45,46,47,48]CD29+ [52,43,53,44,47,45,48]CD44+ [49,46]CD49f+/- [52,42,53,46] CD58+ [48]CD90+/- [43,44,48] [43,46]CD90+ [43,44,48] CD105+/-	[49,50,51,46,47]OCT4+, TERT+ [42,49,46,51]SOX2+ NANOG+ c-Kit- [45]STRO4+ SSEA4+ [49,45]OCT4+ NANOG+ SOX2+ [49] TERT+	[42,51,46]MHC-I+ MHC-II- [49]Survival after allotransplantation in tendon defect	Mesodermal differentiation: [52,42,49,43,51,46,47,48]osteo, [52,43,48]adipo, [47,25,54]chondro, [43]teno, [48]myogenic, [47]endothelial lineage

(Continued)

TABLE 12.2 *(Continued)*

Comparison of Key Features of Domestic Animal Amniotic-Derived Cells

Cell Type	Species	Surface Markers	Expression of Pluripotent/ Embryonic Markers	Immunological Features	Differentiation Potential
		[52,42,53,44,45,46,47]CD166+		[51]Survival after allotransplantation in bone defect	[52]Long-term engraftment into hematopoietic organs of CD34+ sorted ovine AFMSCs after autologous uterus stem cell transplantation in sheep
		[46]Ca-SR+			
		[47]CD11b− [52,42,49,46,51]CD14−			
		[52]CD34+ [52,42,49,51,53,46]CD45−			
		[52,42,49,43,51,53,44,46,48,55,24, 54]CD31−			
	Caprine	[56]CD73+ CD90+ CD105+	[56]ALP+ [56]OCT4+ NANOG+ SOX2+		Mesodermal differentiation: [56]chondro, [56,57]osteo, [56,57]adipogenic, [57]endothelial lineage
		[56]*CD73+ CD90+ CD105+*	[56]SSEA1+ SSEA4+ [57]OCT4+ NANOG+ SOX2+ HES1+		[57]Endodermal differentiation: neurogenic
		[56]*CD34−*			
	Bovine	[58,59]CD44+ [59]CD71+ [58]CD90+ CD105+ [58,59]b integrin+ [59]CD29+ [24,59]*CD29+ CD44+ CD73+* [24]CD105− [24,59]*CD166+*	[58]OCT4− [59]OCT4+ [58]SSEA4+ /−[47]STRO4+ [58]*OCT4−* [25,59]*OCT4+* [25]*Myc+ c-Kit+* [58,47]*NANOG− SOX2−* [47]*STAT3−*	[25]*MHC-I+ MHC-II−*	Mesodermal differentiation: [58]chondro, [58,25,59]osteo, [58,25,59]adipogenic [25,59]Endodermal differentiation: neurogenic

TABLE 12.2 *(Continued)*

Comparison of Key Features of Domestic Animal Amniotic-Derived Cells

Cell Type	Species	Surface Markers	Expression of Pluripotent/ Embryonic Markers	Immunological Features	Differentiation Potential
	Buffalo	[58]CD14- [58]CD34+ [59]CD34- [59]CD45- [24]CD14- [24,59]CD34- CD45-	[60,61,62]ALP+ assay [60,61,62]OCT4+ [61,60]SOX2+ NANOG+ SCF+		[61]Endodermal differentiation: neurogenic

AEC, amniotic epithelial cell; AFMSC, amniotic fluid mesenchymal stem cell; AMSC, amniotic mesenchymal stromal cells; HLA, human leukocyte antigen; MHC, major histocompatibility complex; PBMC, peripheral blood mononuclear cell.

Note: Surface markers are organized into categories: light gray, mesenchymal; dark gray with black writing, hematopoietic; dark gray with white writing, immune cells; black with white writing, plateled markers. The markers in *italics* are genes; the others markers are proteins detected with different techniques.

[1]Rutigliano et al. 2013; [2]Lange-Consiglio et al. 2012; [3]Barboni et al. 2012b; [4]Barboni et al. 2014; [5]Barboni et al. 2012a; [6]Mattioli et al. 2012; [7]Muttini et al. 2013; [8]Muttini et al. 2010; [9]Barboni et al. 2013; [10]Zhu et al. 2013; [11]Mann et al. 2013; [12]Iacono et al. 2012b; [13]Vidane et al. 2014; [14]Park et al. 2012; [15]Filioli Uranio et al. 2011; [16]Filioli Uranio et al. 2014; [17]Corradetti et al. 2014; [18]Seo et al. 2009; [19]Coli et al. 2011; [20]Lange-Consiglio et al. 2013a; [21]Violini et al. 2012; [22] Lange-Consiglio et al. 2013b; [23]Kimura et al. 2012; [24]Mauro et al. 2010; [25]Corradetti et al. 2013; [26]Dev et al. 2012c; [27]Choi et al. 2013; [28]Kim et al. 2014; [29]Fernandes et al. 2012; [30]Gulati et al. 2013; [31]Iacono et al. 2012a; [32]Park et al. 2011b; [33]Lovati et al. 2011; [34]Iacono et al. 2012c; [35]Chen et al. 2011; [36]Baulier et al. 2014; [37]Sartore et al. 2005; [38]Zheng et al. 2011a; [39]Zheng et al. 2009; [40]Zheng et al. 2010; [41]Wagner and Storb 1996; [42]Colosimo et al. 2013b; [43]Klein et al. 2011; [44]Gray et al. 2012; [45]Weber et al. 2012; [46]Di Tomo et al. 2013; [47]Weber et al. 2013; [48]Kunisaki et al. 2007; [49]Colosimo et al. 2013a; [50]Mauro et al. 2010; [51]Berardinelli et al. 2013; [52]Shaw et al. 2015; [53]Shaw et al. 2011; [54]Kunisaki et al. 2006a; [55]Kaviani et al. 2001; [56]Pratheesh et al. 2013; [57]Zheng et al. 2011b; [58]Rossi et al. 2014; [59]Gao et al. 2014; [60]Yadav et al. 2011; [61]Dev et al. 2012b; [62]Dev et al. 2012a.

oAECs, whereas OCT4 is limited to a minority of them (Mattioli et al. 2012; Muttini et al. 2013; Barboni et al. 2014). Even if the presence of these transcription factors seems to be unaffected by cryopreservation (Barboni et al. 2012a; Muttini et al. 2013), they can be modulated by the procedure of *in vitro* expansion. Except for OCT4 that was stable in amplified oAECs (Mattioli et al. 2012), eAECs (Lange-Consiglio et al. 2012), fAECs (Rutigliano et al. 2013), and hAECs (Izumi et al. 2009), *SOX2* mRNA levels progressively decreased in oAECs with culture (Mattioli et al. 2012), whereas *NANOG* displayed a different behavior in cats (Rutigliano et al. 2013) and oAECs (Mattioli et al. 2012). *NANOG* and *SOX2* are also affected by the gestational stage of the cells, since they are overexpressed in AECs at the beginning of pregnancy (Barboni et al. 2014). Similarly, TERT activity is reported to be progressively reduced during subculture, even if in long-term expanded oAECs, signs of abnormal karyotype and telomeres size shortening have not been described (Barboni et al. 2012a; Mattioli et al. 2012). In addition, *TERT* mRNA levels decreased with gestation and when the oAECs were exposed to tissue-inductive culture conditions, regardless of the cells' capacity to complete the process of *in vitro* differentiation (Barboni et al. 2014). TERT downregulation through gestation appears to be a highly conserved mechanism: it has been documented in rat (Nakajima et al. 2001) and sheep (Barboni et al. 2014) and may explain the absence of TERT in term hAECs (Miki et al. 2005).

The modulation of oAEC gene expression could be induced by the epigenetic transformation observed in term and expanded cells (Mattioli et al. 2012; Barboni et al. 2014). Global DNA methylation is indeed lower in freshly isolated and *in vitro*–expanded oAECs collected during the early and middle phase of gestation. Considering that mouse ESCs, hESCs, and oAECs are less methylated than differentiated cells (Zvetkova et al. 2005; Allegrucci et al. 2007; Barboni et al. 2014), the low DNA methylation may retain gene expression and provide a condition to preserve cell stemness.

Few studies to date have analyzed the plasticity of AECs during gestation, except in human (Izumi et al. 2009; Lim et al. 2013) and oAECs (Barboni et al. 2014). A greater oAEC aptitude toward the adipogenic lineage was recorded in cells isolated from early gestational amnia, whereas the osteogenic and neurogenic differentiation appeared to be stronger in cells collected at late gestational stages (Barboni et al. 2014) according to the results described in Lim et al. (2013). In this study, term cells, which are less able to express adipose and neural markers *in vitro*, resulted more prone to express lung lineage molecules than preterm cells, and also exerted greater protection against lung inflammation and fibrosis *in vivo*. Altogether, these results could suggest a stage-specific clinical impact of AECs that should be considered, especially in veterinary regenerative medicine where AECs can be easily harvested at different stages of gestation.

Freshly isolated fAECs (Rutigliano et al. 2013) and oAECs (Barboni et al. 2012a), like hAECs (Terada et al. 2000), express very low levels of the major histocompatibility complex (MHC)-I and lack MHC-II. Expansion *in vitro* (Barboni et al. 2012a) as well as gestation (Barboni et al. 2014) in oAECs increased the mRNA content of *MHC-I*, but did not modify *MHC-II*. Probably attributable to their immunoprivileged state, AECs were able to survive for a long time in allo/xenotransplanted

Applications of Placenta-Derived Cells in Veterinary Medicine

immunocompetent hosts (Tables 12.4 and 12.5). This conserved property is seen across human- and large animal–derived AECs that have an inherited immunomodulatory activity confirmed through the lymphocyte *in vitro* proliferation assay. As such, human AM (hAMs) (Rossi et al. 2012), hAECs, human amniotic mesenchymal stromal cells (hAMSCs) (Bailo et al. 2004; Li et al. 2005; Wolbank et al. 2007; Banas et al. 2008), equine amniotic mesenchymal stem cells (eAMSCs) (Lange-Consiglio et al. 2013b), and oAECs (Barboni et al. 2014) are able to suppress mitogen-induced T lymphocyte proliferation at high dilutions both in cell-to-cell contact and in transwell settings. This result reinforces the hypothesis extensively proposed in humans (Rossi et al. 2012; Cargnoni et al. 2014) that the bioactive molecules released from AECs may be responsible for their paracrine immunomodulatory ability that is fundamental in facilitating tissue repair processes. Interestingly, oAECs maintained this property during the entire gestation, whereas cell immunomodulatory activity is reduced during expansion (Barboni et al. 2014). This consequence of cell culture must be carefully considered before the clinical application of oAECs, but, in our opinion, should not restrict the use of these cells. The opportunity to reduce the *in vitro* amplification to preserve their biological properties can be easily proposed for oAECs since a great amount of fresh cells can be collected from a single amnion. AECs can also be easily isolated from cats (Rutigliano et al. 2013), even if at the beginning of culture freshly isolated cells are heterogeneous (polygonal epithelial and fibroblastoid cells). fAECs are able to acquire a high plating efficiency and a stable phenotype only after *in vitro* amplification when they reached doubling time values similar to those described for hAECs (Parolini et al. 2008; Miki et al. 2010; Lange-Consiglio et al. 2012) and oAECs (Mattioli et al. 2012). This preliminary phase of fAEC selection *in vitro* seems to be essential to obtaining a more homogenous population of cells with the typical polygonal epithelial cell shape and cells expresses CD166 and CD44.

In AECs, regardless of the species, CD49, CD105, and CD166 expression progressively reduce with culture. In agreement with previous findings (Bilic et al. 2008; Stadler et al. 2008), vimentin and α-smooth muscle actin positivity increases in oAECs during expansion. This phenotype shift *in vitro* could be the consequence of the prevalence of MSCs, or it could be induced by the process of the epithelial-mesenchymal transition that has also been documented in AM (Bilic et al. 2008).

Regardless of the mechanisms involved, the strategies for preserving the original pheno/geno/epigenotype of AECs remains a crucial biological and methodological challenge that involves different sources of stem cells both in humans and in domestic animals.

12.2.2.1.3 *Amniotic Fluid Mesenchymal Stem Cells*

Human amniotic fluid mesenchymal stem cells (hAFMSCs) are usually collected during the second trimester of pregnancy through amniocentesis. On the contrary, in domestic animals where amniocentesis is not practiced, AFMSCs are mainly collected at term in pets and horses or at different phase of gestation in ovine, porcine, and bovine where amniotic fluid (AF) can be occasionally collected from pregnant animals at the slaughterhouse. In these latter cases, the AFMSCs can be collected in large amounts even if cell yield appears to be, at least for bovine, significantly higher in the first trimester (Rossi et al. 2014). As a consequence, the gestational stage is a frequent variable among groups that must be taken into account to compare the phenotype of AFMSCs from domestic animals (Perin et al. 2008).

The population of cells isolated from different domestic animals appears to be initially heterogeneous, as in humans (Parolini et al. 2008). Based on their morphology, AFMSCs can be classified into epithelioid-like, fibroblastic-like, and round-shaped cells. In domestic animals, the subpopulation of cells that becomes prevalent early in culture belongs to the fibroblast-like morphology (Corradetti et al. 2013; da Cunha et al. 2014; Gao et al. 2014). Differently from hAFMSCs, the round cells decrease rapidly during the culture with the exception of one study on bovine (Rossi et al. 2014) and one study on porcine (Sartore et al. 2005), where the persistence of both fibroblast-like and round-shaped cells has been described late in culture.

Typical MSC markers (Table 12.2) characterize the immunophenotype of these cells. Some discrepancies are however reported, such as the disappearance of CD29 over culture in canine (Filioli Uranio et al. 2011), conflicting reports on CD73 in equine (Iacono et al. 2012a), and the absence of CD105 in bovine (Corradetti et al. 2013) and porcine (Sartore et al. 2005) AFMSCs. Even if a species-specific phenotype cannot be excluded, the differences observed in MSC markers of domestic animals appears to be frequently ascribable to the low cross-reactivity of the commercial antibodies (Iacono et al. 2012a, 2012b; Gulati et al. 2013; Rossi et al. 2014). Indeed, the absence of a surface protein antigen can be, in contrast, frequently accompanied by a positive check of the expression of the relative gene (Chen et al. 2011; Filioli Uranio et al. 2011, 2014; Corradetti et al. 2013; Gulati et al. 2013; Gao et al. 2014). Although a complete MSC immunophenotype is not currently available for all species, AFMSCs may still represent a valid alternative source of MSCs for veterinary regenerative medicine, as suggested by their great mesenchymal *in vitro* lineage differentiation ability (Table 12.2).

Another controversial aspect regarding the phenotype of AFMSCs from large animals is immunopositivity for the hematopoietic marker CD34. Several reports indicate that AFMSCs from domestic animals are negative for CD34, like other hematopoietic antigens (Table 12.2). Recently, Shaw et al. (2015) refuted this result by sorting, at least in ovine, a subpopulation of CD34$^+$ cells that were autologously transplanted *in utero*. The discrepancy regarding CD34 seems ascribable to the different cross-reactivity of the antibodies. Shaw et al. (2015) developed a sheep-specific antibody previously validated by identifying hematopoietic stem cells in the adult ovine BM (Porada et al. 2008). In addition, unlike other reports where the hematopoietic immunophenotype can appear after several passages, Shaw et al. (2015) sorted the CD34$^+$ cells immediately after AF collection, thus isolating a subpopulation of cells that is similar to those obtained from a BM aspirate. The CD34$^+$ subpopulation, homogenous for CD44$^-$, CD58$^-$, CD14$^-$, CD31$^-$, and CD45$^+$ functionally confirms its hematopoietic origin through a long-term engraftment in the peripheral blood and hematopoietic organs, thus supporting the hypothesis of its utility for treating congenital hematopoietic disease.

Besides mesenchymal and hematopoietic potency, pluripotent properties have also been proposed in AFMSCs from humans and animals. In particular, pluripotency has been attributed to mouse and hAFMSC cKit$^+$ subpopulation: they displayed long-term self-renewal, clonal properties, and high differentiation capabilities (De Coppi et al. 2007;

Applications of Placenta-Derived Cells in Veterinary Medicine

Ditadi et al. 2009; Piccoli et al. 2012), and they shared the expression of specific markers (e.g., OCT4 and SSEA4) with ESCs, even if they are not tumorigenic unless reprogrammed as iPSCs (Galende et al. 2010). In contrast, the pluripotency of AFMSCs from domestic animals seems to be supported more by the expression of transcription factors and by TERT activity (Table 12.2) than by cKit+, which is absent in ovine and remains debatable in pAFMSCs (Sartore et al. 2005; Chen et al. 2011; Colosimo et al. 2013b; Baulier et al. 2014). However, the expression of pluripotency markers appears to be quite unstable during expansion, in contrast to the expression of surface antigens. *NANOG* and *SOX2* declined over culture and TERT activity decreased as a consequence of the enzyme cytoplasmic translocation (Colosimo et al. 2013b) in oAFMSCs. Conversely, *NANOG* and *SOX2* were under- and overexpressed, respectively, in late passages of bubaline (b)AFMSCs (Yadav et al. 2011). In addition, the fibroblast-like bAFMSCs subpopulation progressively reduced *SSEA4* expression, while maintaining a stable but weak expression of *NANOG*, and did not express *OCT4* and *SOX2* (Rossi et al. 2014). The expression of *OCT4* remains, however, contradictory since positive results have been reported by other groups in bovine (Corradetti et al. 2013; Gao et al. 2014), human (Prusa et al. 2003; Tsai et al. 2004; Kim et al. 2007; Siegel et al. 2008; You et al. 2008, 2009), and in other animal-derived AFMSCs (Chen et al. 2011; Filioli Uranio et al. 2011; Lovati et al. 2011; Dev et al. 2012a; Colosimo et al. 2013b).

Similar to evaluations performed with human cells, a direct evaluation of TERT expression or activity or, alternatively, its biological effects, has been carried out to confirm the pluripotent nature of AFMSCs from domestic animals and to predict their clinical relevance after prolonged amplification (Mosquera et al. 1999; Kim et al. 2007; Poloni et al. 2011). In agreement with a long-term expression of TERT, expanded ovine (o)AFMSCs did not show any significant reduction in telomere size over 20 passages (Colosimo et al. 2013a). In contrast, TERT activity seems to decline quickly during culture in canine (c)AFMSCs (Filioli Uranio et al. 2011). In addition, however, a normal diploid karyotype after *in vitro* expansion seems to also confirm a long-term effect of TERT in bubaline (Dev et al. 2012a), canine (Filioli Uranio et al. 2011), ovine (Colosimo et al. 2013a), and hAFMSCs (Miranda-Sayago et al. 2011; Poloni et al. 2011). However, subtle *in vitro* chromosomal rearrangements cannot be excluded since all of the data concerning the karyotype of domestic animals and hAFMSCs have not been gathered using advanced techniques (Carpenter 2001).

The pluripotency of AFMSCs from domestic animals also seems to be confirmed by their ability to differentiate into cell lines of the three germ layers (Table 12.2). The *in vitro* plasticity of AFMSCs, however, is influenced by the process of expansion (Colosimo et al. 2013b; Rossi et al. 2014) and indirectly by the global DNA methylation status. Indeed, global DNA methylation of oAFMSCs revealed that the high level of methylation recorded in freshly isolated cells declined over culture (Colosimo et al. 2013b). A functional change of expanded AFMSCs is, in parallel, confirmed by *in vitro* differentiation experiments. Colosimo et al. (2013b) demonstrated a negative effect of *in vitro* expansion on the osteogenetic ability of oAFMSCs when the process of differentiation was tested adopting advanced and quantitative functional assays. These data are in agreement with the results obtained by Schellenberg et al. (2012) who showed that clonogenic subpopulations of MSCs with the highest adipogenic or osteogenic differentiation capacity significantly decreased their plasticity over passages. However, the cultural selection of cell subpopulations with different epigenetic and functional profiles, as recently demonstrated by Rossi et al. (2014) on bAFMSCs and by Roubelakis et al. (2012) on hAFMSCs, cannot be excluded. The evidence gathered so far not only confirms the high differentiation potential of AFMSCs from domestic animals but also focuses the attention on the negative role exerted by culture conditions, which must be taken into careful consideration when AFMSCs are proposed for cell-based therapies.

12.2.2.1.4 Amniotic Mesenchymal Stromal Cells

The majority of animal AMSCs have been isolated at term by using a two-step procedure validated in humans (Parolini et al. 2008). However, two alternative isolation protocols have been proposed: a direct digestion of AM with collagenase in feline (Iacono et al. 2012a; Vidane et al. 2014) and *in vitro* culture-based selection of cells starting from AM fragments in ovine (Mauro et al. 2010) and canine (Filioli Uranio et al. 2011, 2014).

Regardless of the protocols, AMSC from domestic animals have not yet been tested for the minimal criteria identified by the "International Society for Cellular Therapy" (ISCT) (Horwitz et al. 2005). Except for bovine and eAMSC where a complete MSC immunophenotype has been described, canine (Park et al. 2012) and pig (Kimura et al. 2012) AMSC are positive for CD105 and CD90 or CD29 and CD44 (Table 12.2). In addition, AMSC from domestic animals lack hematopoietic surface antigens and MHC-II, express embryonic and pluripotency genes, and display mesenchymal and neurogenic differentiation potential (Table 12.2).

Interestingly, cMSCs derived either from AM or from UC appeared to be highly stable *in vitro* and weakly influenced by the gestational stage of collection (Filioli Uranio et al. 2011, 2014), differently from oAECs (Mattioli et al. 2012); ovine, bubaline, and bAFMSCs (Yadav et al. 2011; Colosimo et al. 2013b; Rossi et al. 2014); and hAMSCs (Evangelista et al. 2008). The great versatility of cAMSCs could facilitate the development of cell-based therapies in dogs, a species that has an emerging role as a model for studying human hereditary diseases.

12.2.2.2 UC-Derived Cells

During the past 15 years, umbilical cord blood (UCB) has been exploited as a rich source of MSCs (Erices et al. 2000; Zarrabi et al. 2014). Although BM and UCB are traditionally considered as the two main sources of MSCs, more recently, stem/progenitor cells derived from UC outer regions, defined below as umbilical cord matrix (UCM) have been extensively considered (Weiss and Deryl 2006; Harris 2013). These cells are a primitive stromal population that display the characteristics of MSCs: they grow as adherent cells with fibroblast-like morphology; are self-renewing even if for limited passages; express cell surface markers typical of MSCs; and can be differentiated *in vitro* at least into bone, cartilage, and ATs (Horwitz et al. 2005; Dominici et al. 2006).

12.2.2.3 Umbilical Cord Blood Mesenchymal Stem Cells

Umbilical cord blood mesenchymal stem cells (UCBMSCs) from domestic animals (Table 12.3) can be isolated by adopting the protocols validated for humans (Campagnoli et al. 2001; Yang et al. 2004), with minor modifications. The isolation

TABLE 12.3

Comparison of Key Features of Domestic Animal UCB and UCM Mesenchymal-Derived Cells

Cell Type	Species	Surface Markers	Expression of Pluripotent/Embryonic Markers	Immunological Features	Differentiation Potential
UCB	Cat	[1]CD9+ CD44+ [1]CD18- CD45-			[1]Endodermal differentiation: neurogenic
	Dog	[2]CD29+ [2,3,4]CD44+ [2]CD73- [3,4]CD73+ [2]CD90- [3,4]CD90+ [3,4]CD105+ CD184+ [2]CD4- CD8a- CD10- [2,3,4]CD14- [2]CD24- CD31- [2,3,4]CD34- [2]CD38- CD41a- [2,3,4] CD45- [2]CD49b- CD41/61- [2] CD62p-	[2]OCT4+	[2]HLA-DR II-	Mesodermal differentiation: [2,3,4,5]osteo, [2]chondro, [3]adipogenic [2,3]Endodermal differentiation: neurogenic
	Horse	[6,7,8,9,10,11,12]CD29+ [8,11,9,14,6,10,7,12]CD44+ [6,7]CD73- [9]CD73+/- [8,11]CD73+ [17,14,6,8,9,10,11,12]CD90+ [7]CD90-	[13,10]OCT4+ [10]OCT4- [10]SSEA3- [13]SSEA1+ [10]SSEA1- [10]SSEA4- [13]SSEA4+/- [13]TRA-1-60+ [10]TRA-1-60- [13]TRA-1-81+	[10,11,12]MHC-I+ [10]MHC-I+/- [6,7,9,10,11,12]MHC-II-	Mesodermal differentiation: [16,17,14,6,15,18,9,13,10,11,19,20,21]osteo, [17,14,6,15,9,18,13,10,11,19,22,21]chondro, [17,14,6,15,18,9,13,10,11,21]adipo, [13]myogenic [13]Ectodermal differentiation: hepatogenic

(Continued)

TABLE 12.3 *(Continued)*

Comparison of Key Features of Domestic Animal UCB and UCM Mesenchymal-Derived Cells

Cell type	Species	Surface Markers	Expression of Pluripotent Embryonic Markers	Immunological Features	Differentiation Potential
		[17,14,9,10,11]CD105[+] [7]CD105[−]	[8,23]*OCT4[+] NANOG[+] SOX2[+]*		
		[8]*CD73+ CD90[+] CD105[+]*			
		[14,7,10]CD14[−] [15]CD18[−]	[8,23]*Klf4[+] cMyc[+]*		
		[17,6,7,8]CD34[−] [7,8,9,10,11]CD45[−]			
		[6,9,10,7,11]CD79a[−] [15]vWF[−]			
		[8]*CD14[−] CD34[−] CD45[−]*			
		[17]CD20[−] CD28[−] CD38[−]			
		[17]CD62L[−] CD62P[−] CD200[−]			
		[6,9,11]Monocyte marker[−]			
		[11]*CD40[+] CD80[+/−]* [11]*CD86[−]*			
		[17]CD31[−] CD41a[−]			
	Pig	[24]CD29[+]CD105[+]	[24]*SCF[+] LIF[+] G-CSF[+]*		[24]Mesodermal differentiation: osteo, chondro, adipogenic
		[24]CD45[−]			
	Sheep	[25]CD44[+]			[26]Mesodermal differentiation: osteo, adipo, chondrogenic
		[25]CD38[−] CD45[−]			
	Bovine	[27]*CD73[+]*	[27]*OCT4[+]*		[27]Mesodermal differentiation: osteo, chondro, adipogenic

(Continued)

TABLE 12.3 *(Continued)*

Comparison of Key Features of Domestic Animal UCB and UCM Mesenchymal-Derived Cells

Cell type	Species	Surface Markers	Expression of Pluripotent Embryonic Markers	Immunological Features	Differentiation Potential
UCM	Dog	[28,29] CD29− [28,3,4,29]CD44+ [3,4]CD73+ [3,4,29,31]CD90+ [3,4,32,29,31]CD105+ [28,29]CD184+ [30,29]CD29− [30,32,29]CD44+ [32]CD54+ CD61+ [29]CD90+ CD105+ [30]CD184+ [32]Flk-1+ [32]CD80+ [32,29]CD90+ CD105+ [30]CD184+ [3,4]CD14− [29]CD33− [30,28,3,4,32,29,31]CD34− [30,28,3,4,29,31]CD45− [32]CD8A− CD25− CD33− [29]CD34 CD45− [31]CD3− CD11c− CD 28− CD38− CD62L− [31]CD41a−	[30,28,29]OCT4+ [30,29]NANOG+ SOX2+ SSEA4+ [30,32,29]OCT4+ NANOG SOX2++ [32]HMGA2+ cKit−	[30,28]DLA-DQA1− DLA79− DLA88−	Mesodermal differentiation: [16,4,29,31]osteo, [30,32,29,31] chondro, [30,3,29,31]adipogenic [30,3,29,31]Endodermal differentiation: neurogenic

(Continued)

TABLE 12.3 *(Continued)*

Comparison of Key Features of Domestic Animal UCB and UCM Mesenchymal-Derived Cells

Cell type	Species	Surface Markers	Expression of Pluripotent Embryonic Markers	Immunological Features	Differentiation Potential
	Horse	[7,33,11,12]CD29− [14,7,33,11,34,12]CD44+ [37]CD54+ [14,7,33]CD73+ [7,11]CD73−	[35,36,37,38]OCT4+ [37]ALP+ [35,37,38]SSEA4+ [36]SSEA4− [37]SSEA3+ [36]SSEA3− [37,38]TRA-1-60+ [36]TRA-1-60− [35]TRA-1-60+/− [36]cMyc+ c-Kit+	[37]HLA-ABC− [37]HLA-1AG− [11,35,33,11,36,12]MHC-I+ [11,35,7,33,11,34,12,37]MHC-II− [36,38] *MHC-I+* *MHC-II−* [11] Mild inflammation induced by autologous, related allogenic and allogeneic intra-articular injection of UCMMSCs	Mesodermal differentiation: [16,14,33,39,19,36,34,37]osteo, [14,33,19,39,36,34,37,40]chondro, [14,33,39,34,36,37]adipogenic [36]Endodermal differentiation: neurogenic
		[14,7,33,11,34,37]CD90+ [7]CD90−			
		[14,33,36,34,37]CD105+ [7]CD105− [37]CD146+ [35,36,38]*CD29−* [35,36,38]*CD44+* [35,36,38]*CD105+* [36,38]*CD166+*			
		[14,7]CD14− [16]CD18− [16]CD31− [14,35,7,36,34]CD34− [14,7,33,11]CD45− [7,33,11]CD79a− [36,38]*CD14−* [36,38]*CD34−*			
		[16,38]*CD86−* [33,11]Monocyte marker− [33]*CD40+ CD80+ CD86−*			
		[41]CaR+			

(Continued)

TABLE 12.3 *(Continued)*

Comparison of Key Features of Domestic Animal UCB and UCM Mesenchymal-Derived Cells

Cell type	Species	Surface Markers	Expression of Pluripotent Embryonic Markers	Immunological Features	Differentiation Potential
	Pig	[42]CD29+ CD 44+ [42]CD45−	[43,44]c-Kit+ TERT activity+ [45]ALP+ [45]OCT4+ NANOG+ SOX2+ [45]*OCT4+ SOX2+ NANOG+*	[44]Survival after xenotransplantation in rat brain [45]Allogenic engraftment into neonatal intestine	[43]Endodermal differentiation: neurogenic [44]*In vivo* neurodifferentiation
	Caprine	[46]CD 44+ [48]CD73+ CD105+ STRO1+ [46,48]*CD34−*	[47,46,48]ALP+ [47]*NANOG+/−*		[48]Mesodermal differentiation: chondro, adipo, osteogenic
	Bovine	[49,50]CD29+ CD73+ CD90+ [49,50]CD105− [49,50]CD34+ CD45−	[49]OCT4+ TERT activity+		[49]Mesodermal differentiation: chondro, adipo, osteogenic [49,50]Endodermal differentiation: neurogenic

(Continued)

TABLE 12.3 *(Continued)*

Comparison of Key Features of Domestic Animal UCB and UCM Mesenchymal-Derived Cells

Cell type	Species	Surface Markers	Expression of Pluripotent Embryonic Markers	Immunological Features	Differentiation Potential
	Buffalo	[51,52]CD29+ CD73+ CD90+ CD105+ [51,52]*CD73+ CD90+ CD105+* [51]CD34- CD45-	[52] ALP+ OCT4+ NANOG+ SOX2+[51,52]*OCT4+ SOX2+ NANOG+*		[51,52]Mesodermal differentiation: chondro, adipo, osteogenic

HLA, human leukocyte antigen; MHC, major histocompatibility complex; UCB, umbilical cord blood; UCM, umbilical cord matrix; UCMMSC, umbilical cord matrix mesenchymal stem cell.

Note: The surface markers are organized into categories: light gray, mesenchymal; dark gray with black writing, hematopoietic; dark gray with white writing, immune cell; black with white writing, plateled markers. The markers in *italics* are genes; the others markers are proteins detected with different techniques.

[1]Jin et al. 2008;[2]Seo et al. 2009; [3]Ryu et al. 2012; [4]Kang et al. 2012; [5]Jang et al. 2008; [6]De Schauwer et al. 2011; [7]Paebst et al. 2014; [8]Mohanty et al. 2014; [9]De Schauwer et al. 2012; [10] Guest et al. 2008; [11]De Schauwer et al. 2013; [12]Carrade et al. 2012; [13]Reed and Johnson 2008; [14]Iacono et al. 2012a; [15]Schuh et al. 2009; [16]Toupadakis et al. 2010; [17]Kang et al. 2013; [18]Eini et al. 2012; [19]Burk et al. 2013; [20]Figueroa et al. 2011; [21]Koch et al. 2007; [22]Berg et al. 2009; [23]Reed and Johnson 2014; [24]Kumar et al. 2007; [25]Fadel et al. 2011; [26]Jager et al. 2006; [27]Raoufi et al. 2011; [28]Filioli Uranio et al. 2014; [29]Lee et al. 2013b; [30]Filioli Uranio et al. 2011; [31]Seo et al. 2012; [32]Lee et al. 2013a; [33]De Schauwer et al. 2014; [34]Barberini et al. 2014; [35]Lovati et al. 2011; [36]Corradetti et al. 2011; [37]Hoynowski et al. 2007; [38]Lange-Consiglio et al. 2011; [39]Passeri et al. 2009; [40]Co et al. 2014; [41]Martino et al. 2011; [42]Kimura et al. 2012; [43]Mitchell et al. 2003; [44]Weiss et al. 2003; [45]Carlin et al. 2006; [46]Azari et al. 2011; [47]Babaei et al. 2008; [48]Pratheesh et al. 2014; [49]Cardoso et al. 2012a; [50]Cardoso et al. 2012b; [51]Singh et al. 2013; [52]Sreekumar et al. 2014.

Applications of Placenta-Derived Cells in Veterinary Medicine

of UCBMSCs is quite critical in domestic animals and requires careful sterilization procedures (Cremonesi et al. 2011) and adequate gradient separation methods (De Schauwer et al. 2011) as well as automated processing procedures and temperature-controlled shipping to reduce red blood cell contamination and support nucleated cell enrichment (Schuh et al. 2009). The refinement of MSC culture techniques is another phase that can improve the efficiency of UCBMSC expansion. Mild hypoxic conditions and fibronectin as a substrate to coat culture plates have been useful in equine (e)UCBMSCs (Schuh et al. 2009) to enhance and extend cell proliferation, as also described for human UCBMSCs (Grayson et al. 2007). A linear increase of proliferation over approximately 10 passages has been documented in canine (Seo et al. 2009), equine (Schuh et al. 2009; Eini et al. 2012; Kang et al. 2013), and porcine (Kumar et al. 2007) UCBMSCs, along with a rapid ability to form colonies in cells from domestic animals (Schuh et al. 2009; Eini et al. 2012).

In addition to the different culture conditions used, the immunophenotype of UCBMSCs has revealed some differences among domestic animals (Table 12.3). Specifically, the typical MSC marker CD73 is expressed in canine (Kang et al. 2012; Ryu et al. 2012) and bovine (b)UCBMSC (Raoufi et al. 2011), whereas it was not always detected in eUCBMSCs (Iacono et al. 2012a; De Schauwer et al. 2014; Paebst et al. 2014). A discrepancy was also observed in UCBMSCs for CD90 and CD105 (Seo et al. 2009; Paebst et al. 2014). The reasons for these differences could be ascribable to the species but also to cell manipulation and low cross-reactivity of commercial antibodies (De Schauwer et al. 2012; Mohanty et al. 2014; Paebst et al. 2014). For all these reasons, MSC immunophenotyping in large animals remains a challenge and a general characterization of cell typology for various species is lacking, in contrast to that of humans and small laboratory mammals.

UCBMSCs express multiple ESC markers (Tondreau et al. 2005). OCT4, essential for inhibiting tissue-specific genes and enhancing self-renewal, is stably expressed in cUCBMSCs (Seo et al. 2009), eUCBMSCs (Guest et al. 2008; Reed and Johnson 2008; Mohanty et al. 2014), and bUCBMSCs (Raoufi et al. 2011) (Table 12.3). In addition to OCT4, several others pluripotency markers have been reported in eUCBMSCs: *NANOG, SOX2, Klf4*, and *cMyc* (Mohanty et al. 2014; Reed and Johnson 2014), whereas positivity for SSEA1, SSEA4, TRA-1-60, and TRA-1-81 (Guest et al. 2008; Reed and Johnson 2008) remains debated. Moreover, evidence of the ectodermal, mesodermal, and endodermal *in vitro* tissue lineage differentiation (Table 12.3) is supported the multipotency of UCBMSCs from domestic animals.

Comparative studies of MSC plasticity *in vitro* have demonstrated that eUCBMSC display a greater adipogenic potential than eAFMSCs and equine umbilical cord matrix mesenchymal stem cells (eUCMMSCs) (Iacono et al. 2012a). In addition, Reed and Johnson (2008) suggested that foal UCBMSCs are more plastic in terms of mesodermal lineage (adipogenic and myogenic) than adult MSCs derived from AT. Indeed, even if UCBMSCs displayed a low adipogenic potential, they are always able to differentiate sporadically into fat cells containing limited amounts of lipid droplets, in contrast to ATMSCs, which did not display any Oil Red O-positive cells. The greater differentiation potential of UCBMSCs was also confirmed under myocyte-inductive conditions (Reed and Johnson 2008). Even if only a few studies have demonstrated the immunoprivilege of UCBMSCs from domestic animals through the absence of the

MHC-II antigen canine (Seo et al. 2009), equine (Guest et al. 2008; De Schauwer et al. 2011, 2013), the successful use of these cells in allogeneic settings in immunocompetent racehorses (Kang et al. 2013; Van Loon et al. 2014) and dogs (Lim et al. 2007; Jang et al. 2008; Park et al. 2011a) certainly supports this finding.

12.2.2.4 Umbilical Cord Matrix Mesenchymal Stem Cells

The noninvasive and inexpensive isolation of MSCs from tissues discarded at birth has attracted great attention in veterinary medicine. UC is routinely collected at parturition and due to its extracorporeal elimination, it can be rapidly manipulated. However, since labor in domestic animals does not occur in a sterile environment, UC can be highly contaminated by bacteria and fungi (Bartholomew et al. 2009; Passeri et al. 2009; Cremonesi et al. 2011). With proper precautions, however, UC from domestic animals may offer a large volume of tissue from which a large amount of cells can be isolated. Like in humans (Weiss and Deryl 2006), most of the cells collected from a domestic animal UCM are MSCs (Burk et al. 2013). They can be frozen-thawed canine (Zucconi et al. 2010), equine (Corradetti et al. 2011), buffalo (Singh et al. 2013), clonally expanded, and amplified in culture bovine (Cardoso et al. 2012a), equine (Corradetti et al. 2011), canine (Yu et al. 2013), with an efficiency greater than that described for other adult MSC sources, including BM (Troyer and Weiss 2008).

In domestic animals, spindle-shaped cells represent the most common class of cells isolated from UCM, even if heterogeneous cell populations are frequently described in porcine, caprine, and equine umbilical cord matrix mesenchymal stem cells (UCMMSCs) (Mitchell et al. 2003; Babaei et al. 2008; Corradetti et al. 2011), like in humans (Hung et al. 2002; Smith et al. 2004). The presence of different subpopulations is not surprising when one considers the high level of compartmentalization of the UC structure (Can and Karahuseyinoglu 2007).

A systematic study on the characterization of cells isolated from intervascular and perivascular portions of UC has been carried out in equine by adopting size-sieving methods (Corradetti et al. 2011) previously developed in humans (Hung et al. 2002). The sieving procedure yielded two relatively homogeneous subpopulations. Unlike humans (Hung et al. 2002; Majore et al. 2009), large intervascular and small perivascular cells were the most rapidly replicating cells, displaying higher fibroblast colony-forming unit frequency compared to the unsieved cell populations collected from intervascular and perivascular areas (Corradetti et al. 2011). Cells from both portions expressed MSCs and pluripotent-specific markers and were able to differentiate into mesodermic and ectodermic lineages (Corradetti et al. 2011). Regardless of the selection methods used, UCMMSCs derived from other domestic animals showed linear *in vitro* expansion, reaching a plateau after approximately 10 passages in porcine (Mitchell et al. 2003; Weiss et al. 2003), caprine (Babaei et al. 2008), and canine (Lee et al. 2013a, 2013b) even though Filioli Uranio et al. (2011) and Cardoso et al. (2012a) have documented a more long-term expansion in canine and bUCMMSC, respectively. A similar result was described in bovine by da Cunha et al. (2014). In that study, bUCMMSCs showed a more rapid growth than that of bAFMSCs and were able to conserve a normal karyotype and a low degree of DNA fragmentation over culture.

Applications of Placenta-Derived Cells in Veterinary Medicine

Basic characterization of the typical surface markers of adult MSCs demonstrated that UCMMSCs mostly expressed CD44, CD73, CD90, and CD105, and they lacked CD45, CD34, and CD14. In contrast, conflicting data have been reported for pluripotency markers (Table 12.3). Despite the large number of studies regarding the selection methods and the molecular characterization of UCMMSCs from domestic animals, there are a few reports describing their differentiation potential *in vitro* and *in vivo* (Tables 12.3 and 12.4).

Like other MSCs, cUCMMSCs (Filioli Uranio et al. 2011, 2014) and equine Umbilical Cord Matrix Mesenchymal Stem Cells (eUCMMSCs) (Hoynowski et al. 2007; Carrade et al. 2011, 2012; Corradetti et al. 2011; Barberini et al. 2014; De Schauwer et al. 2014; Paebst et al. 2014) are immunoprivileged cells, and at least eUCMMSCs display immunosuppressive effects (Carrade et al. 2012). In contrast to AC human (Bailo et al. 2004; Soncini et al. 2007; Magatti et al. 2008), ovine (Barboni et al. 2014), equine (Lange-Consiglio et al. 2013b), human and human (hUCMMSCs) and eUCMMSCs are not intrinsically able to immunostimulate *in vitro*, as shown by their failure to induce lymphocyte proliferation. Indeed, Carrade et al. (2012) demonstrated that eMSCs derived from BM, AT, UCB, and UCM do not alter lymphocyte proliferation or growth factor secretion, except for transforming growth factor beta (TGF-β1). However, when stimulated, all MSC types decreased lymphocyte proliferation and increased prostaglandin and interleukin-6 production. The interactions between MSCs and lymphocytes have provided some of the rationale for the use of MSCs in treating T cell–mediated diseases, but no studies are available to determine whether the *in vitro* functions correlate with the *in vivo* properties of these MSC sources.

12.3 PLACENTA CELL–BASED REGENERATIVE MEDICINE IN DOMESTIC ANIMALS

An urgent current need in regenerative veterinary medicine is that of identifying an easily accessible, plentiful, and safe source for the development of therapeutic strategies to restore functionality in damaged or diseased organs and tissues. In this context, PCs are becoming a prime candidate in veterinary practice, as they are available in a nearly unlimited supply, possess high plasticity, and are immunoprivileged, thus allowing their use in immunocompetent patients without adopting pharmacological treatments or genetic matching tests before transplantation. These latter properties are strategic when animal models are used for preclinical purposes, since PCs can be used under allogeneic and xenogeneic settings.

Although promising data have been reported to date, further studies are required to fully characterize the *in vivo* regenerative potential of PCs, either to move results into translation to human clinical practice or to identify their possible applications in veterinary medicine.

12.3.1 PRECLINICAL STUDIES IN DOMESTIC ANIMAL MODELS

In Table 12.4 we provide a snapshot of the current knowledge regarding the potential of cells from the AM and UC to address current shortcomings in regenerative medicine field in preclinical settings.

TABLE 12.4
Placenta Cell–Based Regenerative Medicine in Domestic Animals

	Cell Type	Animal Model and Disease Model	Procedures of Cell Manipulation	Detection Time[a] and *In Vivo* Results	Demonstrated Mechanisms	
					Paracrine Mechanisms	Tissue-Specific Differentiation
Spinal cord injury	cUCBMSC (Lim et al. 2007)	Dog spinal cord injury model (first lumbar vertebra)	**MIV**: culture **CT**: 1 × 10^6 cells into spinal cord 7 days after injury **TO**: allogeneic **IC**: yes	**DT**: 8 weeks **RIV**: nerve conduction velocity and functional recovery of the hind limbs		
	cUCBMSC (Park et al. 2011c)	Dog spinal cord injury model (first lumbar vertebra)	**MIV**: culture **CT**: 1 × 10^6 cells into spinal cord 12 h, 1 week, and 2 weeks after injury **TO**: allogeneic **IC**: yes	**DT**: 8 weeks **RIV**: better window of cell transplantation is 1 week; in this frame, improved neuronal regeneration, functional recovery of the hind limbs and increased expression of markers for neuronal regeneration (Tuj-1, nestin, MAP2, NF-M) and signal molecules for actin cytoskeleton (Cdc42 and Rac1) were observed	Reduced inflammatory response (low COX2) and higher neurotrophins (NT-3) expression	

(Continued)

TABLE 12.4 *(Continued)*

Placenta Cell–Based Regenerative Medicine in Domestic Animals

| | Cell Type | Animal Model and Disease Model | Procedures of Cell Manipulation | Detection Time[a] and *In Vivo* Results | Demonstrated Mechanisms | |
					Paracrine Mechanisms	Tissue-Specific Differentiation
	cUCBMSC and cUCMMSC (Ryu et al. 2012)	Dog spinal cord injury model (first lumbar vertebra)	**MIV**: culture, labeling with NEO-STEMTM **CT**: 1×10^5 cells/μl in 60 μl of Matrigel into spinal cord 7 days after injury **TO**: allogeneic **IC**: yes	**DT**: 8 weeks **RIV**: functional recovery of the hind limbs, nerve regeneration, and higher neuroprotection in than UCBMSCs, AT, BM, UCM, MSCs	Lower inflammation with UCBMSCs than other MSCs; reduced interleukin-6 and COX2 than in control	
Myocardial infarction	pAFMSC (Sartore et al. 2005)	Porcine model of acute myocardial ischemia	**MIV**: culture, expansion (5–6 P), labeling with CMFDA **CT**: 7.5×10^6 cells in 1 ml into ischemic region (three to four sites) 1 week after induction **TO**: autogeneic **IC**: yes	**DT**: 30 days **RIV**: low percentage of cells (5 ± 1%) survived displayed a downregulation of Oct4, SSEA4, and MSC markers		pAFMSCs differentiated into vascular cell lineages, but not into cardiomyocytes

(Continued)

TABLE 12.4 (Continued)
Placenta Cell–Based Regenerative Medicine in Domestic Animals

Cell Type	Animal Model and Disease Model	Procedures of Cell Manipulation	Detection Time[a] and *In Vivo* Results	Demonstrated Mechanisms	
				Paracrine Mechanisms	Tissue-Specific Differentiation
pAMSC (Kimura et al. 2012)	Transgenic porcine model of chronic myocardial ischemia	**MIV**: culture, expansion (3–10 P) collected from GFP-transgenic Jinhua pigs **CT**: 1×10^6 cells injected epicardially close to the ischemic region (16–20 sites) 4 weeks after induction **TO**: allogeneic **IC**: yes	**DT**: 4 weeks **RIV**: GFP-pAMSCs survived for 4 weeks colocalizing cardiac troponin T and cardiac troponin I; reduced fibrosis		GFP-pAMSCs gained cardiac phenotype *in situ*
Wound healing gUCMMSC (Azari et al. 2011)	Goat model of first intention cutaneous wound healing	**MIV**: culture, labeling with 5-bromo-2[prime]-deoxyuridine **CT**: 3×10^6 cells in 0.6 ml of PBS on left wounds **TO**: allogeneic **IC**: yes	**DT**: 7 and 12 days **RIV**: cells survived within wound zone that showed a complete re-epithelialization and a thinner granulation with minimum scar at day 7	Minimum inflammation	

(Continued)

TABLE 12.4 *(Continued)*
Placenta Cell–Based Regenerative Medicine in Domestic Animals

Cell Type	Animal Model and Disease Model	Procedures of Cell Manipulation	Detection Time[a] and *In Vivo* Results	Demonstrated Mechanisms	
				Paracrine Mechanisms	Tissue-Speci c Differentiation
oAMSC (Klein et al. 2011)	Sheep model of fetal wound healing	**MIV**: culture of cells collected during fetal life, expansion, and GFP viral transfection **CT**: 3–4 × 10^9 cells, instillation into the amniotic cavity on two symmetrical, size-matched skin wounds: one wound with an open titanium chamber and the other wound covered with a semipermeable membrane that allowed passage of molecules, but not any cells **TO**: autogeneic **IC**: yes	**DT**: 5–6, 9, 14, 20, and 32 days **RIV**: labeled oAMSCs migrated into the injured area; covered wounds showed a significantly slower healing rate and lower elastin levels at the mid–time points; no significant differences in collagen, hyaluronic acid, and substance P levels were observed		

(Continued)

TABLE 12.4 (Continued)

Placenta Cell–Based Regenerative Medicine in Domestic Animals

	Cell Type	Animal Model and Disease Model	Procedures of Cell Manipulation	Detection Time[a] and *In Vivo* Results	Demonstrated Mechanisms	
					Paracrine Mechanisms	Tissue-Specific Differentiation
Tendon injuries	oAEC (Barboni et al. 2012b)	Sheep mechanical model of acute Achilles tendon defect	**MIV**: culture, expansion (3 P), labeling with PKH26 **CT**: 4×10^6 cells in 30 µl of fibrin glue inserted into one limb immediately after the defect induction **TO**: allogeneic **IC**: yes	**DT**: 7, 14, and 28 days **RIV**: cells survived until day 28 migrating close to the healing portion of tendon; greater structural and biomechanical recovery than control; microarchitecture similar to a healthy tissue after 28 days from treatment; higher levels of COL1, fiber organization with physiological orientation along the longitudinal axis of healthy tendon	Modulation of VEGF and TGF-β1, higher recruitment of host progenitor cells, complete remodeling of newly formed blood vessels, leukocyte infiltration and total vascular area were bought to normal values in 28 days	Proliferating oAECs are mainly found at the edge of the repairing area at all time points, and some of them transdifferentiated into tenocyte-like cells switching on the expression of COL1

(Continued)

Applications of Placenta-Derived Cells in Veterinary Medicine

TABLE 12.4 (*Continued*)
Placenta Cell–Based Regenerative Medicine in Domestic Animals

Cell Type	Animal Model and Disease Model	Procedures of Cell Manipulation	Detection Time[a] and *In Vivo* Results	Demonstrated Mechanisms — Paracrine Mechanisms	Demonstrated Mechanisms — Tissue-Specific Differentiation
oAFMSC (Colosimo et al. 2013a)	Sheep enzymatic model of acute Achilles tendon defect	**MIV**: culture, expansion (12 P), and GFP nucleofection. **CT**: 2×10^6 cells in 0.5 ml of saline solution injected under US guidance 20 days after collagenase induced tendon defects. **TO**: allogeneic **IC**: yes	**DT**: 30 days. **RIV**: cells survived and engrafted into the repairing area; enhanced tissue regeneration and recovery of biomechanical properties; microarchitecture similar to healthy tissue with oriented extracellular matrix, characterized by COL1 fibers filled with fusiform-aligned cells		Some GFP-expressing cells, in particular those displaying a tenocyte-like fusiform shape, were able to colocalize COL1 protein
oAMSC (Turner et al. 2011)	Sheep model of congenital diaphragmatic hernia	**MIV**: culture, expansion (3 P), cryopreservation, and expansion before engineering. **CT**: engineered graft was transplanted into the defect ($\sim 7.5 \times 4.5$ cm) of the left hemidiaphragm. **TO**: autogeneic **IC**: yes	**DT**: 7, 10, and 14 months. **RIV**: safety, and superior morphology compared with control; higher COL content and lower α-elastin and glycosaminoglycan levels in the engineered group than in control		

(Continued)

TABLE 12.4 *(Continued)*

Placenta Cell–Based Regenerative Medicine in Domestic Animals

	Cell Type	Animal Model and Disease Model	Procedures of Cell Manipulation	Detection Time[a] and *In Vivo* Results	Demonstrated Mechanisms	
					Paracrine Mechanisms	Tissue-Specific Differentiation
Bone defects	cUCBMSC (Jang et al. 2008)	Dog model of diaphyseal radius defect and one case report of a dog with nonunion fracture	**MIV**: culture **CT**: 1×10^6 cells in 0.5 ml of saline mixed with 700 mg of β-TCP, implanted into the diaphyseal defect, and wrapped with PLGC membrane **TO**: allogeneic **IC**: yes	**DT**: 6 and 12 weeks **RIV**: increased bone formation around β-TCP than in control; when applied to the nonunion fracture, tissue healing was observed in 6 weeks, and a complete radiolucency was found at 12 weeks after cells injection		
	cUCBMSC and cUCMMSC (Kang et al. 2012)	Dog model of diaphyseal radius defect	**MIV**: culture, expansion (3 P), and cryopreservation **CT**: 1×10^6 cells in 700 μL of saline mixed with 700 mg of β-TCP before implantation; site of implantation covered with a TCP/PLGC composite membrane **TO**: allogeneic **IC**: yes	**DT**: 20 weeks **RIV**: similar osteogenic potential for cMSCs derived from AT, BM, UCB, and UCM mixed with β-TCP compared to control		

(Continued)

TABLE 12.4 *(Continued)*

Placenta Cell–Based Regenerative Medicine in Domestic Animals

	Cell Type	Animal Model and Disease Model	Procedures of Cell Manipulation	Detection Time[a] and *In Vivo* Results	Demonstrated Mechanisms	
					Paracrine Mechanisms	Tissue-Specific Differentiation
	oAEC (Mattioli et al. 2012)	Sheep model of proximal tibia epiphysis defect	**MIV**: culture, expansion (12 P), and labeling with PKH26 **CT**: 2×10^6 cells in 10 μl of fibrin glue injected immediately into the lesion **TO**: allogeneic **IC**: yes	**DT**: 45 days **RIV**: cells survived and supported consistent bone neoformation; osteogenic regenerative property of oAECs resulted similar after 6 and 12 P	Little if any infiltration of leukocytes and absence of abnormal blood vessel formation	Some cells enclosed within the newly deposited trabecular bone matrix–like host osteoblasts
Oral bone defects	oAFMSC (Berardinelli et al. 2013)	Sheep model of bilateral sinus lift augmentation	**MIV**: culture, expansion (3 P), cryopreservation, and labeling with PKH26 **CT**: 10×10^6 cells seeded on a commercial magnesium-enriched hydroxyapatite (RegenOss®) **TO**: allogeneic **IC**: yes	**DT**: 45 and 90 days **RIV**: cells survived up to 90 days by increasing bone deposition	Reduced inflammatory reaction; at 45 days oAFMSCs induced a higher angiogenic response with a more intense expression of VEGF	

(Continued)

TABLE 12.4 (Continued)

Placenta Cell–Based Regenerative Medicine in Domestic Animals

Cell Type	Animal Model and Disease Model	Procedures of Cell Manipulation	Detection Time[a] and *In Vivo* Results	Demonstrated Mechanisms	
				Paracrine Mechanisms	Tissue-Specific Differentiation
oAEC (Barboni et al. 2013)	Sheep model of bilateral sinus lift augmentation	**MIV**: culture, expansion (3 P), cryopreservation, and labeling with PKH26 **CT**: 1×10^6 cells seeded on synthetic bone substitute scaffold (hydroxyapatite/β-TCP) **TO**: allogeneic **IC**: yes	**DT**: 45 and 90 days **RIV**: cells survived until 90 days; Micro-computed tomography analysis demonstrated a greater scaffold integration and bone deposition than control; histology showed a reduced fibrosis and a widespread new bone deposition within the whole grafted area	Reduced inflammatory reaction and a specific influence on angiogenesis mediated by VEGF	Direct contribution to osteogenesis: oAEC osteocalcin expression switch-on and their presence and entrapment within the newly deposited ECM
Cartilage defects	oAM (Garcia et al. 2014)	Sheep model of a full-thickness cartilage defect	**MIV**: cryopreserved (2 months) and fresh (24 h) oAMs **CT**: lesion filled with fresh, cryopreserved AM previously cultured with BMMSCs, and cryopreserved AM. AM cut in a proper size, was fit into the defect and applied in a multilayered disposition **TO**: allogeneic **IC**: yes	**DT**: 2 months **RIV**: all AM groups showed similar regenerative properties; cryopreserved AM showed higher chondrocyte clustering within the defect area and a proteoglycan content	

(Continued)

TABLE 12.4 *(Continued)*
Placenta Cell–Based Regenerative Medicine in Domestic Animals

	Cell Type	Animal Model and Disease Model	Procedures of Cell Manipulation	Detection Time[a] and *In Vivo* Results	Demonstrated Mechanisms	
					Paracrine Mechanisms	Tissue-Specific Differentiation
Joint preclinical safety study	eUCBMSC (Carrade et al. 2011)	Horse model of *in vivo* safety intra-articular cell injection	**MIV**: culture and cryopreservation **CT**: 7.5 × 10⁶ cells/joint; one foal received autologous MSCs and allogeneic MSCs; dams received related allogeneic MSCs **TO**: autologous and allogeneic comparison **IC**: yes	**DT**: synovial fluid collected at 0, 6, 24, 48, and 72 h **RIV**: a single intra-articular injection elicited mild inflammation independently of their origin (self or nonself); inflammation with gait abnormalities and self-limiting was not observed		
Prenatal congenital diseases	oAFMSC (Shaw et al. 2011)	Sheep safety model of IUSCT	**MIV**: culture, expansion, and GFP transfection **CT**: 1–2 ml of cells injected under US guidance into the peritoneal cavity of each cell donor fetus **TO**: autologous **IC**: yes	**DT**: 3 weeks **RIV**: GFP-cells were detected in fetal tissues including liver, heart, placenta, membrane, umbilical cord, adrenal gland, and muscle		Rare injected cells in the fetal liver coexpressed GFP with CK18 or AFP cytoplasmic hepatic markers

(Continued)

TABLE 12.4 *(Continued)*
Placenta Cell–Based Regenerative Medicine in Domestic Animals

Cell Type	Animal Model and Disease Model	Procedures of Cell Manipulation	Detection Time[a] and *In Vivo* Results	Demonstrated Mechanisms	
				Paracrine Mechanisms	Tissue-Specic Differentiation
oAFMSC (Shaw et al. 2015)	Sheep model of IUSCT for prenatal congenital hematopoietic disease	**MIV**: culture, expansion, selection of CD34+ cells, cryopreservation, GFP transfection of fresh or thawed CD34+ **CT**: 2×10^4 cells of freshly isolated or transduced CD34+ cells in 1–2 ml injected under US guidance into the peritoneal cavity of each cell donor fetus **TO**: autologous **IC**: yes	**DT**: 9 months post-IUSCT (6 months after birth) **RIV**: long-term engraftment in the peripheral blood and hematopoietic organs; freshly cells had a slightly higher rate of engraftment compared to frozen CD34+ cells		
oAMSC (Gray et al. 2012)	Sheep model of IUSCT for congenital high airway obstruction syndrome	**MIV**: culture, expansion (5 to 8 P), and transfection with GFP **CT**: engineered graft (decellularized leporine tracheal segment) with 1×10^6 cells/ml under static condition and then 7.5×10^6 cells under dynamic condition **TO**: autologous **IC**: yes	**DT**: from 30 to 44 post-IUSCT (7 days after birth) **RIV**: engineered grafts showed a greater diameter, although variable stenosis was present in all implants; engineered constructs exhibited full epithelialization and had a significantly greater degree of elastin; no such differences were noted in COL and glycosaminoglycan contents		Donor cells were detected in engineered grafts

TABLE 12.4 (Continued)
Placenta Cell–Based Regenerative Medicine in Domestic Animals

Cell Type	Animal Model and Disease Model	Procedures of Cell Manipulation	Detection Time[a] and *In Vivo* Results	Demonstrated Mechanisms	
				Paracrine Mechanisms	Tissue-Specific Differentiation
oAFMSC (Weber et al. 2012)	Sheep model of IUSCT for congenital cardiac malformations	**MIV**: culture and labeling with the CellTrace™ CFSE Cell Proliferation kit **CT**: engineered graft (stented trileaflet heart valves PGA-P4HB) with $4.0 \pm 3.1 \times 10^6$ cells/cm^2 using fibrin as a cell carrier and implanted orthotopically into the pulmonary position **TO**: autologous **IC**: yes	**DT**: from 30 to 44 post-IUSCT (7 days after birth) **RIV**: engineered heart valves showed *in vivo* functionality with intact structures and without any thrombus formation; a high cellularity with profound phagocytic accumulation as well as a distinct macrophage infiltration was found in explants		

CT, cell transplantation; **DT**, detection time; **IC**, immunocompetence; **MIV**, manipulation *in vitro*; **RIV**, results *in vivo*; **TO**, transplant option. AT, adipose tissue; β-TCP, β-tricalcium phosphate; BM, bone marrow; CMFDA, 5-chloromethylfluorescein diacetate, cUCBMSC, canine umbilical cord blood mesenchymal stem cell; eUCBMSC, equine umbilical cord blood mesenchymal stem cell; eUCMMSC, equine umbilical cord matrix mesenchymal stem cell; gUCMMSC, goat umbilical cord matrix mesenchymal stem cell; GFP, green fluorescent protein; IUSCT, *in utero* placental stem cell transplantation; MSC, mesenchymal stem cell; oAEC, ovine amniotic epithelial cell; oAFMSC, ovine amniotic fluid mesenchymal stem cell; oAM, ovine amniotic membrane; oAMSC, amniotic mesenchymal stromal cells; P, passage, pAFMSC, porcine amniotic fluid mesenchymal stem cell; pAMSC, porcine amniotic mesenchymal stromal cells; PBS, phosphate-buffered saline; TGF, transforming growth factor; UCM, umbilical cord matrix; US, ultrasound; VEGF, vascular endothelial growth factor.

[a] Time of detection of transplanted cells in host tissues.

12.3.1.1 Spinal Cord Injury

Spinal cord injury (SCI) produces progressive cell death, axonal degeneration, and functional loss of neurons (Meng et al. 2008). All preclinical research on PC-based SCI regenerative medicine has been carried out in canine models using UCBMSCs. Lim et al. (2007) first conducted a study in which transplantation resulted in recovery of nerve function, whereas Park et al. (2011c), identified the best interval between SCI and cell injection. Interestingly, Ryu et al. (2012) demonstrated that UCBMSCs induced a more effective nerve regeneration and a greater anti-inflammatory activity than AT, BM, UCM, and UCBMSCs.

12.3.1.2 Myocardial Infarction

The ability to replace the lost myocardial tissue induced by myocardial infarction (MI) remains an unsolved challenge (Murphy and Atala 2013). Two preclinical studies have suggested a role for AM-derived MSCs: pAFMSC autotransplantation during the acute phase (Sartore et al. 2005) as well pAMSC allogeneic injection during the chronic phase (Kimura et al. 2012) of experimentally induced MI were both able to improve the functional recovery of the tissue. Even if the mechanisms involved remain unclear, signs of transdifferentiation toward the vascular cell lineages but not to cardiomyocytes have been shown (Sartore et al. 2005), whereas pAMSCs differentiate toward the cardiac phenotype (Kimura et al. 2012).

12.3.1.3 Wound Healing

Optimum healing of cutaneous wound requires a well-orchestrated integration of complex biological and molecular events involving cell migration and proliferation as well extracellular matrix deposition, angiogenesis, and remodeling (Wu et al. 2007). In domestic animals, a complete re-epithelialization of a wound defect was described by Azari et al. (2011) after goat UCMMSC allotransplantation. An interesting experiment performed by Klein et al. (2011) in fetal lambs clarified a direct role of oAMSCs that were more capable of accelerating wound closure than the released soluble factors.

12.3.1.4 Tendon Injuries

Tendon injuries are a common cause of disease in both humans and domestic animals. These injuries are difficult to manage because injured tendons cause a fibrotic scar with poor tissue quality and reduced mechanical properties (Sharma and Maffulli 2006). Given the relatively poor results obtained through surgical intervention, in recent years, stem-based therapy, including PCs, have attracted increasing attention (Muttini et al. 2010; Barboni et al. 2012b; Colosimo et al. 2013a; Lange-Consiglio et al. 2013c; Muttini et al. 2013). Our group has carried out several preclinical studies using oAECs to test the regenerative properties after allotransplantation into experimentally injured calcaneal tendons. These studies demonstrated the capability of oAECs to support tendon regeneration and early biomechanical recovery and thus started to clarify the mechanisms involved. Tendon regeneration was in part attributable to paracrine stimulation by oAECs, which induced an adequate release of growth factors (i.e., vascular endothelial growth factor and TGF-β1) and of immunomodulatory cytokines (Russo et al. 2014) in the repairing tissue, as well as a greater

Applications of Placenta-Derived Cells in Veterinary Medicine 261

recruitment of tenocytes involved in the deposition and organization of the extracellular matrix. Interestingly, a direct role of oAECs cannot be excluded. The cells that survived within the host tissue for 28 days, indeed, differentiated into tenocyte-like cells producing COL1 (Barboni et al. 2012b), thus contributing to tissue regeneration through the release of extracellular matrix proteins (Muttini et al. 2010; Barboni et al. 2012b). Similarly, oAFMSCs survived during the early phase of tendon healing (1 month posttransplantation) by supporting a rapid and prompt recovery of tissue microarchitecture (Colosimo et al. 2013a). Turner et al. (2011) also demonstrated tendon regenerative properties for oAMSCs.

12.3.1.5 Bone Defects

Cell-based therapy for bone regeneration is an emerging strategy. Preclinical studies have established a potential role for PCs in orthopedic and maxillo-facial regenerative medicine. In particular, Jang et al. (2008) have demonstrated that the implantation of cUCBMSCs mixed with β-tricalcium phosphate (β-TCP) was able to enhance osteogenesis in a dog diaphyseal radius defect. cMSCs derived from different sources (AT, BM, UCB, and UCM) seem to have similar osteogenic aptitude (Kang et al. 2012), even if clinical application is, however, more feasible for the MSC source that can be most easily and noninvasively collected. Similarly, oAECs successfully supported bone regeneration after allotransplantation into a tibia defect (Mattioli et al. 2012). The labeled oAECs survived in host tissue for 45 days and accelerated the process of new bone deposition by reducing the infiltration of inflammatory cells.

Domestic animal models have also gained notable advantages to study cell-based approaches in oral bone regeneration. All preclinical studies performed to test the regenerative properties of ACs have been carried out on sheep by mimicking the sinus augmentation lift procedure. Both oAFMSCs and oAECs, when engineered on commercial magnesium-enriched hydroxyapatite (MgHA)/collagen (Berardinelli et al. 2013) or custom-made hydroxyapatite (HA)/β-TCP (Barboni et al. 2013) scaffolds, respectively, were able to accelerate the process of bone deposition 90 days after transplantation, suggesting the possibility of anticipating the procedure of dental implant. In the presence of oAECs, bone regeneration is accompanied by a reduced fibrotic reaction, a limited inflammatory response, and an accelerated process of angiogenesis (Barboni et al. 2013). Interestingly, both cell types survived in the host tissue and apparently actively participated in the process of bone regeneration: some oAECs recorded within the injured area began to produce osteocalcin and remained entrapped, like host osteoblasts, within the newly deposited (Barboni et al. 2013) bone extracellular matrix.

12.3.1.6 Cartilage Defects

A unique report demonstrated the synthesis of new cartilage tissue in a full-thickness femoral cartilage defect in a sheep model treated with fresh or cryopreserved AM (Garcia et al. 2014).

12.3.1.7 Joints

A study performed by Carrade et al. (2011) demonstrated the safety of intrasinovial injection of UCMs and UCBMSCs in horses.

12.3.1.8 Prenatal Diseases

Many rigorous preclinical studies have focused the attention on the *in utero* placental stem cell transplantation (IUSCT) to ameliorate prenatal congenital diseases using sheep as a model. After demonstrating the safety of autologous oAFMSCs, Shaw et al. (2015) injected a subpopulation of CD34+ cells. The hematopoietic aptitude of these cells has been thus confirmed by recovering the green fluorescent protein-CD34+-AFMSCs in ovine fetuses from several hematopoietic organs and peripheral blood. This result confirmed that the IUSCT can be adopted to treat fetuses with autologous cells and that oAFMSCs offer an alternative source of stem/progenitor cells to treat congenital hematopoietic disease (Shaw et al. 2015). Prenatal studies also been conducted to verify the possibility of using ACs to treat congenital airway pathologies (Gray et al. 2012) and valve congenital cardiac malformations (Weber et al. 2012).

12.3.2 Clinical Application of PCs in Veterinary Regenerative Medicine

Domestic animal PC therapies have been used mainly to treat equine tendinopathies and equine and dog ocular surface reconstruction (Table 12.5).

12.3.2.1 Tendinopathy

Tendinopathies of the superficial digital flexor tendon (SDFT) are a significant cause of lameness and often a career-ending event in thoroughbred horses (Smith et al. 2003; Richardson et al. 2007; Nixon et al. 2008; de Mattos Carvalho et al. 2011; Muttini et al. 2012). Afflicted horses are prone to SDFT injury due to hyperextension of the metacarpal joint during racing or riding (Kasashima et al. 2004). After injury, the equine SDFT is repaired via a process of fibrosis, and the resulting scar tissue is functionally inferior to a normal tendon (Williams et al. 1980; Crevier-Denoix et al. 1997; Dahlgren et al. 2005). Recently, PCs started to be used to treat tendon injuries in equine medicine. The most widely used cells for this purpose are UC and ACs. The first clinical trial was performed with eAMSCs in spontaneous tendinopathies to investigate cell tolerance after allogeneic transplantation (Lange-Consiglio et al. 2012). A subsequent report by the same group confirmed the efficacy of eAMSCs for treating SDFT tendinopathies by demonstrating their greater clinical outcomes when these cells were applied in parallel with BMMSCs (Lange-Consiglio et al. 2013c). The authors attributed most of the curative effect of eAMSCs to their paracrine actions since similar clinical results were obtained by injecting the conditioned media obtained from cultured eAMSCs (Lange-Consiglio et al. 2013b). Alternatively, oAECs were successfully used in the xenogeneic setting to treat spontaneous lesions of equine SDFT (Muttini et al. 2013). The positive clinical outcomes obtained on 15 horses were substantiated by histological data collected from a repaired tendon recovered from a deceased animal. Interestingly, these results demonstrated the direct contribution of oAECs to tendon healing. Indeed, the recovered oAECs were able to deposit ovine COL1 in the repaired area, as revealed by analysis using ovine-specific primers and antibodies that did not cross-react with equine COL1. eUCBMSCs were also efficient in healing tendinopathies affecting

Applications of Placenta-Derived Cells in Veterinary Medicine 263

TABLE 12.5
Clinical Application of Placental-Derived Cells in Veterinary Regenerative Medicine

Disease	Cell Typology	Pathology	Procedures of Cell Manipulation	Follow-Up and Clinical Outcome
Tendinopathy	eAMSC (Lange-Consiglio et al. 2012)	**Horse:** superficial digital flexor tendon and accessory ligament of the deep digital flexor tendinopathies **CC:** 3	**MIV:** culture, expansion (10 P) and cryopreservation **CT:** 1×10^6 cells in 800 μl of autologous plasma injected under US guidance into the injured tendon/ligament during the acute phase **TO:** allogeneic	**FU:** 90 days **CO:** cells are well tolerated and supported a quick reduction in gross tendon size, palpation sensitivity, and US cross-sectional area; no relapse of the tendon defect was observed
	eAMSC (Lange-Consiglio et al. 2013a)	**Horse:** superficial digital flexor tendinopathies **CC:** 51 treated with eAMSCs and 44 with eBMMSCs	**MIV:** culture, expansion (3 P), and cryopreservation **CT:** 5×10^6 cells compared with autologous fresh eBMMSCs injected intralesionally under US guidance; interval lesion/implantation was of 6–15 days for the eAMSCs and 16–35 days for the eBMMSCs **TO:** allogeneic	**FU:** 2 years **CO:** no significant adverse effects after MSC treatments; all animals resumed their activities; rate of reinjury was lower in horses treated with eAMSCs
	eAMSC CM (Lange-Consiglio et al. 2013b)	**Horse:** different types of ligament/tendinopathies **CC:** 13	**MIV:** CM preparation with 3 P cells and cryopreservation **CT:** 2 ml of CM injected intralesionally under US guidance; interval between lesion/implantation ranged from 8 to 30 days **TO:** allogeneic	**FU:** 2 years **CO:** no significant adverse effect; detection of neovessels 1 month after CM injection that disappeared in 4 months; CM promoted tendon repair by improving echogenicity and fiber alignment; favorable clinical outcomes with lower rate of reinjuries than untreated animals

(Continued)

TABLE 12.5 (Continued)
Clinical Application of Placental-Derived Cells in Veterinary Regenerative Medicine

Disease	Cell Typology	Pathology	Procedures of Cell Manipulation	Follow-Up and Clinical Outcome
	oAEC (Muttini et al. 2013)	**Horse:** superficial digital flexor tendinopathies **CC:** 15 (1 deceased for unrelated causes)	**MIV:** culture, expansion (3 P), labeling with PKH26, and cryopreservation **CT:** 7×10^6 cells in 500 µl of αMEM injected under US guidance into the acute phase of tendon injury **TO:** xenogeneic	**FU:** clinical examination every 30 days for the first 4 months and every 3 months thereafter; histological tendon analysis of one horse deceased 60 days after cell transplantation **CO:** accelerated defect-infilling with fibers; horses returned to work on the same or a higher level; histological examinations of the explanted tendon demonstrated oAEC survival, good microarchitecture recovery with low leucocyte infiltration. *In vivo* transdifferentiation demonstrated by the presence of ovine COL1 into equine host tissue
	eUCBMSC (Kang et al. 2013)	**Horse:** superficial digital flexor tendinopathies **CC:** 6	**MIV:** culture **CT:** 2×10^7 cells in 10 ml of PBS injected into the tendon lesions under US guidance into standardized points; interval between diagnosis and implantation was ~2–4 weeks **TO:** allogeneic	**FU:** 3–6 months **CO:** volume reduction of the core lesion

(Continued)

TABLE 12.5 (Continued)
Clinical Application of Placental-Derived Cells in Veterinary Regenerative Medicine

Disease	Cell Typology	Pathology	Procedures of Cell Manipulation	Follow-Up and Clinical Outcome
	eUCBMSC (Van Loon et al. 2014)	**Horse:** tendinitis of superficial digital flexor and deep digital flexor tendons, and desmitis of the suspensory ligament, and of the inferior check ligament CC: 52	**MIV:** culture, expansion, and cryopreservation **CT:** $2–10 \times 10^6$ cells in 3 ml were injected into the lesions under US guidance; number of injections was not standardized and depended on lesion size and position **TO:** allogeneic	FU: 6 months CO: no adverse reactions to cell transplantation; horses returned to work on the same or a higher level; neither the injured structure nor the age of the horses had a statistically significant influence on the result of treatment
Ophthalmology	eAM (Barros et al. 1998)	**Dogs:** full-thickness corneal defects CC: 18	**MIV:** 5-mm^2 pieces of glycerol-preserved membranes **CT:** membranes were sutured in the corneal defect in a single interrupted pattern **TO:** xenogeneic	FU: 2, 7, 15, 30, 60, and 180 days CO: after 180 days, corneal architecture was restored, with a correct layering of the epithelium, and pigmentation and vascularization present in the deep layers of the cornea
	cAM (Barros et al. 2005)	**Dogs:** keratomalacia and fibrous histiocytoma **Cat:** symblepharon CC: 2 dogs and 1 cat	**MIV:** glycerol stored or cryopreservation cAM **CT:** cAM was sutured in the defects in an interrupted pattern **TO:** allogeneic and xenogeneic	FU:30 and 60 days and 6 months CO: corneal opacity; cornea conjunctivalization with acceptable cosmetics in a fibrous histiocytoma; no ocular pain in keratomalacia and symblepharon, with a little corneal scarring and vascularization

(Continued)

TABLE 12.5 *(Continued)*
Clinical Application of Placental-Derived Cells in Veterinary Regenerative Medicine

Disease	Cell Typology	Pathology	Procedures of Cell Manipulation	Follow-Up and Clinical Outcome
	eAM (Lassaline et al. 2005)	**Horse**: corneal ulceration and severe keratomalacia **CC**: 3	**MIV**: glycerol stored or cryopreservation, orientation of eAM on a nitrocellulose membrane **CT**: eAM was sutured in place **TO**: allogeneic	**FU**: 2 weeks to 6 months **CO**: eAM sloughed over a 4- to 6-week period; last follow-up demonstrated good cosmetic, light perception, and an inconsistent positive menace response; all horses returned to their prior work
	eAM (Plummer et al. 2009)	**Horse**: ulcers or bullous keratopathy, corneal and conjunctival defects **CC**: 58	**MIV**: fresh eAM harvested aseptically **CT**: different procedures for AM transplantation **TO**: allogeneic	**FU**: 6 months **CO**: integrity of the globe, good visual outcome with minimal scarring in severely diseased corneas; improvement of cosmetic
Wound healing	eAFMSC (Iacono et al. 2012c)	**Foal** severe decubitus ulcers **CC**: 1	**MIV**: culture, gelation in 10 ml of PRP thawed gel with calcium gluconate **CT**: 5×10^6 cells transferred over sterile gauzes soaked in sterile saline solution and applied on washed ulcers **TO**: allogeneic	**FU**: 7 months **CO**: faster healing

CC, clinical cases; **CO**, clinical outcome; **CT**, cell transplantation; **FU**, follow-up; **MIV**, manipulation *in vitro*; **TO**, transplant option. CM, conditioned medium; cAM, canine amniotic membrane; eAFMSC, equine amniotic fluid mesenchymal stem cell; eAM, equine amniotic membrane; eAMSC, equine amniotic mesenchymal stromal cells; eBM-MSC, equine bone marrow mesenchymal stem cell; eUCB-MSC, equine umbilical cord blood mesenchymal stem cell; MSC, mesenchymal stem cell; oAEC, ovine amniotic epithelial cell; PBS, phosphate-buffered saline; PRP, Platelet-rich plasma; US, ultrasound.

Applications of Placenta-Derived Cells in Veterinary Medicine

SDFT (Kang et al. 2013; Van Loon et al. 2014) and the deep digital flexor tendon, as well desmitis of the suspensory ligament and of the inferior check ligament (Van Loon et al. 2014).

12.3.2.2 Ophthalmology

The positive experience in humans in treating a continuously widening spectrum of ophthalmic disorders (Azuara-Blanco et al. 1999; Parolini et al. 2009) with AM has also led to its use in veterinary medicine. AM transplantation (AMT), through a combination of mechanical and biological factors, can preserve the integrity of the globe, optimize the visual outcome, and minimize scarring in severely diseased corneas. Based on these assumptions, Barros et al. (1998) carried out a first study in which they demonstrated a curative role in dogs of xenogeneic, preserved equine amniotic membrane (eAM) on full-thickness corneal defects. These authors demonstrated similar positive effects by using frozen canine AMT on different types of ocular diseases in pets (Barros et al. 2005). In the same year, Lassaline et al. (2005) successfully treated corneal ulceration and severe keratomalacia of three horses with cryopreserved eAM. A more complete clinical study was performed by Plummer et al. (2009) who published a retrospective study on 58 equine clinical cases, thus strongly confirming the validity of AMT for ocular surface reconstruction.

12.4 CONCLUDING REMARKS AND FUTURE PERSPECTIVES

PC-based therapies represent an emerging challenge in veterinary medicine. Currently, several PC-based approaches for domestic animals are being developed and some of them, previously verified in preclinical settings, are now being gradually implemented in the clinic for the treatment of equine tendinopathies and canine and equine reconstruction of ocular defects.

Recently, our knowledge of the biology and properties of stem/progenitor cells isolated from the placenta of domestic animals has considerably advanced by starting to identify, at least for the AC, which protocols for isolating, expanding, and transplanting can preserve their original stemness state. To further progress, several issues need to be addressed, such as a higher degree of standardization of the *in vitro* procedures, common guidelines between different laboratories, development and validation of species-specific reagents, and the identification of biomarkers to characterize and test cell states. All these prerequisites are essential to create stable and well-characterized PC lines and to make them available to the biomedical veterinary community for research or clinical applications. This ambitious target could be easier achieved in domestic animals where the high availability of PCs is not accompanied by any gestational age–related tissue limitation. Indeed, animal models can play a great role in the study of the properties and the potential of PCs, thereby contributing to the bridging of knowledge gaps and safety concerns that still limit their future application in human medicine. The immunoprivilege status and the inherited immunomodulatory activity of human and domestic animal ACs represent a unique experimental and clinical opportunity to respond to the unsolved and high-priority challenges related to the efficient and safe use of these stem cells. The behavior and the fate of different animal PCs and their derivatives upon introduction into specific host tissues are now being clarified. Moreover, criteria for selecting the type/subpopulation of PCs for a particular application and for increasing cell survival after grafting are better understood. In this context, a powerful tool to analyze the crosstalk between transplanted cells and host tissues or whole organism is offered by the molecular chimeras that can be easily reproduced experimentally by using PCs in domestic animals of high translational value in xenotransplantation settings. The huge amount of data that has been and will be generated from such integrated research, operating in a context of "one health vision," will not only improve the therapeutic opportunities addressed for humans but also produce a secondary positive benefit for cutting-edge veterinary therapies.

ACKNOWLEDGMENTS

This work was supported by Tercas Foundation grants and from PRIN 2010-2011. We thank Dr. Lisa Di Marcantonio for valuable technical assistance in formatting the manuscript, Dr. Oriana Di Giacinto for figure and table preparation, and Dr. Nicola Bernabò for scientometric analysis.

REFERENCES

Ahern, B. J., J. Parvizi, R. Boston, and T. P. Schaer. 2009. Preclinical animal models in single site cartilage defect testing: A systematic review. *Osteoarthritis Cartilage* 17 (6): 705–13. doi: 10.1016/j.joca.2008.11.008.

Ahern, B. J., T. P. Schaer, S. P. Terkhorn, et al. 2011. Evaluation of equine peripheral blood apheresis product, bone marrow, and adipose tissue as sources of mesenchymal stem cells and their differentiation potential. *Am J Vet Res* 72 (1): 127–33. doi: 10.2460/ajvr.72.1.127.

Allegrucci, C., Y. Z. Wu, A. Thurston, et al. 2007. Restriction landmark genome scanning identifies culture-induced DNA methylation instability in the human embryonic stem cell epigenome. *Hum Mol Genet* 16 (10): 1253–68. doi: 10.1093/hmg/ddm074.

Arufe, M. C., A. De la Fuente, I. Fuentes, F. J. Toro, and F. J. Blanco. 2011. Umbilical cord as a mesenchymal stem cell source for treating joint pathologies. *World J Orthop* 2 (6): 43–50. doi: 10.5312/wjo.v2.i6.43.

Azari, O., H. Babaei, A. Derakhshanfar, et al. 2011. Effects of transplanted mesenchymal stem cells isolated from Wharton's jelly of caprine umbilical cord on cutaneous wound healing; histopathological evaluation. *Vet Res Commun* 35 (4): 211–22. doi: 10.1007/s11259-011-9464-z.

Azuara-Blanco, A., C. T. Pillai, and H. S. Dua. 1999. Amniotic membrane transplantation for ocular surface reconstruction. *Br J Ophthalmol* 83 (4): 399–402.

Babaei, H., M. Moshrefi, M. Golchin, and S. N. Nematollahi-Mahani. 2008. Assess the pluripotency of caprine umbilical cord Wharton's jelly mesenchymal cells by RT-PCR analysis of early transcription factor nanog. *Iran J Vet Surg* 3: 57–65.

Bailo, M., M. Soncini, E. Vertua, et al. 2004. Engraftment potential of human amnion and chorion cells derived from term placenta. *Transplantation* 78 (10): 1439–48.

Banas, R. A., C. Trumpower, C. Bentlejewski, et al. 2008. Immunogenicity and immunomodulatory effects of amnion-derived multipotent progenitor cells. *Hum Immunol* 69 (6): 321–8. doi: 10.1016/j.humimm.2008.04.007.

Barberini, D. J., N. P. Freitas, M. S. Magnoni, et al. 2014. Equine mesenchymal stem cells from bone marrow, adipose tissue and umbilical cord: Immunophenotypic characterization and differentiation potential. *Stem Cell Res Ther* 5 (1): 25. doi: 10.1186/scrt414.

Applications of Placenta-Derived Cells in Veterinary Medicine 269

Barboni, B., V. Curini, V. Russo, et al. 2012a. Indirect co-culture with tendons or tenocytes can program amniotic epithelial cells towards stepwise tenogenic differentiation. *PLoS One* 7 (2): e30974. doi: 10.1371/journal.pone.0030974.

Barboni, B., C. Mangano, L. Valbonetti, et al. 2013. Synthetic bone substitute engineered with amniotic epithelial cells enhances bone regeneration after maxillary sinus augmentation. *PLoS One* 8 (5): e63256. doi: 10.1371/journal.pone.0063256.

Barboni, B., V. Russo, V. Curini, et al. 2012b. Achilles tendon regeneration can be improved by amniotic epithelial cell allotransplantation. *Cell Transplant* 21 (11): 2377–95. doi: 10.3727/096368912x638892.

Barboni, B., V. Russo, V. Curini, et al. 2014. Gestational stage affects amniotic epithelial cells phenotype, methylation status, immunomodulatory and stemness properties. *Stem Cell Rev* 10 (5): 725–41. doi: 10.1007/s12015-014-9519-y.

Barone, R. 2003. *Trattato di anatomia comparata dei mammiferi domestici. Vol. 4: Splancnologia. Apparecchio uro-genitale. Feto e i suoi annessi. Edagricole-New Business Media.*

Barros, P. S., J. A. Garcia, J. L. Laus, A. L. Ferreira, and T. L. Salles Gomes. 1998. The use of xenologous amniotic membrane to repair canine corneal perforation created by penetrating keratectomy. *Vet Ophthalmol* 1 (2–3): 119–23.

Barros, P. S., A. M. Safatle, C. A. Godoy, et al. 2005. Amniotic membrane transplantation for the reconstruction of the ocular surface in three cases. *Vet Ophthalmol* 8 (3): 189–92. doi: 10.1111/j.1463-5224.2005.00391.x.

Bartholomew, S., S. D. Owens, G. L. Ferraro, et al. 2009. Collection of equine cord blood and placental tissues in 40 thoroughbred mares. *Equine Vet J* 41 (8): 724–8.

Baulier, E., F. Favreau, A. Le Corf, et al. 2014. Amniotic fluid-derived mesenchymal stem cells prevent fibrosis and preserve renal function in a preclinical porcine model of kidney transplantation. *Stem Cells Transl Med* 3 (7): 809–20. doi: 10.5966/sctm.2013-0186.

Berardinelli, P., L. Valbonetti, A. Muttini, et al. 2013. Role of amniotic fluid mesenchymal cells engineered on MgHA/collagen-based scaffold allotransplanted on an experimental animal study of sinus augmentation. *Clin Oral Investig* 17 (7): 1661–75. doi: 10.1007/s00784-012-0857-3.

Berg, L., T. Koch, T. Heerkens, et al. 2009. Chondrogenic potential of mesenchymal stromal cells derived from equine bone marrow and umbilical cord blood. *Vet Comp Orthop Traumatol* 22 (5): 363–70. doi: 10.3415/vcot-08-10-0107.

Bilic, G., S. M. Zeisberger, A. S. Mallik, R. Zimmermann, and A. H. Zisch. 2008. Comparative characterization of cultured human term amnion epithelial and mesenchymal stromal cells for application in cell therapy. *Cell Transplant* 17 (8): 955–68.

Biologics, Medivet. Available at http://www.medivet-america.com

Black, L. L., J. Gaynor, C. Adams, et al. 2008. Effect of intraarticular injection of autologous adipose-derived mesenchymal stem and regenerative cells on clinical signs of chronic osteoarthritis of the elbow joint in dogs. *Vet Ther* 9 (3): 192–200.

Black, L. L., J. Gaynor, D. Gahring, et al. 2007. Effect of adipose-derived mesenchymal stem and regenerative cells on lameness in dogs with chronic osteoarthritis of the coxofemoral joints: A randomized, double-blinded, multicenter, controlled trial. *Vet Ther* 8 (4): 272–84.

Brehm, W., B. Aklin, T. Yamashita, et al. 2006. Repair of superficial osteochondral defects with an autologous scaffold-free cartilage construct in a caprine model: Implantation method and short-term results. *Osteoarthritis Cartilage* 14 (12): 1214–26. doi: 10.1016/j.joca.2006.05.002.

Bruns, J., J. Kampen, J. Kahrs, and W. Plitz. 2000. Achilles tendon rupture: Experimental results on spontaneous repair in a sheep-model. *Knee Surg Sports Traumatol Arthrosc* 8 (6): 364–9.

Burk, J., I. Ribitsch, C. Gittel, et al. 2013. Growth and differentiation characteristics of equine mesenchymal stromal cells derived from different sources. *Vet J* 195 (1): 98–106. doi: 10.1016/j.tvjl.2012.06.004.

Campagnoli, C., I. A. Roberts, S. Kumar, et al. 2001. Identification of mesenchymal stem/ progenitor cells in human first-trimester fetal blood, liver, and bone marrow. *Blood* 98 (8): 2396–402.

Can, A., and S. Karahuseyinoglu. 2007. Concise review: Human umbilical cord stroma with regard to the source of fetus-derived stem cells. *Stem Cells* 25 (11): 2886–95. doi: 10.1634/stemcells.2007-0417.

Cardoso, T. C., H. F. Ferrari, A. F. Garcia, et al. 2012a. Isolation and characterization of Wharton's jelly-derived multipotent mesenchymal stromal cells obtained from bovine umbilical cord and maintained in a defined serum-free three-dimensional system. *BMC Biotechnol* 12: 18. doi: 10.1186/1472-6750-12-18.

Cardoso, T. C., J. B. Novais, T. F. Antello, et al. 2012b. Susceptibility of neuron-like cells derived from bovine Wharton's jelly to bovine herpesvirus type 5 infections. *BMC Vet Res* 8: 242. doi: 10.1186/1746-6148-8-242.

Cargnoni, A., E. C. Piccinelli, L. Ressel, et al. 2014. Conditioned medium from amniotic membrane-derived cells prevents lung fibrosis and preserves blood gas exchanges in bleomycin-injured mice-specificity of the effects and insights into possible mechanisms. *Cytotherapy* 16 (1): 17–32. doi: 10.1016/j.jcyt.2013.07.002.

Carlin, R., D. Davis, M. Weiss, B. Schultz, and D. Troyer. 2006. Expression of early transcription factors Oct-4, Sox-2 and Nanog by porcine umbilical cord (PUC) matrix cells. *Reprod Biol Endocrinol* 4: 8. doi: 10.1186/1477-7827-4-8.

Carpenter, N. J. 2001. Molecular cytogenetics. *Semin Pediatr Neurol* 8 (3): 135–46.

Carrade, D. D., M. W. Lame, M. S. Kent, et al. 2012. Comparative analysis of the immunomodulatory properties of equine adult-derived mesenchymal stem cells(). *Cell Med* 4 (1): 1–11. doi: 10.3727/215517912x647217.

Carrade, D. D., S. D. Owens, L. D. Galuppo, et al. 2011. Clinicopathologic findings following intra-articular injection of autologous and allogeneic placentally derived equine mesenchymal stem cells in horses. *Cytotherapy* 13 (4): 419–30. doi: 10.3109/14653249.2010.536213.

Cebrian-Serrano, A., T. Stout, and A. Dinnyes. 2013. Veterinary applications of induced pluripotent stem cells: Regenerative medicine and models for disease? *Vet J* 198 (1): 34–42. doi: 10.1016/j.tvjl.2013.03.028.

Celavet. Available at http://www.celavet.com

Chen, J., Z. Lu, D. Cheng, S. Peng, and H. Wang. 2011. Isolation and characterization of porcine amniotic fluid-derived multipotent stem cells. *PLoS One* 6 (5): e19964. doi: 10.1371/journal.pone.0019964.

Chiang, H., and C. C. Jiang. 2009. Repair of articular cartilage defects: Review and perspectives. *J Formos Med Assoc* 108 (2): 87–101. doi: 10.1016/s0929-6646(09)60039-5.

Choi, S. A., H. S. Choi, K. J. Kim, et al. 2013. Isolation of canine mesenchymal stem cells from amniotic fluid and differentiation into hepatocyte-like cells. *In Vitro Cell Dev Biol Anim* 49 (1): 42–51. doi: 10.1007/s11626-012-9569-x.

Cibelli, J., M. E. Emborg, D. J. Prockop, et al. 2013. Strategies for improving animal models for regenerative medicine. *Cell Stem Cell* 12 (3): 271–4. doi: 10.1016/j.stem.2013.01.004.

Co, C., M. K. Vickaryous, and T. G. Koch. 2014. Membrane culture and reduced oxygen tension enhances cartilage matrix formation from equine cord blood mesenchymal stromal cells in vitro. *Osteoarthritis Cartilage* 22 (3): 472–80. doi: 10.1016/j.joca.2013.12.021.

Coli, A., F. Nocchi, R. Lamanna, et al. 2011. Isolation and characterization of equine amnion mesenchymal stem cells. *Cell Biol Int Rep (2010)* 18 (1): e00011. doi: 10.1042/cbr20110004.

Colocho, G., W. P. Graham, 3rd, A. E. Greene, D. W. Matheson, and D. Lynch. 1974. Human amniotic membrane as a physiologic wound dressing. *Arch Surg* 109 (3): 370–3.

Colosimo, A., V. Curini, V. Russo, et al. 2013a. Characterization, GFP gene Nucleofection, and allotransplantation in injured tendons of ovine amniotic fluid-derived stem cells. *Cell Transplant* 22 (1): 99–117. doi: 10.3727/096368912x638883.

Colosimo, A., V. Russo, A. Mauro, et al. 2013b. Prolonged in vitro expansion partially affects phenotypic features and osteogenic potential of ovine amniotic fluid-derived mesenchymal stromal cells. *Cytotherapy* 15 (8): 930–50. doi: 10.1016/j.jcyt.2013.03.014.

Corradetti, B., A. Correani, A. Romaldini, et al. 2014. Amniotic membrane-derived mesenchymal cells and their conditioned media: Potential candidates for uterine regenerative therapy in the horse. *PLoS One* 9 (10): e111324. doi: 10.1371/journal.pone.0111324.

Corradetti, B., A. Lange-Consiglio, M. Barucca, F. Cremonesi, and D. Bizzaro. 2011. Size-sieved subpopulations of mesenchymal stem cells from intervascular and perivascular equine umbilical cord matrix. *Cell Prolif* 44 (4): 330–42. doi: 10.1111/j.1365-2184.2011.00759.x.

Corradetti, B., A. Meucci, D. Bizzaro, F. Cremonesi, and A. Lange Consiglio. 2013. Mesenchymal stem cells from amnion and amniotic fluid in the bovine. *Reproduction* 145 (4): 391–400. doi: 10.1530/rep-12-0437.

Cremonesi, F., B. Corradetti, and A. Lange Consiglio. 2011. Fetal adnexa derived stem cells from domestic animal: Progress and perspectives. *Theriogenology* 75 (8): 1400–15. doi: 10.1016/j.theriogenology.2010.12.032.

Crevier-Denoix, N., C. Collobert, P. Pourcelot, et al. 1997. Mechanical properties of pathological equine superficial digital flexor tendons. *Equine Vet J Suppl* (23): 23–6.

Crovace, A., L. Lacitignola, G. Rossi, and E. Francioso. 2010. Histological and immunohistochemical evaluation of autologous cultured bone marrow mesenchymal stem cells and bone marrow mononucleated cells in collagenase-induced tendinitis of equine superficial digital flexor tendon. *Vet Med Int* 2010: 250978. doi: 10.4061/2010/250978.

Cuervo, B., M. Rubio, J. Sopena, et al. 2014. Hip osteoarthritis in dogs: A randomized study using mesenchymal stem cells from adipose tissue and plasma rich in growth factors. *Int J Mol Sci* 15 (8): 13437–60. doi: 10.3390/ijms150813437.

Cyranoski, D. 2013. Stem cells boom in vet clinics. *Nature* 496 (7444): 148–9. doi: 10.1038/496148a.

da Cunha, E. R., C. F. Martins, C. G. Silva, H. C. Bessler, and S. N. Bao. 2014. Effects of prolonged in vitro culture and cryopreservation on viability, DNA fragmentation, chromosome stability and ultrastructure of bovine cells from amniotic fluid and umbilical cord. *Reprod Domest Anim* 49 (5): 806–12. doi: 10.1111/rda.12372.

Dahlgren, L. A., H. O. Mohammed, and A. J. Nixon. 2005. Temporal expression of growth factors and matrix molecules in healing tendon lesions. *J Orthop Res* 23 (1): 84–92. doi: 10.1016/j.orthres.2004.05.007.

De Coppi, P., G. Bartsch, Jr., M. M. Siddiqui, et al. 2007. Isolation of amniotic stem cell lines with potential for therapy. *Nat Biotechnol* 25 (1): 100–6. doi: 10.1038/nbt1274.

de Mattos Carvalho, A., A. L. Alves, P. G. G. de Oliveira, et al. 2011. Use of adipose tissue-derived mesenchymal stem cells for experimental tendinitis therapy in equines. *J Equine Vet Sci* 31 (1): 26–34. doi: 10.1016/j.jevs.2010.11.014.

De Schauwer, C., K. Goossens, S. Piepers, et al. 2014. Characterization and profiling of immunomodulatory genes of equine mesenchymal stromal cells from non-invasive sources. *Stem Cell Res Ther* 5 (1): 6. doi: 10.1186/scrt395.

De Schauwer, C., E. Meyer, P. Cornillie, et al. 2011. Optimization of the isolation, culture, and characterization of equine umbilical cord blood mesenchymal stromal cells. *Tissue Eng Part C Methods* 17 (11): 1061–70. doi: 10.1089/ten.tec.2011.0052.

De Schauwer, C., S. Piepers, G. R. Van de Walle, et al. 2012. In search for cross-reactivity to immunophenotype equine mesenchymal stromal cells by multicolor flow cytometry. *Cytometry A* 81 (4): 312–23. doi: 10.1002/cyto.a.22026.

De Schauwer, C., G. R. van de Walle, S. Piepers, et al. 2013. Successful isolation of equine mesenchymal stromal cells from cryopreserved umbilical cord blood-derived mononuclear cell fractions. *Equine Vet J* 45 (4): 518–22. doi: 10.1111/evj.12003.

Deeg, H. J., and R. Storb. 1994. Canine marrow transplantation models. *Curr Topics Vet Res* 1: 103–14.

Dehoux, J. P., and P. Gianello. 2007. The importance of large animal models in transplantation. *Front Biosci* 12: 4864–80.

DeRoth, A. 1940. Plastic repair of conjunctival defects with fetal membranes. *Arch Ophthalmol* 23: 522–55.

Dev, K., S. K. Gautam, S. K. Giri, et al. 2012a. Isolation, culturing and characterization of feeder-independent amniotic fluid stem cells in buffalo (*Bubalus bubalis*). *Res Vet Sci* 93 (2): 743–8. doi: 10.1016/j.rvsc.2011.09.007.

Dev, K., S. K. Giri, A. Kumar, et al. 2012b. Derivation, characterization and differentiation of buffalo (Bubalus bubalis) amniotic fluid derived stem cells. *Reprod Domest Anim* 47 (5): 704–11. doi: 10.1111/j.1439-0531.2011.01947.x.

Dev, K., S. K. Giri, A. Kumar, et al. 2012c. Expression of transcriptional factor genes (Oct-4, Nanog, and Sox-2) and embryonic stem cell-like characters in placental membrane of Buffalo (Bubalus bubalis). *J Membr Biol* 245 (4): 177–83. doi: 10.1007/s00232-012-9427-5.

Di Tomo, P., C. Pipino, P. Lanuti, et al. 2013. Calcium sensing receptor expression in ovine amniotic fluid mesenchymal stem cells and the potential role of R-568 during osteogenic differentiation. *PLoS One* 8 (9): e73816. doi: 10.1371/journal.pone.0073816.

Ditadi, A., P. de Coppi, O. Picone, et al. 2009. Human and murine amniotic fluid c-Kit+Lin- cells display hematopoietic activity. *Blood* 113 (17): 3953–60. doi: 10.1182/blood-2008-10-182105.

Dominici, M., K. Le Blanc, I. Mueller, et al. 2006. Minimal criteria for defining multipotent mesenchymal stromal cells. The International Society for Cellular Therapy position statement. *Cytotherapy* 8 (4): 315–17. doi: 10.1080/14653240600855905.

Eini, F., T. Foroutan, A. Bidadkosh, A. Barin, M. M. Dehghan, and P. Tajik. 2012. The effects of freeze/thawing process on cryopreserved equine umbilical cord blood-derived mesenchymal stem cells. *Comp Clin Pathol* 21 (6): 1713–18. doi: 10.1007/s00580-011-1355-8.

Erices, A., P. Conget, and J. J. Minguell. 2000. Mesenchymal progenitor cells in human umbilical cord blood. *Br J Haematol* 109 (1): 235–42.

Evangelista, M., M. Soncini, and O. Parolini. 2008. Placenta-derived stem cells: New hope for cell therapy? *Cytotechnology* 58 (1): 33–42. doi: 10.1007/s10616-008-9162-z.

Evans, H. E., and A. de Lahunta. 2013. *Miller's Anatomy of the Dog*, 4th ed. Elsevier, Saunders, PA: Philadelphia.

Fadel, L., B. R. Viana, M. L. Feitosa, et al. 2011. Protocols for obtainment and isolation of two mesenchymal stem cell sources in sheep. *Acta Cir Bras* 26 (4): 267–73.

Faulk, W. P., R. Matthews, P. J. Stevens, et al. 1980. Human amnion as an adjunct in wound healing. *Lancet* 1 (8179): 1156–8.

Fernandes, R. A., C. V. Wenceslau, A. L. Reginato, I. Kerkis, and M. A. Migli. 2012. Derivation and characterization of progenitor stem cells from canine allantois and amniotic fluids at the third trimester of gestation. *Placenta* 33 (8): 640–4. doi: 10.1016/j.placenta.2012.03.009.

Figueroa, R. J., T. G. Koch, and D. H. Betts. 2011. Osteogenic differentiation of equine cord blood multipotent mesenchymal stromal cells within coralline hydroxyapatite scaffolds in vitro. *Vet Comp Orthop Traumatol* 24 (5): 354–62. doi: 10.3415/vcot-10-10-0142.

Filioli Uranio, M., M. E. Dell'Aquila, M. Caira, et al. 2014. Characterization and in vitro differentiation potency of early-passage canine amnion- and umbilical cord-derived mesenchymal stem cells as related to gestational age. *Mol Reprod Dev* 81 (6): 539–51. doi: 10.1002/mrd.22322.

Applications of Placenta-Derived Cells in Veterinary Medicine 273

Filioli Uranio, M., L. Valentini, A. Lange-Consiglio, et al. 2011. Isolation, proliferation, cytogenetic, and molecular characterization and in vitro differentiation potency of canine stem cells from foetal adnexa: A comparative study of amniotic fluid, amnion, and umbilical cord matrix. *Mol Reprod Dev* 78 (5): 361–73. doi: 10.1002/mrd.21311.

Fortier, L. A., H. G. Potter, E. J. Rickey, et al. 2010. Concentrated bone marrow aspirate improves full-thickness cartilage repair compared with microfracture in the equine model. *J Bone Joint Surg Am* 92 (10): 1927–37. doi: 10.2106/jbjs.i.01284.

Fortier, L. A., and A. J. Travis. 2011. Stem cells in veterinary medicine. *Stem Cell Res Ther* 2 (1): 9. doi: 10.1186/scrt50.

Frey-Vasconcells, J., K. J. Whittlesey, E. Baum, and E. G. Feigal. 2012. Translation of stem cell research: Points to consider in designing preclinical animal studies. *Stem Cells Transl Med* 1 (5): 353–8. doi: 10.5966/sctm.2012-0018.

Frisbie, D. D., J. D. Kisiday, C. E. Kawcak, N. M. Werpy, and C. W. McIlwraith. 2009. Evaluation of adipose-derived stromal vascular fraction or bone marrow-derived mesenchymal stem cells for treatment of osteoarthritis. *J Orthop Res* 27 (12): 1675–80. doi: 10.1002/jor.20933.

Gade, N. E., M. D. Pratheesh, A. Nath, et al. 2012. Therapeutic potential of stem cells in veterinary practice. *Vet World* 5 (8): 499–507. doi: 10.5455/vetworld.2012.499-507.

Galende, E., I. Karakikes, L. Edelmann, et al. 2010. Amniotic fluid cells are more efficiently reprogrammed to pluripotency than adult cells. *Cell Reprogram* 12 (2): 117–25. doi: 10.1089/cell.2009.0077.

Gao, Y., Z. Zhu, Y. Zhao, et al. 2014. Multilineage potential research of bovine amniotic fluid mesenchymal stem cells. *Int J Mol Sci* 15 (3): 3698–710. doi: 10.3390/ijms15033698.

Garcia, D., U. G. Longo, J. Vaquero, et al. 2014. Amniotic membrane transplant for articular cartilage repair: An experimental study in sheep. *Curr Stem Cell Res Ther* 10 (1): 77–83.

Gray, F. L., C. G. Turner, A. Ahmed, et al. 2012. Prenatal tracheal reconstruction with a hybrid amniotic mesenchymal stem cells-engineered construct derived from decellularized airway. *J Pediatr Surg* 47 (6): 1072–9. doi: 10.1016/j.jpedsurg.2012.03.006.

Grayson, W. L., F. Zhao, B. Bunnell, and T. Ma. 2007. Hypoxia enhances proliferation and tissue formation of human mesenchymal stem cells. *Biochem Biophys Res Commun* 358 (3): 948–53. doi: 10.1016/j.bbrc.2007.05.054.

Gruss, J. S., and D. W. Jirsch. 1978. Human amniotic membrane: A versatile wound dressing. *Can Med Assoc J* 118 (10): 1237–46.

Guercio, A., P. Di Marco, S. Casella, et al. 2012. Production of canine mesenchymal stem cells from adipose tissue and their application in dogs with chronic osteoarthritis of the humeroradial joints. *Cell Biol Int* 36 (2): 189–94. doi: 10.1042/cbi20110304.

Guest, D. J., J. C. Ousey, and M. R. Smith. 2008. Defining the expression of marker genes in equine mesenchymal stromal cells. *Stem Cells Cloning* 1: 1–9.

Guest, D. J., M. R. Smith, and W. R. Allen. 2010. Equine embryonic stem-like cells and mesenchymal stromal cells have different survival rates and migration patterns following their injection into damaged superficial digital flexor tendon. *Equine Vet J* 42 (7): 636–42. doi: 10.1111/j.2042-3306.2010.00112.x.

Gulati, B. R., R. Kumar, N. Mohanty, et al. 2013. Bone morphogenetic protein-12 induces tenogenic differentiation of mesenchymal stem cells derived from equine amniotic fluid. *Cells Tissues Organs* 198 (5): 377–89. doi: 10.1159/000358231.

Harris, D. T. 2013. Umbilical cord tissue mesenchymal stem cells: Characterization and clinical applications. *Curr Stem Cell Res Ther* 8 (5): 394–9.

Hendrickson, D. A., A. J. Nixon, D. A. Grande, et al. 1994. Chondrocyte-fibrin matrix transplants for resurfacing extensive articular cartilage defects. *J Orthop Res* 12 (4): 485–97. doi: 10.1002/jor.1100120405.

Horwitz, E. M., K. Le Blanc, M. Dominici, et al. 2005. Clarification of the nomenclature for MSC: The International Society for Cellular Therapy position statement. *Cytotherapy* 7 (5): 393–5. doi: 10.1080/14653240500319234.

Hoynowski, S. M., M. M. Fry, B. M. Gardner, et al. 2007. Characterization and differentiation of equine umbilical cord-derived matrix cells. *Biochem Biophys Res Commun* 362 (2): 347–53. doi: 10.1016/j.bbrc.2007.07.182.

Hung, S. C., N. J. Chen, S. L. Hsieh, et al. 2002. Isolation and characterization of size-sieved stem cells from human bone marrow. *Stem Cells* 20 (3): 249–58. doi: 10.1634/stemcells.20-3-249.

Iacono, E., L. Brunori, A. Pirrone, et al. 2012a. Isolation, characterization and differentiation of mesenchymal stem cells from amniotic fluid, umbilical cord blood and Wharton's jelly in the horse. *Reproduction* 143 (4): 455–68. doi: 10.1530/rep-10-0408.

Iacono, E., M. Cunto, D. Zambelli, et al. 2012b. Could fetal fluid and membranes be an alternative source for mesenchymal stem cells (MSCs) in the feline species? A preliminary study. *Vet Res Commun* 36 (2): 107–18. doi: 10.1007/s11259-012-9520-3.

Iacono, E., B. Merlo, A. Pirrone, et al. 2012c. Effects of mesenchymal stem cells isolated from amniotic fluid and platelet-rich plasma gel on severe decubitus ulcers in a septic neonatal foal. *Res Vet Sci* 93 (3): 1439–40. doi: 10.1016/j.rvsc.2012.04.008.

Izumi, M., B. J. Pazin, C. F. Minervini, et al. 2009. Quantitative comparison of stem cell marker-positive cells in fetal and term human amnion. *J Reprod Immunol* 81 (1): 39–43. doi: 10.1016/j.jri.2009.02.007.

Jager, M., R. Bachmann, A. Scharfstadt, and R. Krauspe. 2006. Ovine cord blood accommodates multipotent mesenchymal progenitor cells. *In Vivo* 20 (2): 205–14.

Jang, B. J., Y. E. Byeon, J. H. Lim, et al. 2008. Implantation of canine umbilical cord blood-derived mesenchymal stem cells mixed with beta-tricalcium phosphate enhances osteogenesis in bone defect model dogs. *J Vet Sci* 9 (4): 387–93.

Jin, G. Z., X. J. Yin, X. F. Yu, et al. 2008. Generation of neuronal-like cells from umbilical cord blood-derived mesenchymal stem cells of a RFP-transgenic cloned cat. *J Vet Med Sci* 70 (7): 723–6.

Kang, B. J., H. H. Ryu, S. S. Park, et al. 2012. Comparing the osteogenic potential of canine mesenchymal stem cells derived from adipose tissues, bone marrow, umbilical cord blood, and Wharton's jelly for treating bone defects. *J Vet Sci* 13 (3): 299–310.

Kang, J. G., S. B. Park, M. S. Seo, et al. 2013. Characterization and clinical application of mesenchymal stem cells from equine umbilical cord blood. *J Vet Sci* 14 (3): 367–71.

Kasashima, Y., T. Takahashi, R. K. Smith, et al. 2004. Prevalence of superficial digital flexor tendonitis and suspensory desmitis in Japanese Thoroughbred flat racehorses in 1999. *Equine Vet J* 36 (4): 346–50.

Kaviani, A., T. E. Perry, A. Dzakovic, et al. 2001. The amniotic fluid as a source of cells for fetal tissue engineering. *J Pediatr Surg* 36 (11): 1662–5.

Kim, E. Y., K. B. Lee, J. Yu, et al. 2014. Neuronal cell differentiation of mesenchymal stem cells originating from canine amniotic fluid. *Hum Cell* 27 (2): 51–8. doi: 10.1007/s13577-013-0080-9.

Kim, J., Y. Lee, H. Kim, et al. 2007. Human amniotic fluid-derived stem cells have characteristics of multipotent stem cells. *Cell Prolif* 40 (1): 75–90. doi: 10.1111/j.1365-2184.2007.00414.x.

Kim, J. Y., Y. M. Choi, S. W. Jeong, and D. L. Williams. 2009. Effect of bovine freeze-dried amniotic membrane (Amnisite-BA) on uncomplicated canine corneal erosion. *Vet Ophthalmol* 12 (1): 36–42. doi: 10.1111/j.1463-5224.2009.00671.x.

Kimura, M., M. Toyoda, S. Gojo, et al. 2012. Allogeneic amniotic membrane-derived mesenchymal stromal cell transplantation in a porcine model of chronic myocardial ischemia. *J Stem Cells Regen Med* 8 (3): 171–80.

Kisiel, A. H., L. A. McDuffee, E. Masaoud, et al. 2012. Isolation, characterization, and in vitro proliferation of canine mesenchymal stem cells derived from bone marrow, adipose tissue, muscle, and periosteum. *Am J Vet Res* 73 (8): 1305–17. doi: 10.2460/ajvr.73.8.1305.

Klein, J. D., C. G. Turner, S. A. Steigman, et al. 2011. Amniotic mesenchymal stem cells enhance normal fetal wound healing. *Stem Cells Dev* 20 (6): 969–76. doi: 10.1089/scd.2010.0379.

Koch, T. G., T. Heerkens, P. D. Thomsen, and D. H. Betts. 2007. Isolation of mesenchymal stem cells from equine umbilical cord blood. *BMC Biotechnol* 7: 26. doi: 10.1186/1472-6750-7-26.

Koerner, J., D. Nesic, J. D. Romero, et al. 2006. Equine peripheral blood-derived progenitors in comparison to bone marrow-derived mesenchymal stem cells. *Stem Cells* 24 (6): 1613–19. doi: 10.1634/stemcells.2005-0264.

Kraus, K. H., and C. Kirker-Head. 2006. Mesenchymal stem cells and bone regeneration. *Vet Surg* 35 (3): 232–42. doi: 10.1111/j.1532-950X.2006.00142.x.

Kumar, B. M., J. G. Yoo, S. A. Ock, et al. 2007. In vitro differentiation of mesenchymal progenitor cells derived from porcine umbilical cord blood. *Mol Cells* 24 (3): 343–50.

Kunisaki, S. M., D. A. Freedman, and D. O. Fauza. 2006a. Fetal tracheal reconstruction with cartilaginous grafts engineered from mesenchymal amniocytes. *J Pediatr Surg* 41 (4): 675–82; discussion 675–82. doi: 10.1016/j.jpedsurg.2005.12.008.

Kunisaki, S. M., J. R. Fuchs, S. A. Steigman, and D. O. Fauza. 2007. A comparative analysis of cartilage engineered from different perinatal mesenchymal progenitor cells. *Tissue Eng* 13 (11): 2633–44. doi: 10.1089/ten.2006.0407.

Kuzmuk, K. N., and L. B. Schook. 2011. Pigs as a model for biomedical sciences. In *The Genetics of the Pig*, M. F. Eothschild and A. Ruvinsky (eds.) 2nd ed., pp. 426–44. doi: 10.1079/9781845937560.0014.

Laboratories, Fat-Stem. Available at http://fat-stem.be

Lange-Consiglio, A., B. Corradetti, D. Bizzaro, et al. 2012. Characterization and potential applications of progenitor-like cells isolated from horse amniotic membrane. *J Tissue Eng Regen Med* 6 (8): 622–35. doi: 10.1002/term.465.

Lange-Consiglio, A., B. Corradetti, A. Meucci, et al. 2013a. Characteristics of equine mesenchymal stem cells derived from amnion and bone marrow: In vitro proliferative and multilineage potential assessment. *Equine Vet J* 45 (6): 737–44. doi: 10.1111/evj.12052.

Lange-Consiglio, A., B. Corradetti, L. Rutigliano, F. Cremonesi, and D. Bizzarro. 2011. In vitro studies of horse umbilical cord matrix-derived cells: From characterization to labeling for magnetic resonance imaging. *Open Tissue Eng Regen Med J* 4: 120–33.

Lange-Consiglio, A., D. Rossi, S. Tassan, et al. 2013b. Conditioned medium from horse amniotic membrane-derived multipotent progenitor cells: Immunomodulatory activity in vitro and first clinical application in tendon and ligament injuries in vivo. *Stem Cells Dev* 22 (22): 3015–24. doi: 10.1089/scd.2013.0214.

Lange-Consiglio, A., S. Tassan, B. Corradetti, et al. 2013c. Investigating the efficacy of amnion-derived compared with bone marrow-derived mesenchymal stromal cells in equine tendon and ligament injuries. *Cytotherapy* 15 (8): 1011–20. doi: 10.1016/j.jcyt.2013.03.002.

Lassaline, M. E., D. E. Brooks, F. J. Ollivier, et al. 2005. Equine amniotic membrane transplantation for corneal ulceration and keratomalacia in three horses. *Vet Ophthalmol* 8 (5): 311–17. doi: 10.1111/j.1463-5224.2005.00405.x.

Lee, K. S., S. H. Cha, H. W. Kang, et al. 2013a. Effects of serial passage on the characteristics and chondrogenic differentiation of canine umbilical cord matrix derived mesenchymal stem cells. *Asian Aust J Anim Sci* 26 (4): 588–95. doi: 10.5713/ajas.2012.12488.

Lee, K. S., J. J. Nah, B. C. Lee, et al. 2013b. Maintenance and characterization of multipotent mesenchymal stem cells isolated from canine umbilical cord matrix by collagenase digestion. *Res Vet Sci* 94 (1): 144–51. doi: 10.1016/j.rvsc.2012.07.033.

Li, H., J. Y. Niederkorn, S. Neelam, et al. 2005. Immunosuppressive factors secreted by human amniotic epithelial cells. *Invest Ophthalmol Vis Sci* 46 (3): 900–7. doi: 10.1167/iovs.04-0495.

Lim, J. H., Y. E. Byeon, H. H. Ryu, et al. 2007. Transplantation of canine umbilical cord blood-derived mesenchymal stem cells in experimentally induced spinal cord injured dogs. *J Vet Sci* 8 (3): 275–82.

Lim, R., S. T. Chan, J. L. Tan, et al. 2013. Preterm human amnion epithelial cells have limited reparative potential. *Placenta* 34 (6): 486–92. doi: 10.1016/j.placenta.2013.03.010.

Lovati, A. B., B. Corradetti, A. Lange Consiglio, et al. 2011. Comparison of equine bone marrow-, umbilical cord matrix and amniotic fluid-derived progenitor cells. *Vet Res Commun* 35 (2): 103–21. doi: 10.1007/s11259-010-9457-3.

Lunney, J. K. 2007. Advances in swine biomedical model genomics. *Int J Biol Sci* 3 (3): 179–84.

Magatti, M., S. De Munari, E. Vertua, et al. 2008. Human amnion mesenchyme harbors cells with allogeneic T-cell suppression and stimulation capabilities. *Stem Cells* 26 (1): 182–92. doi: 10.1634/stemcells.2007-0491.

Majore, I., P. Moretti, R. Hass, and C. Kasper. 2009. Identification of subpopulations in mesenchymal stem cell-like cultures from human umbilical cord. *Cell Commun Signal* 7: 6. doi: 10.1186/1478-811x-7-6.

Mann, A., R. P. Yadav, J. Singh, et al. 2013. Culture, characterization and differentiation of cells from buffalo (Bubalus bubalis) amnion. *Cytotechnology* 65 (1): 23–30. doi: 10.1007/s10616-012-9464-z.

Martin, D. R., N. R. Cox, T. L. Hathcock, G. P. Niemeyer, and H. J. Baker. 2002. Isolation and characterization of multipotential mesenchymal stem cells from feline bone marrow. *Exp Hematol* 30 (8): 879–86.

Martino, N. A., A. Lange-Consiglio, F. Cremonesi, et al. 2011. Functional expression of the extracellular calcium sensing receptor (CaSR) in equine umbilical cord matrix size-sieved stem cells. *PLoS One* 6 (3): e17714. doi: 10.1371/journal.pone.0017714.

Mattioli, M., A. Gloria, M. Turriani, et al. 2012. Stemness characteristics and osteogenic potential of sheep amniotic epithelial cells. *Cell Biol Int* 36 (1): 7–19. doi: 10.1042/cbi20100720.

Mauro, A., M. Turriani, A. Ioannoni, et al. 2010. Isolation, characterization, and in vitro differentiation of ovine amniotic stem cells. *Vet Res Commun* 34 (Suppl 1): S25–8. doi: 10.1007/s11259-010-9393-2.

McCarty, R. C., S. Gronthos, A. C. Zannettino, B. K. Foster, and C. J. Xian. 2009. Characterisation and developmental potential of ovine bone marrow derived mesenchymal stem cells. *J Cell Physiol* 219 (2): 324–33. doi: 10.1002/jcp.21670.

Meng, X. T., C. Li, Z. Y. Dong, et al. 2008. Co-transplantation of bFGF-expressing amniotic epithelial cells and neural stem cells promotes functional recovery in spinal cord-injured rats. *Cell Biol Int* 32 (12): 1546–58. doi: 10.1016/j.cellbi.2008.09.001.

Mermet, I., N. Pottier, J. M. Sainthillier, et al. 2007. Use of amniotic membrane transplantation in the treatment of venous leg ulcers. *Wound Repair Regen* 15 (4): 459–64. doi: 10.1111/j.1524-475X.2007.00252.x.

Miki, T., T. Lehmann, H. Cai, D. B. Stolz, and S. C. Strom. 2005. Stem cell characteristics of amniotic epithelial cells. *Stem Cells* 23 (10): 1549–59. doi: 10.1634/stemcells.2004-0357.

Miki, T., F. Marongiu, K. Dorko, E. C. Ellis, and S. C. Strom. 2010. Isolation of amniotic epithelial stem cells. *Curr Protoc Stem Cell Biol* Chapter 1: Unit 1E.3. doi: 10.1002/9780470151808.sc01e03s12.

Miranda-Sayago, J. M., N. Fernandez-Arcas, C. Benito, et al. 2011. Lifespan of human amniotic fluid-derived multipotent mesenchymal stromal cells. *Cytotherapy* 13 (5): 572–81. doi: 10.3109/14653249.2010.547466.

Mitchell, K. E., M. L. Weiss, B. M. Mitchell, et al. 2003. Matrix cells from Wharton's jelly form neurons and glia. *Stem Cells* 21 (1): 50–60. doi: 10.1634/stemcells.21-1-50.

Applications of Placenta-Derived Cells in Veterinary Medicine

Mohanty, N., B. R. Gulati, R. Kumar, et al. 2014. Immunophenotypic characterization and tenogenic differentiation of mesenchymal stromal cells isolated from equine umbilical cord blood. *In Vitro Cell Dev Biol Anim* 50 (6): 538–48. doi: 10.1007/s11626-013-9729-7.

Mosquera, A., J. L. Fernandez, A. Campos, et al. 1999. Simultaneous decrease of telomere length and telomerase activity with ageing of human amniotic fluid cells. *J Med Genet* 36 (6): 494–6.

Murphy, S. V., and A. Atala. 2013. Amniotic fluid and placental membranes: Unexpected sources of highly multipotent cells. *Semin Reprod Med* 31 (1): 62–8. doi: 10.1055/s-0032-1331799.

Muttini, A., M. Mattioli, L. Petrizzi, et al. 2010. Experimental study on allografts of amniotic epithelial cells in calcaneal tendon lesions of sheep. *Vet Res Commun* 34 (Suppl 1): S117–20. doi: 10.1007/s11259-010-9396-z.

Muttini, A., V. Salini, L. Valbonetti, and M. Abate. 2012. Stem cell therapy of tendinopathies: Suggestions from veterinary medicine. *Muscles Ligaments Tendons J* 2 (3): 187–92.

Muttini, A., L. Valbonetti, M. Abate, et al. 2013. Ovine amniotic epithelial cells: In vitro characterization and transplantation into equine superficial digital flexor tendon spontaneous defects. *Res Vet Sci* 94 (1): 158–69. doi: 10.1016/j.rvsc.2012.07.028.

Nakajima, T., S. Enosawa, T. Mitani, et al. 2001. Cytological examination of rat amniotic epithelial cells and cell transplantation to the liver. *Cell Transplant* 10 (4–5): 423–7.

Neupane, M., C. C. Chang, M. Kiupel, and V. Yuzbasiyan-Gurkan. 2008. Isolation and characterization of canine adipose-derived mesenchymal stem cells. *Tissue Eng Part A* 14 (6): 1007–15. doi: 10.1089/tea.2007.0207.

Nixon, A. J., L. A. Dahlgren, J. L. Haupt, A. E. Yeager, and D. L. Ward. 2008. Effect of adipose-derived nucleated cell fractions on tendon repair in horses with collagenase-induced tendinitis. *Am J Vet Res* 69 (7): 928–37. doi: 10.2460/ajvr.69.7.928.

Nixon, A. J., G. Lust, and M. Vernier-Singer. 1992. Isolation, propagation, and cryopreservation of equine articular chondrocytes. *Am J Vet Res* 53 (12): 2364–70.

Ollivier, F. J., M. E. Kallberg, C. E. Plummer, et al. 2006. Amniotic membrane transplantation for corneal surface reconstruction after excision of corneolimbal squamous cell carcinomas in nine horses. *Vet Ophthalmol* 9 (6): 404–13. doi: 10.1111/j.1463-5224.2006.00480.x.

Pacini, S., S. Spinabella, L. Trombi, et al. 2007. Suspension of bone marrow-derived undifferentiated mesenchymal stromal cells for repair of superficial digital flexor tendon in race horses. *Tissue Eng* 13 (12): 2949–55. doi: 10.1089/ten.2007.0108.

Paebst, F., D. Piehler, W. Brehm, et al. 2014. Comparative immunophenotyping of equine multipotent mesenchymal stromal cells: An approach toward a standardized definition. *Cytometry A* 85 (8): 678–87. doi: 10.1002/cyto.a.22491.

Park, D. H., J. H. Lee, C. V. Borlongan, et al. 2011a. Transplantation of umbilical cord blood stem cells for treating spinal cord injury. *Stem Cell Rev* 7 (1): 181–94. doi: 10.1007/s12015-010-9163-0.

Park, S. B., M. S. Seo, J. G. Kang, J. S. Chae, and K. S. Kang. 2011b. Isolation and characterization of equine amniotic fluid-derived multipotent stem cells. *Cytotherapy* 13 (3): 341–9. doi: 10.3109/14653249.2010.520312.

Park, S. B., M. S. Seo, H. S. Kim, and K. S. Kang. 2012. Isolation and characterization of canine amniotic membrane-derived multipotent stem cells. *PLoS One* 7 (9): e44693. doi: 10.1371/journal.pone.0044693.

Park, S. S., Y. E. Byeon, H. H. Ryu, et al. 2011c. Comparison of canine umbilical cord blood-derived mesenchymal stem cell transplantation times: Involvement of astrogliosis, inflammation, intracellular actin cytoskeleton pathways, and neurotrophin-3. *Cell Transplant* 20 (11–12): 1867–80. doi: 10.3727/096368911x566163.

Parker, H. G., A. L. Shearin, and E. A. Ostrander. 2010. Man's best friend becomes biology's best in show: Genome analyses in the domestic dog. *Annu Rev Genet* 44: 309–36. doi: 10.1146/annurev-genet-102808-115200.

Parolini, O., F. Alviano, G. P. Bagnara, et al. 2008. Concise review: Isolation and characterization of cells from human term placenta: Outcome of the first international Workshop on Placenta Derived Stem Cells. *Stem Cells* 26 (2): 300–11. doi: 10.1634/stemcells.2007-0594.

Parolini, O., and M. Caruso. 2011. Review: Preclinical studies on placenta-derived cells and amniotic membrane: An update. *Placenta* 32 (Suppl 2): S186–95. doi: 10.1016/j.placenta.2010.12.016.

Parolini, O., M. Soncini, M. Evangelista, and D. Schmidt. 2009. Amniotic membrane and amniotic fluid-derived cells: Potential tools for regenerative medicine? *Regen Med* 4 (2): 275–91. doi: 10.2217/17460751.4.2.275.

Passeri, S., F. Nocchi, R. Lamanna, et al. 2009. Isolation and expansion of equine umbilical cord-derived matrix cells (EUCMCs). *Cell Biol Int* 33 (1): 100–5. doi: 10.1016/j.cellbi.2008.10.012.

Perin, L., S. Sedrakyan, S. Da Sacco, and R. De Filippo. 2008. Characterization of human amniotic fluid stem cells and their pluripotential capability. *Methods Cell Biol* 86: 85–99. doi: 10.1016/s0091-679x(08)00005-8.

Piccoli, M., C. Franzin, E. Bertin, et al. 2012. Amniotic fluid stem cells restore the muscle cell niche in a HSA-Cre, Smn(F7/F7) mouse model. *Stem Cells* 30 (8): 1675–84. doi: 10.1002/stem.1134.

Plummer, C. E., F. Ollivier, M. Kallberg, et al. 2009. The use of amniotic membrane transplantation for ocular surface reconstruction: A review and series of 58 equine clinical cases (2002–2008). *Vet Ophthalmol* 12 (Suppl 1): 17–24. doi: 10.1111/j.1463-5224.2009.00741.x.

Poloni, A., G. Maurizi, L. Babini, et al. 2011. Human mesenchymal stem cells from chorionic villi and amniotic fluid are not susceptible to transformation after extensive in vitro expansion. *Cell Transplant* 20 (5): 643–54. doi: 10.3727/096368910x536518.

Porada, C. D., D. D. Harrison-Findik, C. Sanada, et al. 2008. Development and characterization of a novel CD34 monoclonal antibody that identifies sheep hematopoietic stem/progenitor cells. *Exp Hematol* 36 (12): 1739–49. doi: 10.1016/j.exphem.2008.09.003.

Pratheesh, M. D., N. E. Gade, P. K. Dubey, et al. 2014. Molecular characterization and xenogenic application of Wharton's jelly derived caprine mesenchymal stem cells. *Vet Res Commun* 38 (2): 139–48. doi: 10.1007/s11259-014-9597-y.

Pratheesh, M. D., N. E. Gade, A. N. Katiyar, et al. 2013. Isolation, culture and characterization of caprine mesenchymal stem cells derived from amniotic fluid. *Res Vet Sci* 94 (2): 313–19. doi: 10.1016/j.rvsc.2012.08.002.

Prusa, A. R., E. Marton, M. Rosner, G. Bernaschek, and M. Hengstschlager. 2003. Oct-4-expressing cells in human amniotic fluid: A new source for stem cell research? *Hum Reprod* 18 (7): 1489–93.

Raoufi, M. F., P. Tajik, M. M. Dehghan, F. Eini, and A. Barin. 2011. Isolation and differentiation of mesenchymal stem cells from bovine umbilical cord blood. *Reprod Domest Anim* 46 (1): 95–9. doi: 10.1111/j.1439-0531.2010.01594.x.

Ravishanker, R., A. S. Bath, and R. Roy. 2003. "Amnion Bank"—The use of long term glycerol preserved amniotic membranes in the management of superficial and superficial partial thickness burns. *Burns* 29 (4): 369–74.

Reed, S. A., and S. E. Johnson. 2008. Equine umbilical cord blood contains a population of stem cells that express Oct4 and differentiate into mesodermal and endodermal cell types. *J Cell Physiol* 215 (2): 329–36. doi: 10.1002/jcp.21312.

Reed, S. A., and S. E. Johnson. 2014. Expression of scleraxis and tenascin C in equine adipose and umbilical cord blood derived stem cells is dependent upon substrata and FGF supplementation. *Cytotechnology* 66 (1): 27–35. doi: 10.1007/s10616-012-9533-3.

RENOVOCyte. Available at http://renovocyte.com

Ribitsch, I., J. Burk, U. Delling, et al. 2010. Basic science and clinical application of stem cells in veterinary medicine. *Adv Biochem Eng Biotechnol* 123: 219–63. doi: 10.1007/10_2010_66.

Richardson, L. E., J. Dudhia, P. D. Clegg, and R. Smith. 2007. Stem cells in veterinary medicine—Attempts at regenerating equine tendon after injury. *Trends Biotechnol* 25 (9): 409–16. doi: 10.1016/j.tibtech.2007.07.009.

Riddle, Rood and. Available from http://www.roodandriddle.com/stemcell.html

Rossi, B., B. Merlo, S. Colleoni, et al. 2014. Isolation and in vitro characterization of bovine amniotic fluid derived stem cells at different trimesters of pregnancy. *Stem Cell Rev* 10 (5): 712–24. doi: 10.1007/s12015-014-9525-0.

Rossi, D., S. Pianta, M. Magatti, P. Sedlmayr, and O. Parolini. 2012. Characterization of the conditioned medium from amniotic membrane cells: Prostaglandins as key effectors of its immunomodulatory activity. *PLoS One* 7 (10): e46956. doi: 10.1371/journal.pone.0046956.

Roubelakis, M. G., O. Trohatou, and N. P. Anagnou. 2012. Amniotic fluid and amniotic membrane stem cells: Marker discovery. *Stem Cells Int* 2012: 107836. doi: 10.1155/2012/107836.

Russo, V., P. Berardinelli, V. Gatta, et al. 2014. Cross-talk between human amniotic derived cells and host tendon supports tissue regeneration. *J Tissue Eng Regen Med* 8: 142.

Rutigliano, L., B. Corradetti, L. Valentini, et al. 2013. Molecular characterization and in vitro differentiation of feline progenitor-like amniotic epithelial cells. *Stem Cell Res Ther* 4 (5): 133. doi: 10.1186/scrt344.

Ryu, H. H., B. J. Kang, S. S. Park, et al. 2012. Comparison of mesenchymal stem cells derived from fat, bone marrow, Wharton's jelly, and umbilical cord blood for treating spinal cord injuries in dogs. *J Vet Med Sci* 74 (12): 1617–30.

Sandoval, J. A., C. López, and J. U. Carmona. 2013. Therapies intended for joint regeneration in the horse. *Arch Med Vet* 45: 229–36.

Sartore, S., M. Lenzi, A. Angelini, et al. 2005. Amniotic mesenchymal cells autotransplanted in a porcine model of cardiac ischemia do not differentiate to cardiogenic phenotypes. *Eur J Cardiothorac Surg* 28 (5): 677–84. doi: 10.1016/j.ejcts.2005.07.019.

Schellenberg, A., T. Stiehl, P. Horn, et al. 2012. Population dynamics of mesenchymal stromal cells during culture expansion. *Cytotherapy* 14 (4): 401–11. doi: 10.3109/14653249.2011.640669.

Schuh, E. M., M. S. Friedman, D. D. Carrade, et al. 2009. Identification of variables that optimize isolation and culture of multipotent mesenchymal stem cells from equine umbilical-cord blood. *Am J Vet Res* 70 (12): 1526–35. doi: 10.2460/ajvr.70.12.1526.

Seo, M. S., Y. H. Jeong, J. R. Park, et al. 2009. Isolation and characterization of canine umbilical cord blood-derived mesenchymal stem cells. *J Vet Sci* 10 (3): 181–7.

Seo, M. S., S. B. Park, and K. S. Kang. 2012. Isolation and characterization of canine Wharton's jelly-derived mesenchymal stem cells. *Cell Transplant* 21 (7): 1493–502. doi: 10.3727/096368912x647207.

Sharma, P., and N. Maffulli. 2006. Biology of tendon injury: Healing, modeling and remodeling. *J Musculoskelet Neuronal Interact* 6 (2): 181–90.

Shaw, S. W., M. P. Blundell, C. Pipino, et al. 2015. Sheep CD34+ amniotic fluid cells have hematopoietic potential and engraft after autologous in utero transplantation. *Stem Cells* 33 (1): 122–32. doi: 10.1002/stem.1839.

Shaw, S. W., S. Bollini, K. A. Nader, et al. 2011. Autologous transplantation of amniotic fluid-derived mesenchymal stem cells into sheep fetuses. *Cell Transplant* 20 (7): 1015–31. doi: 10.3727/096368910x543402.

Siegel, N., M. Rosner, M. Hanneder, A. Freilinger, and M. Hengstschlager. 2008. Human amniotic fluid stem cells: A new perspective. *Amino Acids* 35 (2): 291–3. doi: 10.1007/s00726-007-0593-1.

Singh, J., A. Mann, D. Kumar, J. S. Duhan, and P. S. Yadav. 2013. Cultured buffalo umbilical cord matrix cells exhibit characteristics of multipotent mesenchymal stem cells. *In Vitro Cell Dev Biol Anim* 49 (6): 408–16. doi: 10.1007/s11626-013-9617-1.

Sinowatz, F. 2009. Essentials of Domestic Animal Embryology, In *Essentials of Domestic Animal Embryology*, Hyttel, P., F. Sinowatz, M. Vejlsted, K. Betteridge (eds.), Saunders Ltd, PA: Philadelphia.

Smith, J. R., R. Pochampally, A. Perry, S. C. Hsu, and D. J. Prockop. 2004. Isolation of a highly clonogenic and multipotential subfraction of adult stem cells from bone marrow stroma. *Stem Cells* 22 (5): 823–31. doi: 10.1634/stemcells.22-5-823.

Smith, R. K., M. Korda, G. W. Blunn, and A. E. Goodship. 2003. Isolation and implantation of autologous equine mesenchymal stem cells from bone marrow into the superficial digital flexor tendon as a potential novel treatment. *Equine Vet J* 35 (1): 99–102.

Soncini, M., E. Vertua, L. Gibelli, et al. 2007. Isolation and characterization of mesenchymal cells from human fetal membranes. *J Tissue Eng Regen Med* 1 (4): 296–305. doi: 10.1002/term.40.

Sreekumar, T. R., M. M. Ansari, V. Chandra, and G. T. Sharma. 2014. Isolation and characterization of buffalo Wharton's jelly derived mesenchymal stem cells. *J Stem Cell Res Ther* 4: 207. doi: 10.4172/2157-7633.1000207.

Stadler, G., S. Hennerbichler, A. Lindenmair, et al. 2008. Phenotypic shift of human amniotic epithelial cells in culture is associated with reduced osteogenic differentiation in vitro. *Cytotherapy* 10 (7): 743–52. doi: 10.1080/14653240802345804.

Subrahmanyam, M. 1995. Amniotic membrane as a cover for microskin grafts. *Br J Plast Surg* 48 (7): 477–8.

Switonski, M. 2014. Dog as a model in studies on human hereditary diseases and their gene therapy. *Reprod Biol* 14 (1): 44–50. doi: 10.1016/j.repbio.2013.12.007.

Sykes, M. 2009. Hematopoietic cell transplantation for tolerance induction: Animal models to clinical trials. *Transplantation* 87 (3): 309–16. doi: 10.1097/TP.0b013e31819535c2.

Terada, S., K. Matsuura, S. Enosawa, et al. 2000. Inducing proliferation of human amniotic epithelial (HAE) cells for cell therapy. *Cell Transplant* 9 (5): 701–4.

Therapies, ART Advanced Regenerative. Available at http://www.art4dvm.com

Thomas, E. D. 1999. A history of haemopoietic cell transplantation. *Br J Haematol* 105 (2): 330–9.

Thomas, E. D., and R. Storb. 1999. The development of the scientific foundation of hematopoietic cell transplantation based on animal and human studies. In *Hematopoietic Cell Transplantation*, K. G. Blume, S. J. Forman, and F. R. Appelbaum (eds.), 2nd ed., pp. 1–11.

Tondreau, T., N. Meuleman, A. Delforge, et al. 2005. Mesenchymal stem cells derived from CD133-positive cells in mobilized peripheral blood and cord blood: Proliferation, Oct4 expression, and plasticity. *Stem Cells* 23 (8): 1105–12. doi: 10.1634/stemcells.2004-0330.

Toupadakis, C. A., A. Wong, D. C. Genetos, et al. 2010. Comparison of the osteogenic potential of equine mesenchymal stem cells from bone marrow, adipose tissue, umbilical cord blood, and umbilical cord tissue. *Am J Vet Res* 71 (10): 1237–45. doi: 10.2460/ajvr.71.10.1237.

Trelford, J. D., and M. Trelford-Sauder. 1979. The amnion in surgery, past and present. *Am J Obstet Gynecol* 134 (7): 833–45.

Trelford-Sauder, M., E. J. Dawe, and J. D. Trelford. 1978. Use of allograft amniotic membrane for control of intra-abdominal adhesions. *J Med* 9 (4): 273–84.

Trobridge, G. D., and H. P. Kiem. 2010. Large animal models of hematopoietic stem cell gene therapy. *Gene Ther* 17 (8): 939–48. doi: 10.1038/gt.2010.47.

Troyer, D. L., and M. L. Weiss. 2008. Wharton's jelly-derived cells are a primitive stromal cell population. *Stem Cells* 26 (3): 591–9. doi: 10.1634/stemcells.2007-0439.

Tsai, M. S., J. L. Lee, Y. J. Chang, and S. M. Hwang. 2004. Isolation of human multipotent mesenchymal stem cells from second-trimester amniotic fluid using a novel two-stage culture protocol. *Hum Reprod* 19 (6): 1450–6. doi: 10.1093/humrep/deh279.

Turner, C. G., J. D. Klein, S. A. Steigman, et al. 2011. Preclinical regulatory validation of an engineered diaphragmatic tendon made with amniotic mesenchymal stem cells. *J Pediatr Surg* 46 (1): 57–61. doi: 10.1016/j.jpedsurg.2010.09.063.

UC Davis. Available at http://www.vetmed.ucdavis.edu/vmth/regen_med

Van Loon, V. J., C. J. Scheffer, H. J. Genn, A. C. Hoogendoorn, and J. W. Greve. 2014. Clinical follow-up of horses treated with allogeneic equine mesenchymal stem cells derived from umbilical cord blood for different tendon and ligament disorders. *Vet Q* 34 (2): 92–7. doi: 10.1080/01652176.2014.949390.

VetBiologics. Available at www.stemlogix.com

Vet-Stem. Available at https://www.vet-stem.com

Vidal, M. A., S. O. Robinson, M. J. Lopez, et al. 2008. Comparison of chondrogenic potential in equine mesenchymal stromal cells derived from adipose tissue and bone marrow. *Vet Surg* 37 (8): 713–24. doi: 10.1111/j.1532-950X.2008.00462.x.

Vidal, M. A., N. J. Walker, E. Napoli, and D. L. Borjesson. 2012. Evaluation of senescence in mesenchymal stem cells isolated from equine bone marrow, adipose tissue, and umbilical cord tissue. *Stem Cells Dev* 21 (2): 273–83. doi: 10.1089/scd.2010.0589.

Vidane, A. S., A. F. Souza, R. V. Sampaio, et al. 2014. Cat amniotic membrane multipotent cells are nontumorigenic and are safe for use in cell transplantation. *Stem Cells Cloning* 7: 71–8. doi: 10.2147/sccaa.s67790.

Vilar, J. M., M. Batista, M. Morales, et al. 2014. Assessment of the effect of intraarticular injection of autologous adipose-derived mesenchymal stem cells in osteoarthritic dogs using a double blinded force platform analysis. *BMC Vet Res* 10: 143. doi: 10.1186/1746-6148-10-143.

Violini, S., C. Gorni, L. F. Pisani, et al. 2012. Isolation and differentiation potential of an equine amnion-derived stromal cell line. *Cytotechnology* 64 (1): 1–7. doi: 10.1007/s10616-011-9398-x.

Volk, S. W., D. L. Diefenderfer, S. A. Christopher, M. E. Haskins, and P. S. Leboy. 2005. Effects of osteogenic inducers on cultures of canine mesenchymal stem cells. *Am J Vet Res* 66 (10): 1729–37.

Volk, S. W., and C. Theoret. 2013. Translating stem cell therapies: The role of companion animals in regenerative medicine. *Wound Repair Regen* 21 (3): 382–94. doi: 10.1111/wrr.12044.

Volk, S. W., Y. Wang, and K. D. Hankenson. 2012. Effects of donor characteristics and ex vivo expansion on canine mesenchymal stem cell properties: Implications for MSC-based therapies. *Cell Transplant* 21 (10): 2189–200. doi: 10.3727/096368912x636821.

Wagner, J. L., and R. Storb. 1996. Preclinical large animal models for hematopoietic stem cell transplantation. *Curr Opin Hematol* 3 (6): 410–15.

Wang, J. H. 2006. Mechanobiology of tendon. *J Biomech* 39 (9): 1563–82. doi: 10.1016/j.jbiomech.2005.05.011.

Ward, D. J., and J. P. Bennett. 1984. The long-term results of the use of human amnion in the treatment of leg ulcers. *Br J Plast Surg* 37 (2): 191–3.

Ward, D. J., J. P. Bennett, H. Burgos, and J. Fabre. 1989. The healing of chronic venous leg ulcers with prepared human amnion. *Br J Plast Surg* 42 (4): 463–7.

Watts, A. E., A. E. Yeager, O. V. Kopyov, and A. J. Nixon. 2011. Fetal derived embryonic-like stem cells improve healing in a large animal flexor tendonitis model. *Stem Cell Res Ther* 2 (1): 4. doi: 10.1186/scrt45.

Webb, T. L., J. M. Quimby, and S. W. Dow. 2012. In vitro comparison of feline bone marrow-derived and adipose tissue-derived mesenchymal stem cells. *J Feline Med Surg* 14 (2): 165–8. doi: 10.1177/1098612x11429224.

Weber, B., M. Y. Emmert, L. Behr, et al. 2012. Prenatally engineered autologous amniotic fluid stem cell-based heart valves in the fetal circulation. *Biomaterials* 33 (16): 4031–43. doi: 10.1016/j.biomaterials.2011.11.087.

Weber, B., D. Kehl, U. Bleul, et al. 2013. In vitro fabrication of autologous living tissue-engineered vascular grafts based on prenatally harvested ovine amniotic fluid-derived stem cells. *J Tissue Eng Regen Med.* doi: 10.1002/term.1781.

Weiss, M. L., and D. L. Deryl. 2006. Stem cells in the umbilical cord. *Stem Cell Rev* 2 (2): 155–62. doi: 10.1007/s12015-006-0022-y.

Weiss, M. L., K. E. Mitchell, J. E. Hix, et al. 2003. Transplantation of porcine umbilical cord matrix cells into the rat brain. *Exp Neurol* 182 (2): 288–99.

Wichayacoop, T., P. Briksawan, P. Tuntivanich, and S. Yibchok-Anun. 2009. Anti-inflammatory effects of topical supernatant from human amniotic membrane cell culture on canine deep corneal ulcer after human amniotic membrane transplantation. *Vet Ophthalmol* 12 (1): 28–35. doi: 10.1111/j.1463-5224.2009.00670.x.

Wilke, M. M., D. V. Nydam, and A. J. Nixon. 2007. Enhanced early chondrogenesis in articular defects following arthroscopic mesenchymal stem cell implantation in an equine model. *J Orthop Res* 25 (7): 913–25. doi: 10.1002/jor.20382.

Williams, I. F., A. Heaton, and K. G. McCullagh. 1980. Cell morphology and collagen types in equine tendon scar. *Res Vet Sci* 28 (3): 302–10.

Wolbank, S., A. Peterbauer, M. Fahrner, et al. 2007. Dose-dependent immunomodulatory effect of human stem cells from amniotic membrane: A comparison with human mesenchymal stem cells from adipose tissue. *Tissue Eng* 13 (6): 1173–83. doi: 10.1089/ten.2006.0313.

Wu, Y., L. Chen, P. G. Scott, and E. E. Tredget. 2007. Mesenchymal stem cells enhance wound healing through differentiation and angiogenesis. *Stem Cells* 25 (10): 2648–59. doi: 10.1634/stemcells.2007-0226.

Xiang, Z., W. Hu, Q. Kong, H. Zhou, and X. Zhang. 2006. Preliminary study of mesenchymal stem cells-seeded type I collagen-glycosaminoglycan matrices for cartilage repair. *Zhongguo Xiu Fu Chong Jian Wai Ke Za Zhi* 20 (2): 148–54.

Yadav, P. S., A. Mann, V. Singh, et al. 2011. Expression of pluripotency genes in buffalo (Bubalus bubalis) amniotic fluid cells. *Reprod Domest Anim* 46 (4): 705–11. doi: 10.1111/j.1439-0531.2010.01733.x.

Yang, S. E., C. W. Ha, M. Jung, et al. 2004. Mesenchymal stem/progenitor cells developed in cultures from UC blood. *Cytotherapy* 6 (5): 476–86. doi: 10.1080/14653240410005041.

You, Q., L. Cai, J. Zheng, et al. 2008. Isolation of human mesenchymal stem cells from third-trimester amniotic fluid. *Int J Gynaecol Obstet* 103 (2): 149–52. doi: 10.1016/j.ijgo.2008.06.012.

You, Q., X. Tong, Y. Guan, et al. 2009. The biological characteristics of human third trimester amniotic fluid stem cells. *J Int Med Res* 37 (1): 105–12.

Young, R. L., J. Cota, G. Zund, B. A. Mason, and J. M. Wheeler. 1991. The use of an amniotic membrane graft to prevent postoperative adhesions. *Fertil Steril* 55 (3): 624–8.

Yu, S., N. Tajiri, N. Franzese, et al. 2013. Stem cell-like dog placenta cells afford neuroprotection against ischemic stroke model via heat shock protein upregulation. *PLoS One* 8 (9): e76329. doi: 10.1371/journal.pone.0076329.

Zarrabi, M., S. H. Mousavi, S. Abroun, and B. Sadeghi. 2014. Potential uses for cord blood mesenchymal stem cells. *Cell J* 15 (4): 274–81.

Zheng, Y. M., Y. H. Dang, Y. P. Xu, W. J. Sai, and Z. X. An. 2011a. Differentiation of AFS cells derived from the EGFP gene transgenic porcine fetuses. *Cell Biol Int* 35 (8): 835–9. doi: 10.1042/cbi20100317.

Zheng, Y. M., X. E. Zhao, and Z. X. An. 2010. Neurogenic differentiation of EGFP gene transfected amniotic fluid-derived stem cells from pigs at intermediate and late gestational ages. *Reprod Domest Anim* 45 (5): e78–82. doi: 10.1111/j.1439-0531.2009.01526.x.

Applications of Placenta-Derived Cells in Veterinary Medicine

Zheng, Y. M., H. Y. Zhao, X. E. Zhao, et al. 2009. Development of cloned embryos from porcine neural stem cells and amniotic fluid-derived stem cells transfected with enhanced green fluorescence protein gene. *Reproduction* 137 (5): 793–801. doi: 10.1530/rep-08-0469.

Zheng, Y. M., Y. L. Zheng, X. Y. He, et al. 2011b. Multipotent differentiation of the EGFP gene transgenic stem cells derived from amniotic fluid of goat at terminal gestational age. *Cell Biol Int* 35 (12): 1243–6. doi: 10.1042/cbi20110128.

Zhu, X., X. Wang, G. Cao, et al. 2013. Stem cell properties and neural differentiation of sheep amniotic epithelial cells. *Neural Regen Res* 8 (13): 1210–19. doi: 10.3969/j.issn.1673-5374.2013.13.006.

Zucconi, E., N. M. Vieira, D. F. Bueno, et al. 2010. Mesenchymal stem cells derived from canine umbilical cord vein—A novel source for cell therapy studies. *Stem Cells Dev* 19 (3): 395–402. doi: 10.1089/scd.2008.0314.

Zvetkova, I., A. Apedaile, B. Ramsahoye, et al. 2005. Global hypomethylation of the genome in XX embryonic stem cells. *Nat Genet* 37 (11): 1274–9. doi: 10.1038/ng1663.

Index

A

Activin A, 152
Adherence prevention, 56
Adnexa embryology, comparative placenta, 223–226
AFP, *see* alpha-fetoprotein
Albumin, 152
Allantois, umbilical cord, 9
Alpha-fetoprotein (AFP), 152
AM, *see* amniotic membrane
Amnion cells
    cardiovascular disease, 131
    cartilage disorders, 133
    hepatic disease, 132
    inflammation, 133–137
    MSC sources, 21
    muscular disorders, 133
    neurological disorders, 131
    overview, 130
    pancreatic disease, 132–133
    pulmonary disease, 132
    stroke, 134–137
    use in human disorders, 130–133
Amnion epithelial cells, liver disorders; *see also* hAECs
    background, 144–145
    cell-based therapy, 145–148
    clinical program, 154–155
    hepatocyte transplantation, 146–148
    overview, 143–144, 148–151
    preclinical models, 152–153
    stem-cell–based therapies, 148
    summary, 155–156
    in vitro hepatic differentiation, 151–152
Amniotic epithelial cells
    cell characterization in animals, 226–235
    lung injury, 183–185
    regenerative veterinary application, 227–228
Amniotic fluid and amniotic fluid stem cells
    bronchopulmonary dysplasia-like injury, 185–186
    cell characterization in animals, 235–238
    lung injury, 185–186
    MSC sources, 21
    regenerative veterinary application, 230–233, 235–238
Amniotic membrane (AM)
    boiled and dehydrated forms, 57–58
    cell characterization in animals, 226
    cell populations, 70–78

    cells comprising, 72
    cells isolated from, 72–78
    chronic wounds, 57–59
    cryopreserved form, 59
    fetal membrane, 6
    fresh form, 58
    hAECs, 72–75
    histology of tissue, 70–71
    HLA molecule expression, 93, 96
    immunogenicity, 92–99
    immunomodulatory properties, 92, 99–111
    immunosuppression mechanisms, 92, 111–116
    irradiated form, 59
    mesenchymal stromal cells, 75–76
    skin re-epithelialization, 59–60
    source of, 57
    T cell immunological tolerance, 100–101
    trophic functions, AM-derived cells, 76–78
Amniotic MSCs, 228–229, 238
Anatomy changes during gestation, 9–10
Anecdotal reports, 220
Angiogenesis, 15, 52–53
Animal models; *see also specific issue*; veterinary medicine applications
    cardiovascular disease, 131
    CSMC therapeutic potential, 25
    liver disease, 152–153
    liver fibrosis, 190
    MSCs, 20–24
    mural trophectoderm, 3
    neurological disorders, 131
    in vivo studies, 99
Antigen-presenting cells (APCs), 92, 105–107
Antigen processing, 41
Antiproliferative effect; *see also* proliferative phase
    hAMSCs, 101–102, 235
    hUCMSCs, 98, 102
    Wharton's jelly, 101–102, 116
APCs, *see* antigen-presenting cells (APCs)
Arterial malperfusion, maternal, 18
Arthritis, collagen-induced, 103
Autoimmune disorders
    autoimmune myocarditis, 170–171
    autoimmunity, 162–163
    clinical trials, 171–172
    current treatment, 165–166
    disease induction mechanisms, 164–165
    environmental factors, 163–164
    genetic predisposition, 163
    inflammatory bowel disease, 166–167

285

286 Index

multiple sclerosis, 168–169
overview, 161–162
pathogenesis, 163–164
preclinical models, cell use, 166–171
published clinical trials, 172
rheumatoid arthritis, 169–170
summary, 172
SLE, 167–168
Autoimmune myocarditis, 170–171
Autoreactivity, 162; *see also* autoimmune
disorders

B

Banks, *see* strategy, regulations, and GMP
B cells
costimulatory family molecule
expression, 96–97
immune system, 40–41
MSC immunomodulatory features, 104
SLE, 167
B7 costimulatory family molecule
expression, 96–97
Bedsores, 59
Blastocyst chimera experiments, 3
Blastomeres, placenta development, 2–3
Boiled amniotic membrane, 57–58
Bone defects, animals, 254–255, 261
Brain injury, 77, 133
Branched-chain keto-acid dehydrogenase
(BCKDH) enzyme complex, 153
Bromodeoxyuridine (BrdU) label, 26–27
Burns and wounds, 56, 57, 111

C

Cancer, 84
Cardiovascular disease, 131, 170–171;
see also stroke
Cartilage defects and disorders
amnion cells, 133
amniotic membrane, 226
regenerative veterinary application,
256, 261
Case reports, 220
Cavitation, 223
CD40/CD40L, 97–98
Cell-based regenerative veterinary medicine,
218–221; *see also* veterinary medicine
applications
Cell-based therapy, liver disorders, 145–148
Cell characterization in animals
amniotic-derived tissue and cells, 226–228
overview, 226
umbilical cord blood MSCs, 238–246
umbilical cord-derived cells, 238
umbilical cord matrix MSCs, 246–247

Cell populations
amniotic membrane, 70–78
chorionic membrane, 78–80
overview, 69–70
summary, 84
Wharton's jelly, 81–84
Cell replacement therapy, 77
Central intolerance, 164
Chemical exposures, 164
Chorangioma, 18
Chorioamnionitis, 17
Chorion, MSC sources, 21–22
Chorionic membrane
cell populations, 78–80
fetal membrane, 6–8
histology of tissue, 79
mesenchymal stromal cells, 79–80
Chorionic placenta, 26–27
Chorionic villi
immunogenicity, 92–99
immunomodulatory properties, 92, 99–111
immunosuppression mechanisms, 92, 111–116
Chorionic villi MSCs (CMSCs/CVMSCs)
antigen-presenting cells, 106–108
antiproliferative effect, 101–102
engraftment, 103–104
hematopoietic marker expression, 98
HLA molecule expression, 93, 96
HLA-G, 114
IDO, 114
inflammatory cytokines, 115
IL-10, 111–112
overview, 24–25
PGE2, 113
role in preeclampsia, 29–30
sources, 22
T cell immunological tolerance, 100–101
TGF-β, 112
Chorionic villous cells, 24–25
Chorion laevae, 78
Chronic intervillositis, 17
CIA, *see* collagen-induced arthritis
Circumvallate placenta, 17
Cirrhosis, *see* liver disorders
Clinical trials
autoimmune disorders, 171–172
lung injury, 188–189
CMSCs, *see* chorionic villi MSCs
Collagen-induced arthritis (CIA), 103, 169–170
Collagen synthesis, 52
Companion and farm animal models, 220
Comparative placenta adnexa embryology, 223–226
Complications during pregnancy, predonation
phase, 209
Concavillin A, 101
Conjunctival epithelium reconstruction, 226
Connecting stalk, 4

Index

Constraints, 199
Contamination, 24–25
Corneal reconstruction, 226; *see also* ophthalmology
Costimulatory family molecule expression, 96–97, 161
COX activities, 113
Crohn's disease, 163, 171; *see also* inflammatory bowel disease
Cryopreservation
    amniotic epithelial cells, 234
    amniotic membrane, 59
    hAECs *vs.* hepatocytes, 155
Current good manufacturing practice (cGMP), 155
CVMSC, *see* chorionic villi MSCs
Cytomegalovirus (CVM) recall antigen, 101
Cytotrophoblasts, 9

D

DCs, *see* dendtritic cells
Decidua and decidualization
    fetal membrane origins, 6
    human development within, 4
    overview, 3–4
    placenta and immunity, 43
Decidua basalis
    chorionic membrane, 78
    CMSCs, 24
    contamination, 24–25
    MSC niche, 27–28
    MSC sources, 23
    stem cell populations, 25–26
    stem cell source, 20
Decidual mesenchymal stem cells (DMSCs), 25–26, 28
Decidua parietalis
    MSC niche, 27
    sources, 20, 22–23
Dehydrated amniotic membrane, 57–58; *see also* amniotic membrane
Dendtritic cells (DCs)
    IL-10, 112
    maternal immune system, 42
    MSC immunomodulatory features, 105–107
    PGE2, 113
Dexamethasone, 151
Diabetes
    dehydrated amniotic membrane, 57
    genetic predisposition, 163
    hWJMSCs, 99, 103
    MSCs, potential role, 32
Differentiation
    cell source, 18
    hAECs, 74–75, 151–152
    hAMSCs, 76
    hCMSCs, 80

hWJMSCs, 84
inner cell mass, 223
regenerative applications, 76–77
regenerative veterinary application, 223
stem cell classification, 18–19
Disease induction mechanisms, 164–165
Donors and donation; *see also* strategy, regulations, and GMP
    asking and communication, 202
    execution, 203–204
    final release, 213
    informed consent, 209–210
    legal requirements, 204–206
    predonation phase, 206–210
    pregnancy complications, 209
    procedure, 210
    production, 212
    quality control, 213
    questionnaire, 208–209
    selection, 207
    tissue donation, 210–211
    transport and reception, 211
    viral testing, 211

E

Early postimplantation placenta, 4
Embryonic stem cells (ESCs), 19
Endometrial cell replenishment, 226
Endovascular trophoblasts, 2
Engraftment, long-term, 99
Environmental factors, 163–164
Epidermal growth factor (EGF), 51, 53
Epithelialization, 53, 61–62; *see also* amnion epithelial cells
Estrogen, 164
Execution, GMP environment, 203–204
Exocoelomic membrane, 5
Experimental autoimmune encephalomyelitis (EAE), 103, 169
Experimental autoimmune myocarditis (EAM), 171
Extracellular matrix (ECM), 52, 182
Extraembryonic endoderm, 5
Extraembryonic mesoderm, 6
Extraembryonic yolk sac, 6
Extravillous trophoblasts (EVT)
    placenta development, 2
    placenta structure and function, 15–17
    spiral aterioles, 28
    trophoblast stem cells, 19
    villous placenta development, 7

F

Farm and companion animal models, 220
Fas ligands
    amnion epithelial cells, 151

288

Index

placenta and immunity, 44
TNF/TNFR superfamily molecule
    expression, 97
Fetal growth restriction (FGR)
    HLAs, 44
    MSCs, potential role, 28, 32
    oxygen levels, 30
    placenta pathology, 18
Fetal immune system, programming, 44–45
Fetal-maternal interface, 15
Fetal-maternal tolerance
    fetal immune programming, 44–45
    immune system, 40–45
    maternal during pregnancy, 41–43
    overview, 39–40
    placenta and immunity, 43–44
    summary, 45
Fetal membrane
    amniotic membrane, 6–7
    chorionic membrane, 6–8
    membrane fusion, 9
    origins, 4
    overview, 6
Fetal thrombotic vasculopathy, 17
Fibroblast migration, 52
Fibrotic diseases, hAECs, 77–78
Final release, GMP environment, 213
Financial matters, 201–202
Folding modality, 223
Foot ulcers, see ulcers
Fresh amniotic membrane, 58; see also
    amniotic membrane
Function, placenta, 14–17
Future directions and perspectives
    neurological disorders, 136–137
    veterinary medicine, 267–268
FZD-9 immunoreactivity, 28

G

Gamma-irradiated AM, 59
Genetic predisposition, autoimmune disorders, 163
Genomic profile, AM, 226
Germ cells, 18
Gestation, 9–10
GMP environment, see strategy, regulations,
    and GMP
Granulation tissue, 53, 55
Granulocytes, inflammation, 51
Graves' disease, 163

H

HaCaT cells, 60, 62
hAECs (human amniotic epithelial cells);
    see also amnion epithelial cells
    amnion cells, 130

antigen-presenting cells, 106–109
antiproliferative effect, 101–102
applications, 77–78
bronchopulmonary dysplasia-like injury,
    184–185
critical parameters, 150
cryopreservation, 155
engraftment, 103–104
fibrotic disorders, 183–185
hematopoietic marker expression, 98
HLA molecule expression, 93, 96
HLA-G, 114
inflammatory cytokines, 115
multiple sclerosis, 168–169
NK cells, 105
neurological disorders, 131
overview, 72–75
pancreatic disease, 131, 132–133
proliferation suppression, 235
PGE2, 113
pulmonary disease, 131, 132
rheumatoid arthritis, 169–170
T cell immunological tolerance, 100–101
TGF-β, 112
in vitro studies, 98
in vivo studies, 99
Half-life, transplanted cells, 154–155
hAMSCs (human amniotic membrane
        mesenchymal stromal cells)
    amnion cells, 130
    antigen-presenting cells, 106–109
    antiproliferative effect, 101–102, 235
    engraftment, 103–104
    hematopoietic marker expression, 98
    HGF, 112–113
    HLA molecule expression, 93, 96
    HLA-G, 114
    IDO, 114
    inflammatory bowel disease, 167
    inflammatory cytokines, 115
    IL-10, 111
    multiple sclerosis, 169
    NK cells, 105
    overview, 72, 75–76
    proliferation suppression, 101–102, 235
    PGE2, 113
    T cell immunological tolerance, 100–101
    TGF-β, 112
    in vitro studies, 98
    in vivo studies, 99
hCMSCs (human chorionic mesenchymal
        stem/stromal cells), 79–80; see also
        mesenchymal stem/stromal cells
Heart disorders, see cardiovascular disease
Heat shock protein 27 (Hsp27), 136
Hematopoietic marker expression, 98, 236
Hemostasis, wound healing, 51

Index

289

Hepatic disease, 131, 132; *see also* liver disorders
Hepatic gene expression, 152
Hepatocyte growth factor (HGF), 112–113
Hepatocyte transplantation (HTx), 146–148, 190–191
Hind limb injury, 25, 111
Histology of tissue, *see* tissue
HLAs, *see* human leukocyte antigens
Hormones
  autoimmune triggers, 164
  steroids, exposure, 151
hPMSCs (human chorionic villous/placenta mesenchymal stem/stromal cells), 79–80; *see also* chorionic villi MSCs
Hsp27, *see* heat shock protein 27
HTx, *see* hepatocyte transplantation
hUCMSCs (human umbilical cord MSCs); *see also* umbilical cord
  antigen-presenting cells, 106–109
  antiproliferation effect, 98, 102
  IDO, 114
  inflammatory cytokines, 115
  NK cells, 105
  PGE2, 113
  published clinical trials, 172
  SLE, 167–168
Human leukocyte antigens (HLAs)
  amnion epithelial cells, 150–151
  inflammatory cytokines, 115
  limited expression, 161
  MSC immunogenicity, 92
  MSC immunomodulatory features, 93, 96, 113–114
  placenta and immunity, 44
Human placenta, structure and development
  amniotic membrane, 6–7
  anatomy changes during gestation, 9–10
  chorionic membrane, 7–8
  connecting stalk origins, 4
  decidua, 3–4
  early postimplantation placenta, 4
  fetal membrane, 4, 6–9
  lineage derivation, 2–3
  membrane fusion, 9
  overview, 1–2
  preimplantation development, 2–3
  summary, 10
  umbilical cord, 9
  villous placenta development, 4–6
  Wharton's jelly, 4, 9
hUVECs (human umbilical vein endothelial cells), 29, 30
Hydrocortisone, 151
Hypertrophic decidual vasculopathy, 17–18; *see also* decidua and decidualization
Hypoblast layer, 5

I

IBD, *see* inflammatory bowel disease
IDO (indoleamine 2,3-dioxygenase), 44, 114
IGF-1, *see* insulin-like growth factor 1
"Immune-privileged" status, 150–151
Immune system
  fetal immune programming, 44–45
  hWJMSCs, 93
  maternal, during pregnancy, 41–43
  overview, 40–41
  placenta and immunity, 43–44
Immunological tolerance, T cells, 100–104
Immunomodulatory features, MSC
  APCs, effects on, 105–110
  B cells, effects on, 104
  B7 costimulatory family molecule expression, 96–97
  DCs, effects on, 105–107
  hematopoietic marker expression, 98
  HGF, 112–113
  HLA-G, 113–114
  HLA molecule expression, 93, 96
  IDO, 114
  immunogenicity, 92–99
  immunomodulatory properties, 99–111
  immunosuppression mechanisms, 111–116
  inflammatory cytokines, 115
  IL-10, 111–112
  macrophages, effects on, 105–107
  neutrophils, effects on, 110–111
  NK cells, effects on, 104–105
  overview, 92
  PGE2, 113
  secreted factors, 115–116
  T cells, effects on, 100–104
  TGF-β, 112–113
  TNF/TNFR superfamily molecule expression, 97–98
  in vitro studies, 98
  in vivo studies, 99
"Implantation window," 4
Infarctions, placenta pathology, 18; *see also* myocardial infarction, animals
Infectious agents, autoimmunity, 163–164
Inflammation
  amnion cells, 130, 133–137
  amniotic membrane, 226
  B cells, 104
  fetal immune programming, 44
  future directions, 136–137
  IDO, 114
  IL-10, 111
  liver, 190
  long-term damage, 135
  placenta pathology, 17, 32
  wound healing, 51–52, 54

Inflammatory and fibrotic disorders
amniotic epithelial cells, 183–185
amniotic fluid stem cells, 185–186
animal models, 190
clinical trials, 188–189
hepatocyte transplantation studies, 190–191
kidney fibrosis, 189
liver fibrosis, 189–193
lung injury, 182–189
mechanisms of actions, proposed, 187–188
mesenchymal stromal cells, 185–186
overview, 181–182
summary, 194
term placenta cells, 186
treatment studies, 187, 191–193
Inflammatory bowel disease (IBD), 166–167;
see also Crohn's disease
Inflammatory cytokines
bronchopulmonary dysplasia-like injury,
184–185
MSC immunomodulatory features, 115
rheumatoid arthritis, 169
Informed consent, 209–210
Insulin-like growth factor 1 (IGF-1), 51
Integration, GMP environment, 202–203
Interleukin 10 (IL-10)
inflammatory bowel disease, 167
MSC immunomodulatory features, 111–112
rheumatoid arthritis, 169–170
Intervillositis, 17
Intervillous spaces, 5
In vitro hepatic differentiation, 151–152
In vitro studies, 98
In vivo studies, 99
Irradiated amniotic membrane, 59; *see also*
amniotic membrane
Issues, GMP environment, 198–199

J

Joint preclinical safety study, 257

K

Keratinocyte migration acceleration, 62
Kidney fibrosis, 189

L

Label-retaining cells, 26–27
Lacunae, 4–5
Legal matters, *see* strategy, regulations, and GMP
Leg ulcers, *see* ulcers
Lineage derivation, human placenta, 2–3
Liver disorders
animal models, 190
background, 144–145

cell-based therapy, 145–148
clinical program, 154–155
hepatic disease, 131, 132
hepatocyte transplantation, 146–148, 190–191
inflammation, 190
overview, 143–144, 148–151
placental cells, 191–193
preclinical models, 152–153
stem-cell–based therapies, 148
summary, 155–156
in vitro hepatic differentiation, 151–152
Liver fibrosis
animal models, 190
hepatocyte transplantation studies, 190–191
overview, 189–190
treatment studies, 191–193
Lung injury and inflammation
amnion cells, 131, 132
amniotic epithelial cells, 183–185
amniotic fluid stem cells, 185–186
clinical trials, 188–189
CSMC therapeutic potential, 25
hAECs/hAMSCs, 110
mechanisms of actions, proposed, 187–188
mesenchymal stromal cells, 185–186
neutrophils, 111
overview, 182–183
term placenta cells, 186
treatment, 187
Lymphocyte proliferation, 112; *see also*
antiproliferative effect; proliferative
phase

M

Macrophages
angiogenesis, 53
inflammation, 52
IL-10, 112
maternal immune system, 43
MSC immunomodulatory features, 105–107
Maori people, 10
Maple syrup urine disease (MSUD), 144, 153–154
Mastocytes, chronic healing, 55
Maternal arterial malperfusion, 18
Maternal blood flow, 10
Maternal immune system
dysregulation of, 44–45
during pregnancy, 41–43
Medawar, P.B., 40
Membrane fusion, 9
Memorizing function, 43
Mesenchymal stem/stromal cells (MSCs)
amniotic membrane, 75–76
APCs, effects on, 105–110
autoimmune disorders, 165
B cells, effects on, 104

Index

B7 costimulatory family molecule
  expression, 96–97
chorionic membrane, 79–80
chorionic placenta, 26–27
chorionic villous type, 24–25
CMSC role in, 29–30
DCs, effects on, 105–107
Decidua basalis, 25–28
definition, lack of, 24
DMSC role in, 30–31
hematopoietic marker expression, 98
HGF, 112–113
HLA-G, 113–114
HLA molecule expression, 93, 96
IDO, 114
immunogenicity, 92–99
immunomodulatory properties, 92, 99–111
immunosuppression mechanisms, 92, 111–116
inflammatory cytokines, 115
IL-10, 111–112
lung injury, 185–186
macrophages, effects on, 105–107
mesenchymal stem cells, 20, 24–26
neutrophils, effects on, 110–111
niche localization in the placenta, 26–28
NK cells, effects on, 104–105
other stem cell types, 20
overview, 14, 28, 92
pathogenesis, 28–29
PGE2, 113
placental pathologies, 17–18
placental structure and function, 14–17
as potential treatment, 31–32
preeclampsia, 28–32
pregnancy pathologies role, 28–32
secreted factors, 115–116
sources, 21–23
stem cell niche, 26
stem cell populations, 19–26
stem cells, 18–26
T cells, effects on, 100–104
TGF-β, 112–113
TNF/TNFR superfamily molecule
  expression, 97–98
trophoblast stem cells, 19–20
umbilical cord blood, 238–246
umbilical cord matrix, 246–247
veterinary applications, 219
in vitro studies, 98
in vivo studies, 99
Wharton's jelly, 81–84
Metabolic liver disorders, *see* liver disorders
MiRs (microRNAs), 30
Miscarriage, 44
MOG (myelin oligodendrocyte glycoprotein), 168
Monocytes, 43
Mononuclear cytotrophoblasts, 9

Mononuclear villous cytotrophoblast cells, 15
Morula stage, 223
Multinucleated syncytiotrophoblasts, 6
Multiple sclerosis (MS), 168–169
Murine blastocyst chimera experiments, 3
Muscular disorders, 133
Myocardial infarction, animals, 249–250, 260
Myocardial injury, *see* cardiovascular disease

N

Nanog
  AFMSCs, 237
  hAECs, 73, 150
  placenta development, 3
  regenerative veterinary application, 226, 234
Natural killer (NK) cells
  fetal immune programming, 44
  HLA-G, 113
  IDO, 114
  immune system, 41
  immunomodulatory properties, 104–105
  inflammatory cytokines, 115
  IL-10, 112
  maternal immune system, 43
  placenta and immunity, 43–44
  PGE2, 113
Nerve growth factor, 98
Neurological disorders
  amnion cells, 130–133
  cardiovascular disease, 131
  cartilage disorders, 133
  future directions, 136–137
  hepatic disease, 132
  inflammation, 133–137
  muscular disorders, 133
  neurological disorders, 131
  overview, 129–130
  pancreatic disease, 132–133
  pulmonary disease, 132
  STEPS guidelines, 137
  stroke, 134–137
  summary, 136–137
  use in human disorders, 130–133
Neutrophils, 110–111
New Zealand Maori people, 10
Niche, MSC localization, 26–28

O

Oct4 expression, 3
Onfaloceles covering, 56
Ophthalmology
  amniotic membrane, 56–57
  placenta cell-based regenerative veterinary
    medicine, 267
  regenerative veterinary applications, 265–266

292 Index

Oral bone defects, animals, 255–256
Ornithine transcarbamylase (OTC), 145
Oxygen levels, 10, 30

P

Pancreatic disease, 132–133
Peripheral blood mononuclear cells (PBMCs)
  HGF, 113
  HLA-G, 113
  IDO, 114
  IL-10, 111–112
  PGE2, 113
  rheumatoid arthritis, 169–170
  T cell immunological tolerance, 101
  TGF-β, 112
  in vitro studies, 98
Peripheral intolerance, 165
Peritoneum substitute, 56
PGE2, *see* prostaglandin E2
Phenylketonuria (PKU), 144, 153–154
Placenta
  cell-based regenerative veterinary
    medicine, 247–267
  cell characteristics, 218
  function, 14–17
  immune system, 43–44
  pathologies, 17–18
  pathologies, MSC role, 17–18
  structure, 14–17
Placenta adnexa embryology, 223–226
Placental supernatants (PSs), 170
Placenta roles, fetal-maternal tolerance
  fetal immune programming, 44–45
  immune system, 40–45
  maternal during pregnancy, 41–43
  overview, 39–40
  placenta and immunity, 43–44
  summary, 45
Planning, GMP environment
  approaches, 200–201
  constraints, 199
  financial matters, 201–202
  integration, 202–203
  overview, 199
  risks, 199–200
  staffing, 202
Platelet-derived growth factor receptor
    (PDGFR), 27, 51
Pluripotent properties
  AFMSCs, 236–237
  amniotic membrane, 72
  hAECs, 73–74, 150–151
  hAMSCs, 75
  hPMSCs, 79
  stem cell-based therapies, 148
  veterinary applications, 219

Polymorphonuclear leukocytes (PMNLs), 51
Preclinical animal models, 220–221, 247–267;
    see also specific disorder
Preclinical human models, 152–153, 166–171;
    see also specific disorder
Predonation phase
  donor questionnaire, 208–209
  informed consent, 209–210
  overview, 206
  pregnancy complications, 209
  selection of donors, 207
Preeclampsia (PE)
  CMSC role in, 29–30
  DMSC role in, 30–31
  HLAs, 44
  memorizing function, 43
  MSCs as potential treatment, 31–32
  overview, 28
  pathogenesis, 28–29
  placenta pathology, 18
Pregnancy complications, predonation phase, 209
Pregnancy pathologies, MSC role
  CMSC role in, 29–30
  DMSC role in, 30–31
  overview, 28
  pathogenesis, 28–29
  preeclampsia, 28–29
Preimplantation development, 2–3
Prenatal congenital diseases, animals,
    257–259, 262
Primary umbilical vesicle, 5
Primary villi, 78
Primitive syncytium, 4
Production, GMP environment, 212
Progenitor cells, veterinary applications,
    221–247, 267
Progesterone, 3
Proliferative phase, 52–53, 54; *see also*
    antiproliferative effect
Prostaglandin E2 (PGE2), 113, 115
PSs, *see* placental supernatants
Published clinical trials, 172; *see also* clinical trials
Pulmonary disease, *see* lung injury and inflammation

Q

Quality control, 213

R

Reception, GMP environment, 211
Reconstructive surgery, 56
Re-epithelialization, skin, 59–60
Regenerative applications; *see also* veterinary
    medicine applications
  amnion cells, human disorders, 130–131
  differentiation, 76–77

Index

Regulations, *see* strategy, regulations, and GMP
Remodeling phase, 53–54
Rheumatoid arthritis (RA), 163, 169–170
Risks, GMP environment, 199–200

S

Secondary villi, 6, 78
Secreted factors, 115–116
Selection of donors, 207
SFlt-1 (fms-like tyrosine kinase), 29
Skin re-epithelialization, 59–60
SLE, *see* systemic lupus erythematosus
Smad family, 61–62
Soluble endoglin, 29
SOX2, 226, 234, 237
Spinal cord injury, animals, 248–249, 260
Spiral arterioles
    DSMC niche, 28
    fetal growth restriction, 32
    preeclampsia, 29, 31
Staffing, GMP environment, 202
Stem-cell–based therapies, 148; *see also*
    mesenchymal stem/stromal cells (MSC)
Stem cell niche, 26
Stem cell populations
    chorionic villous type, 24–25
    Decidua basalis, 25–26
    mesenchymal stem cells, 20, 24–26
    other stem cell types, 20
    overview, 19
    sources, 21–23
    stem cell populations, 19–26
    trophoblast stem cells, 19–20
Stem cells, 18–26; *see also specific type*
Stem Cell Therapeutics as an Emerging Platform
    for Stroke (STEPS), 137
Stem/progenitor cells source, veterinary medicine
    amniotic-derived tissue and cells, 226–228
    comparative placenta adnexa embryology,
        223–226
    overview, 221, 226
    umbilical cord blood MSCs, 238–246
    umbilical cord-derived cells, 238
    umbilical cord matrix MSCs, 246–247
STEPS, *see* Stem Cell Therapeutics as an
    Emerging Platform for Stroke
Steroid hormone exposure, 151
Stillbirth, 18
Strategy, regulations, and GMP
    approaches, 200–201
    constraints, 199
    donation procedure, 210
    donor questionnaire, 208–209
    execution, 203–204
    final release, 213
    financial matters, 201–202

    informed consent, 209–210
    integration, 202–203
    issues, 198–199
    legal requirements, 204–206
    overview, 198
    planning, 199–203
    practicing, 155
    predonation phase, 206–210
    pregnancy complications, 209
    production, 212
    quality control, 213
    risks, 199–200
    selection of donors, 207
    staffing, 202
    summary, 213–214
    tissue donation, 210–211
    transport and reception, 211
    viral testing, 211
Stroke; *see also* cardiovascular disease
    amnion cells, 131, 134–137
    inflammation, 133
Structure, placenta, 14–17
Superficial digital flexor tendon (SDFT) injury, 262
Surgical debridement, 55
Syncytial knots, 9–10
Syncytiotrophoblast layer
    changes, placenta anatomy, 9–10
    placenta structure and function, 15
    trophoblast stem cells, 19
Syncytium, primitive, 4
Synovial inflammation, *see* rheumatoid arthritis
Systemic lupus erythematosus (SLE)
    autoimmune disorders, 167–168
    genetic predisposition, 163
    published clinical trials, 172

T

T cells
    costimulatory family molecule expression,
        96–97
    fetal immune programming, 44
    fibrotic disorders, 184
    IDO, 114
    immune system, 40–41, 44
    immunological tolerance, 100–104
    lack of proliferation, 98
    MSC immunogenicity, 92
    MSC immunomodulatory features, 100–104
    multiple sclerosis, 169
    SLE, 167
Telomerase activities, 151
Tendinopathy, 262–267
Tendon injuries, animals, 218, 252–253, 260–261
Terminal chorionic villi, 15
Term placenta cells, 186
TERT activity, 234, 237

Tertiary villi, 6, 9, 78
Tissue; *see also* strategy, regulations, and GMP
    amniotic membrane, 70–71
    chorionic membrane, 79
    donation, 210–211
    granulation tissue formation, 53
    Wharton's jelly, 81
Tissue plasminogen activator (tPA), 134
TNF/TNFR superfamily molecule expression,
        97–98
Trabeculae, 4
TRAIL (tumor necrosis factor–related
        apoptosis-inducing ligand), 44
Transforming growth factor beta (TGF-β)
    angiogenesis, 52–53
    hemostasis, 51
    MSC immunomodulatory features, 112–113
    multiple sclerosis, 168
    wound healing, 60–62
Transport, GMP environment, 211
Treatment studies
    liver fibrosis, 191–193
    lung injury, 187
Treg (regulatory T) cells
    antiproliferative effect, 102
    autoimmune disorders, 165
    boosting, 116
    costimulatory family molecule expression,
        96–97
    HLA-G, 114
    immune system, general, 41
    maternal immune system, 42–43
    multiple sclerosis, 169
    placenta and immunity, 44
Trophectoderm, 3
Trophic functions, 76–78
Trophoblasts, 2
Trophoblast stem cells, 19–20
Tubaric period, 223

U

Ulcers, 57–60
Umbilical cord (UC)
    blood MSCs, 238–246
    matrix MSCs, 246–247
    regenerative veterinary application, 238–244
    Wharton's jelly, 9, 81–84

V

Vacuum-assisted closure (VAC), 55
Vascular endothelial growth factor (VEGFR), 29
Vasculogenesis, 6, 15
Vasculosyncytial membranes, 10
VCT, *see* villous cytotrophoblast cells
Venous leg ulcers, *see* ulcers

Vessel dilation, reduced, 29
Veterinary medicine applications; *see also*
        animal models
    amniotic-derived tissue and cells, 226–228
    cell-based regenerative medicine, 218–221
    cell characterization in animals, 226–247
    comparative placenta adnexa embryology,
        223–226
    future perspectives, 267–268
    ophthalmology, 267
    overview, 218
    placenta cell-based regenerative medicine,
        247–267
    preclinical animal models, 220–221, 247–267
    stem/progenitor cells source, 221–247
    summary, 267–268
    therapeutic potential, 218–220
    umbilical cord blood MSCs, 238–246
    umbilical cord-derived cells, 238
    umbilical cord matrix MSCs, 246–247
Villous cytotrophoblast (VCT) cells, 15, 19
Villous immaturity, 17
Villous placenta, 4–6
Viral testing, GMP environment, 211

W

Wharton's jelly
    antigen-presenting cells, 106–108
    antiproliferative effect, 101–102, 116
    cell populations, 81–84
    cells isolated from, 81
    diabetes, 103
    engraftment, 103–104
    hematopoietic marker expression, 98
    HGF, 112
    histology of tissue, 81
    HLA molecule expression, 93, 96
    HLA-G, 114
    IDO, 114
    immunogenicity, 92–99
    immunomodulatory properties, 92, 99–111
    immunosuppression mechanisms, 92, 111–116
    inflammatory bowel disease, 167
    inflammatory cytokines, 115
    mesenchymal stromal cells, 81–84
    NK cells, 105
    origins, 4
    PGE2, 113
    T cell immunological tolerance, 100–101
    TGF-β, 112
    umbilical cord, 9
    in vitro studies, 98
    in vivo studies, 99
Whole organ replacement *vs.* cell-based therapy, 145
Witebsky's postulates, 162
Wound dressings/devices, 55–56

Index

Wound healing
- acute wound healing, 51–54
- amniotic membrane, 56–60
- angiogenesis, 52–53
- boiled amniotic membrane, 57–58
- chronic wounds, 54–55, 57–62
- collagen synthesis, 52
- cryopreserved amniotic membrane, 59
- dehydrated amniotic membrane, 57–58
- epithelialization, 53
- fibroblast migration, 52
- fresh amniotic membrane, 58
- granulation tissue formation, 53
- hemostasis, 51
- inflammation, 51–52
- irradiated amniotic membrane, 59
- overview, 50–51
- proliferative phase, 52–53
- regenerative veterinary application, 250–251, 260
- regenerative veterinary applications, 266
- remodeling phase, 53–54
- skin re-epithelialization, 59–60
- summary, 62
- TGF-β, 60–62

X

Xenotransplantation
- hAECs, 99
- hAMSCs, 76, 99
- regenerative veterinary application, 268

Y

Yolk sac, 5, 6, 9

An environmentally friendly book printed and bound in England by www.printondemand-worldwide.com

PEFC Certified

This product is
from sustainably
managed forests
and controlled
sources

www.pefc.org

This book is made of chain-of-custody materials; FSC materials for the cover and PEFC materials for the text pages.